W.J.A.J. Smeets R. Nieuwenhuys
B.L. Roberts

The Central Nervous System of Cartilaginous Fishes

Structure and Functional Correlations

With 51 Figures in the Text and
139 Figures in the Atlas

Springer-Verlag
Berlin Heidelberg NewYork 1983

WILHELMUS J.A.J. SMEETS
Former Assistant of Department of Anatomy, University of Nijmegen, The Netherlands

Prof. Dr. RUDOLF NIEUWENHUYS
Professor of Neuroanatomy, Department of Anatomy, University of Nijmegen, The Netherlands

Dr. BARRY L. ROBERTS
Member of the Staff of the Laboratory of The Marine Biological Association, Plymouth, England

ISBN-13: 978-3-642-68925-3 e-ISBN-13: 978-3-642-68923-9
DOI: 10.1007/978-3-642-68923-9

Library of Congress Cataloging in Publication Data. Smeets, W.J.A.J. (Wilhelmus J.A.J.), 1949– . The central nervous system of cartilaginous fishes.
Bibliography: p. – Includes index. 1. Nervous system—Fishes. 2. Chondrichthyes. 3. Central nervous system. I. Nieuwenhuys, R., 1927– . II. Roberts, B.L. (Barry L.), 1942– . III. Title. [DNLM: 1. Fishes—Anatomy and histology. 2. Fishes—Physiology. 3. Central nervous system—Anatomy and histology. 4. Central nervous system—Physiology. QL 638.6.S637c] . QL638.6.S63 1983 597′.304188 83–336

Softcover reprint of the hardcover 1st edition 1983

Reproduction of the illustrations: G. Dreher GmbH, Stuttgart

Typesetting, printing and bookbinding: Universitätsdruckerei H. Stürtz AG, Würzburg.
2131/3130-543210

Preface

Although a large body of data about the structure of the central nervous system of cartilaginous fishes has now been accumulated, a systematic survey of the neuromorphology of this, in so many respects, highly interesting group of animals is lacking. The present book is an effort to fill this gap. The information provided is derived partly from the literature and partly from original observations based on our own material. We have attempted to present a complete review of the relevant literature of the last 25 years, but the earlier literature has also been thoroughly scrutinized. Wherever possible we have commented on the functional significance of the various structures.

The information available in the literature has been incorporated with our own findings from a detailed study of four species, the sharks *Squalus acanthias* and *Scyliorhinus canicula,* the ray *Raja clavata* and the holocephalian *Hydrolagus collei.* Although these species do not reflect the total range of cartilaginous fishes they do exemplify the main features of the major groups.

The book is divided into two parts: text and atlas. In the first part some brief general introductory chapters are followed by chapters dealing with the major divisions of the central nervous system: spinal cord, brain stem, cerebellum, tectum mesencephali, diencephalon and telencephalon. The text illustrations include photomicrographs and drawings of characteristic structures, some of which have been derived from the literature. A pictorial survey of what is known of the fibre connections in the chondrichthyan brain is provided by a series of diagrams projected onto the brain of *Scyliorhinus.* In the final chapter of the first part we have attempted to foster further research by indicating the numerous gaps in our knowledge of the structural organization of the central nervous system of cartilaginous fishes.

In the second part of the book we present sets of standard illustrations of the four species studied. For each species these include (a) a diagrammatic median section through the brain, (b) a topological chart showing the disposition of the ventricular sulci and the cell masses in the brain stem, (c) topographical reconstructions of the cell masses in brain stem, diencephalon and telencephalon, and (d) sequences of photomicrographs of transversely oriented sections of the brain. It may be expected that these sets of figures will provide initial guidance in experimental studies on the brain of cartilaginous fishes.

Our work has been generously supported by the Medical Faculty of the University of Nijmegen and the Laboratory of the Marine Biological Association, Plymouth, and we are grateful to many of our colleagues in these institutes for their assistance. In particular we have been helped by the excellent histological expertise given by Mrs. Nellie Driessen-Verijdt, Mrs. Carla de Vocht-Poort and Mrs. Tonny Smeulders-van Wezel, who prepared the various series on which the atlas is based. The elasmobranch brains were obtained by us at Plymouth, but the *Hydrolagus* material from North America was collected for us by Dr. Christine Jaeger. The drawings were provided by Mr. Chr. van Huijzen, Mr. W. Maas, Mr. J. Konings and, especially by Mr. J. Russon, whereas the text photomicrographs were prepared by Mr. A.T.A. Reynen. Our special thanks should be given to Mr. Shimon Paniry of the Free University of Amsterdam for all the excellent photomicrographs included in the atlas. We are also very appreciative of the sustained secretarial assistance given by Miss Margaret Sjak Shie throughout the preparation of the book. Furthermore, we wish to thank the following publishers for permission to reproduce illustrations from their journals: Academic Press, Alan R. Liss, Elsevier/North-Holland Biomedical Press, Karger, Springer-Verlag and The Royal Society, London. Finally, the authors thank Dr. K. F. Springer, Mrs. Th. Deigmöller, Dr. D. Czeschlik and Mr. E. Kirchner of the publishing house of Springer for their help during the publication of this book.

Nijmegen/Plymouth, Spring 1983

W.J.A.J. Smeets
R. Nieuwenhuys
B.L. Roberts

Contents

1 Introduction

The fishes are the largest class of vertebrates. They live in a wide range of aquatic habitats and present a great diversity of structure, physiology and behaviour. More than 90% are bony fishes, and the living forms that fall within the scope of this book form a rather small group that includes only about 800 species. Several features obviously separate them from the bony fishes but the most distinctive is their cartilaginous endoskeleton. Fossil remains are consequently rare and the evolution of cartilaginous fishes is presently understood only in broad terms.

The first fishes to appear in the fossil record, dating from the Ordovician, were small-mouthed jawless forms which declined markedly as a group by the Carboniferous and which today are represented only by hagfishes and lampreys. The major line of jawed fishes diverged from the agnathans and developed jaws from the skeleton of the branchial arches. The earliest to appear, the Acanthodi, arose in the Upper Silurian and were followed closely by osteichthyans and placoderms. The Chondrichthyes or cartilaginous fishes appeared later, in the lower Devonian.

Early in this century, the cartilaginous fishes were considered to be generalized, primitive vertebrates and so were important to morphologists for typological comparisons with other vertebrates. As our understanding of these fishes has progressed, we have come to realize that they are less simple in organization than had been supposed and that they possess elegant sensory systems, relatively large brains, sophisticated reproductive mechanisms, specialized osmoregulatory

Table 1-1. Classification of the cartilaginous fishes (largely based on Compagno, 1977)

Class	Subclass	Superorder	Order	Genus
Chondrichthyes	Holocephali			*Chimaera*, *Hydrolagus*
	Elasmobranchii	Squalomorphii	Hexanchiformes	*Chlamydoselachus*, *Hexanchus*, *Heptranchias*
			Squaliformes	*Squalus*, *Deania*, *Centroscymnus*, *Centrophorus*
		Galeomorphii	Heterodontiformes	*Heterodontus*
			Orectolobiformes	*Ginglymostoma*
			Lamniformes	*Carcharodon*, *Cetorhinus*
			Carchariniformes	*Scyliorhinus*, *Galeocerdo*, *Mustelus*, *Carcharhinus*, *Negaprion*, *Sphyrna*, *Galeus*, *Pristiurus*
		Squatinomorphii	Squatiniformes	*Squatina*
		Batoidea	Rajiformes	*Raja*, *Platyrhinoidis*, *Rhinobatos*
			Torpediniformes	*Torpedo*, *Narke*
			Myliobatiformes	*Dasyatis*, *Mobula*

adaptations and even, in some species, thermoregulation.

Early in their history, the cartilaginous fishes separated into two major divisions, the Elasmobranchii and the Holocephali. The living elasmobranchs (sharks and rays) comprise four major superorders (Table 1-1): Squalomorphii, Galeomorphii, Squatinomorphii and Batoidea (Compagno 1977).

The sharks included in the Squalomorphii are principally dwellers of cold deep water and include about 25% of all shark species. The group contains the primitive and relatively uncommon six- and seven-gilled forms (*Chlamydoselachus, Hexanchus, Heptranchias*) as well as several better-known species (*Centrophorus, Deania, Centroscymnus, Squalus*).

The Galeomorphii, constituting 75% of the sharks, differs in many respects from the other groups and includes most of the well-known sharks and dogfishes (e.g. *Carcharodon, Carcharhinus, Negaprion, Sphyrna, Ginglymostoma, Mustelus, Scyliorhinus*).

The Squatinomorphii contains only the single genus, *Squatina.* This genus has some similarities with the Batoidea but resembles the Squalomorphii more closely in cranial and pectoral fin structure.

The Batoidea includes the rajiformes (skates), with large pectoral discs and slender tails, rhinobatiformes (guitar fishes), pristiformes (saw fishes), torpediniformes (electric rays) and myliobatiformes (sting-, butterfly-, eagle-, cow-nosed-, and devil-rays).

The Holocephali, abundant in the Carboniferous, are represented today by only six living genera. These fish, of somewhat bizarre appearance, differ from elasmobranchs in several important ways. The name Holocephali refers to the distinctive autostylic suspension of the jaws of these fishes which contrasts with the hyostylic and amphistylic jaw suspension of elasmobranchs. The gill slits are covered by a large operculum and a spiral valve is absent from the gut. There are also important differences in skeletal organization and in the structure of the brain.

The elasmobranch groups are represented in this book by *Squalus acanthias,* a squalomorph; *Scyliorhinus canicula,* a galeomorph, and *Raja clavata,* a batoid, the general body forms of which are illustrated in Fig. 1-1. Although these species do not represent the total range of cartilaginous fishes they do exemplify the main features of the major groups. Moreover, they have provided the bulk of the data on the central nervous system in the literature, because they are relatively common, are harmless to handle, and are easily maintained in the marine laboratory.

Scyliorhinus is uncommon in the Western Atlantic and its place within the North American laboratory inventory is taken by *Mustelus* in northern waters and by *Ginglymostoma,* or small specimens of *Negaprion* and *Sphyrna,* in warmer waters. Work in North America on Batoidea has been based mostly on *Raja erinacea, Raja eglanteria, Rhinobatos* and *Plathyrhinoidis.*

Squalus acanthias, the spurdog, is worldwide in distribution and lives in open water, feeds voraciously on a great variety of other fish, such as herring, cod and haddock, and performs extensive migrations both in European and North American waters. The roughhound, *Scyliorhinus canicula,* is a non-migratory bottom-dwelling species, living at depths of around 60 m, and feeding primarily on molluscs and crustacea. *Raja clavata,* the thornback ray, is found on the sea bottom at depths of up to 60 m and is non-migratory. It feeds primarily on crustacea. The Holocephali are represented in this book by the ratfish, *Hydrolagus collei,* which lives in the Pacific Ocean. Other common species are *Hydrolagus affinis* from the Gulf of Maine, *Hydrolagus alberto* from the Gulf of Mexico and the south Atlantic and *Chimaera monstrosa* from north European waters. These forms are mostly bottom dwellers, living in deep water with a wide geographic distribution. They have large eyes and a well-developed lateral line system. Most of the teeth are fused into large toothplates adapted for

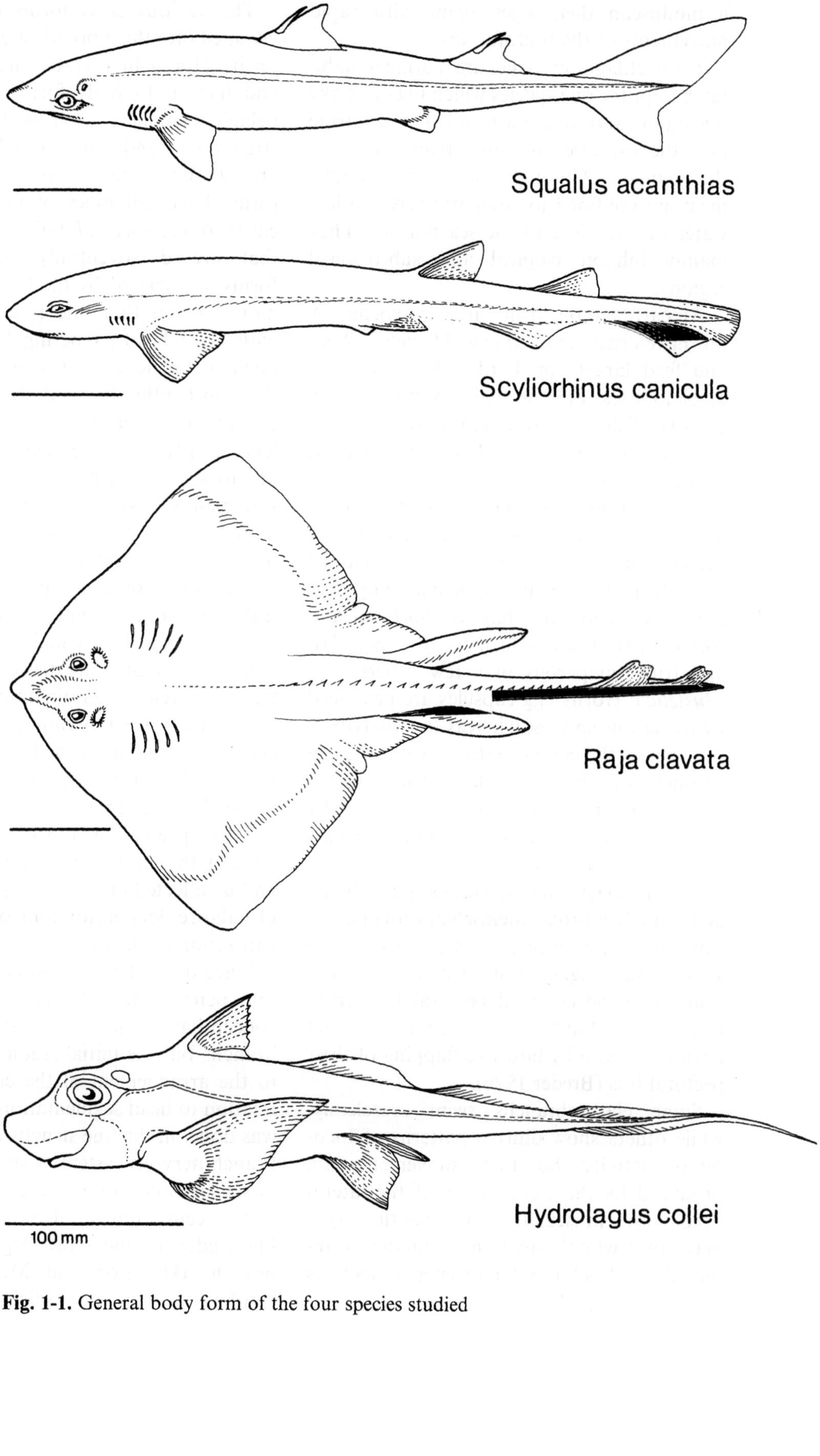

Fig. 1-1. General body form of the four species studied

a molluscan diet. They swim with rapid movements of the pectoral fins.

The cartilaginous fishes are marine in habitat, except for one species which lives in Lake Nicaragua and for a few forms that penetrate into the estuaries of large rivers, such as the Ganges. Although many are oceanic, most are confined to comparatively shallow water or live close to the sea bottom. They mainly inhabit tropical and sub-tropical regions.

All cartilaginous fishes are carnivorous. A few have crushing teeth (e.g. *Mustelus, Raja*) and feed largely on hard-shelled crustacea and molluscs, but the majority prey chiefly on other fishes and squids; the largest forms (*Rhincodon, Cetorhinus, Mobula*), however, feed wholly on plankton.

Fertilization is internal and is achieved by a pair of characteristic male copulatory organs (the claspers) which are developed from the pelvic fins. Embryonic development is oviparous in some (e.g. *Scyliorhinus* and *Raja*), ovoviviparous in most (e.g. *Squalus*) and truly viviparous in a few others (e.g. *Mustelus*). Horny egg capsules are produced by oviparous species, whereas in ovoviviparous species the embryo escapes from the egg capsule early, continues development in the mother's oviduct, and is not born until fully formed. A special yolk-sac placenta is formed in viviparous species.

All the sharks and sawfishes, guitarfishes and torpedoes propel themselves with undulations of the posterior part of the trunk. The skates and stingrays send waves of undulation along the enlarged pectoral fins, from front to rear, but the eagle-, cow-nosed-, and devil-rays swim by bird-like flapping of their pectoral fins (Breder 1926).

Some elasmobranchs swim ceaselessly, while others show diurnal patterns of locomotor activity. Sustained movements are produced by the contractions of the lateral "red" body musculature, whereas the larger mass of "white" myotomal muscle is reserved for brief rapid movements, such as attack or escape (Bone 1978).

The various body forms of sharks were grouped into three broad categories by Klausewitz (1962) in a study that related habit and habitat to body form. The active fast pelagic swimmers, such as the Isuridae, are streamlined and have caudal fin lobes that are of almost equal size. Less streamlined forms have tail lobes of unequal size (the classical *heterocercal* tail) and have heads that are dorsoventrally flattened. These forms, as typified by the Carcharinidae, are more littoral than the first group. Sharks with large heads, tapering bodies and very unequal tail lobes (e.g. *Ginglymostoma*) are sluggish benthic forms that feed on bottom-dwelling fish and invertebrates. This analysis has recently been extended by Thomson and Simanek (1977), who demonstrated the constancy in position of the pectoral and second dorsal fins and the general shape of the body in relation to locomotion.

Although some elasmobranchs are neutrally buoyant, utilizing low-density liver oil as a buoyancy mechanism (Corner et al. 1969), the majority are denser than water and have to develop a certain amount of dynamic lift during level swimming, which they achieve by balancing the lifting action of the tail with that of the large pectoral fins (Harris 1936). The interplay between overall body density, speed and manoeuverability is correlated with habit (Bone and Roberts 1969) and is reflected in the development of organs of balance, locomotor control and in the organization of the nervous system.

Since the end of the last century comparative neuroanatomists have recognized the conceptual advantages of studying elasmobranchs but the initial studies were confined to the arrangement of the cranial nerves in relation to head segmentation and little work was done on the fine structure of the elasmobranch nervous system or on the projections of sensory and motor nerves.

A recent volume of reviews of current knowledge of the sense organs of elasmobranchs (Hodgson and Mathewson 1978) emphasizes that these fish are very well equipped with sensory detectors of several kinds. As well as the senses of taste, smell, sight and hearing, familiar to ourselves, they possess in the lateral line organs and the ampullae of Lorenzini sense organs which are absent from most vertebrates.

There is abundant anecdotal and experimental evidence of the importance of chemical senses to elasmobranchs, particularly of olfaction. The sense of smell is mediated through the receptor cells of the olfactory epithelium, while taste originates in the oral and buccal cavities. Taste is restricted primarily to aspects of food finding and feeding but smell, while certainly utilized in food detection and location, is also important in reproductive and social behaviour.

The eyes of elasmobranchs are prominent, although not particularly large, and are placed laterally on the head (except in Batoidea) to provide some degree of binocular overlap in the visual field. They are thought to be important for predatory behaviour, particularly at close quarters. The eyes can be rotated by the extrinsic eye muscles to maintain a constant visual field during locomotion and turning. The rod-rich retina provides the eye with high sensitivity which can be enhanced in dim light by tapetal mirrors that reflect light from the choroid back to the rods.

It is now clearly recognized that sound plays a significant role in the lives of sharks and other elasmobranchs and is used by them to locate food sources and other objects. Acoustic playback techniques have revealed that sounds below 1000 Hz, particularly low frequencies (around 40 Hz), are attractive to these fishes (Myrberg 1978). The lateral-line sensory system is well developed in cartilaginous fish and was recognized by Lorenzini long ago as comprising two components, canals and gelatinous tubules. In most species the canals are enclosed, opening at intervals to the water, but in *Chimaera* there is simply a groove set into the surface of the skin. The lateral-line canal system is considered to be involved in the detection of water displacements set up close to the surface of the fish.

It is now appreciated that the ampullae of Lorenzini provide an extremely sensitive sensory system that is capable of responding to very small electric currents. This possibility was first demonstrated by Dijkgraaf and Kalmijn (1962) and was then firmly established by Kalmijn's subsequent experiments which revealed the extreme electrosensitivity of elasmobranchs. These animals appear to be able to respond to gradients as low as 0.01 μV/cm. The cartilaginous fish seems to use this sensory system to detect living food and to determine the direction and polarity of electric fields in the ocean (Kalmijn 1978).

Electrosensitivity in teleosts is usually associated with the production of discharges from electric organs, both systems being used for navigation. No shark is known to have an electric organ but electric organs are found in the Batoids. In torpedoids the electric organ is substantial in size and powerful discharges can be produced which are apparently used to stun prey during feeding. A much weaker discharge is emitted from the tail electric organ of rajoids, but the biological significance of this discharge remains uncertain.

Although our knowledge of the sensory systems of cartilaginous fishes has been increasing rapidly in recent years, our understanding of the organization of the brain of these species and of their behaviour remains much more limited. Studies by Springer, Nelson, Myrberg and others (reviewed by Gruber and Myrberg 1977) based on fisheries material and field work have provided information on the structure of shark populations and have demonstrated the presence of relatively complex social interactions in these forms. The brainweight-bodyweight ratio in some elasmobranchs is comparable to that of birds and mammals and exceeds that of most other non-mammalian species. Traditionally, large brains have been thought to reflect complex behaviour. But in the elasmobranchs it is difficult to make any statements about behaviour as so little work has been done.

Behavioural studies are, of course, hampered by the difficulties of keeping large predatory animals in captivity and of observing them safely in the wild. The need for a marine environment makes the cartilaginous fishes less convenient as laboratory material than freshwater animals, but with the development of artificial seawater systems it is now possible to maintain some species inland on a limited scale while marine laboratories are increasingly providing good facilities for experimental work. As these technical difficulties are overcome, the advantages offered to the neurobiologist by elasmobranchs as basic material for comparative work and as model systems in our understanding of brain function will increasingly be recognized. Certain species survive operative procedures well, and compared with the bony fishes the brain is large in size and the nervous system is easily accessible because of the cartilaginous nature of the skeleton.

Until now use of cartilaginous fishes in neurobiology has been seriously restricted because so little is known of their neuroanatomy, despite the widespread use of dogfish in the teaching of gross neuroanatomy in Europe and North America. Consequently, the modern researcher using elasmobranch material for electrophysiological or biophysical studies has often as a first step to unravel the anatomy of the system being investigated. With the application of new experimental methods to the nervous system of these fishes many new data are now being rapidly acquired. It is the purpose of the chapters that follow in this book to provide a survey of our current knowledge of the nervous system of these animals.

2 Materials and Methods

The information provided in this book comes partly from the published work of many authors who since the end of the last century have worked on the nervous system of cartilaginous fishes, and partly from original observations based on our own material.

Brains of the four species selected for this study (*Squalus acanthias, Scyliorhinus canicula, Raja clavata,* and *Hydrolagus collei*) were sectioned in the transverse and sagittal planes at a thickness of 15 µm. Alternate sections were stained with cresyl violet (Nissl) or according to the techniques of Bodian or Klüver and Barrera. For the study of the glial pattern we employed the brain of one specimen of *Squalus,* cut at a thickness of 25 µm, and prepared according to the gold sublimate stain of Ramon y Cajal.

Photomicrographs are provided in the atlas of selected cross sections through the brain of the four species. In each figure the left half is a photomicrograph of a section stained with the method of Klüver and Barrera (or Bodian), while the right half is a photomicrograph of the adjacent Nissl section. Together with the reconstructions of the various parts of the brain (see below) these form an atlas. References to figures of this atlas are indicated in the text by a single number, whereas text figures are indicated by a number preceded by the chapter number.

The size of cells in the various nuclei was estimated in the following way. Twenty cells of each cell mass were drawn at a magnification of 500 diameters, using a Zeiss drawing prism. The size of the soma was determined from these drawings by measuring the diameter in two directions, perpendicular to each other, and computing the average (Chap. 4). This procedure is hardly rigorous, yet it seems preferable to the widespread use of terms such as "small" or "large" when describing cells, without any quantification.

Besides the topographical reconstructions of the brain stem, diencephalon and telencephalon provided in the atlas, we have also made topological reconstructions of the brain stem using the procedure developed by one of us (Nieuwenhuys 1972, 1974) that enables the spatial relations of the various cell masses in the brain stem to be displayed in a single figure (Figs. 2, 35, 67, 101). This procedure has already been applied to the brain stem of *Squalus acanthias* and *Scyliorhinus canicula* (Smeets and Nieuwenhuys 1976) and it is here extended to two more species (i.e. *Raja clavata* and *Hydrolagus collei*). Two additional specimens of *Squalus* and *Scyliorhinus* were also studied to assess further the validity of the method. Full details of the procedure are given in the earlier publications to which the reader is referred, and so we shall give here only a brief description of this procedure.

Drawings of some 50 equidistant sections of the brain stem were made with the aid of a projection apparatus at a magnification of 25 to 40 diameters. The cell picture was studied at higher power and the cell masses were delineated on the drawings on the basis of the usual cytoarchitectonic criteria. The ependymal and meningeal surfaces were connected by a number of curves, termed projection curves, which were derived from the main trajectory of the glial fibres and, to a lesser extent, from the course of the blood vessels. The investigations of Rohon (1878),

Fig. 2-1. A Photograph of a transverse section through the tegmentum mesencephali of the shark *Squalus acanthias,* showing the radial pattern of the processes of the ependymal gliocytes. Gold sublimate stain. **B** Higher power view of part of the same section shown in **A**. In **A** and **B** bar scale = 100 μm. (After Smeets and Nieuwenhuys 1976)

Sauerbeck (1896), Houser (1901), Sterzi (1912) and Horstmann (1954) have revealed that in cartilaginous fishes these structures show a preferentially radial pattern. This pattern could also be recognized in our material, particularly in the gold sublimate series which showed that the ependymal gliocytes possess long basal processes that span the entire width of the wall. Although the processes of these tanycytes (Horstmann 1954) show some degree of ramification, an overall radial pattern can be readily observed (Fig. 2-1).

The outline drawings provided by the projection curves constituted the basis of the topological reconstruction, the preparation of which essentially involved the following steps (Fig. 2-2): (1–5) With the aid of the projection curves the outlines of the cell masses situated at various depths in the brain stem wall were projected upon the ventricular surface. (6) With the aid of a curvimeter, the distances from the zero point (defined as the deepest point of the ventral ventricular midline groove) to the deepest point of the other sulci and to the projections of the outlines of the nuclei upon the ventricular surface were determined on both sides. All the distances were measured along the ventricular surface and plotted graphically on a line. (7) The lines

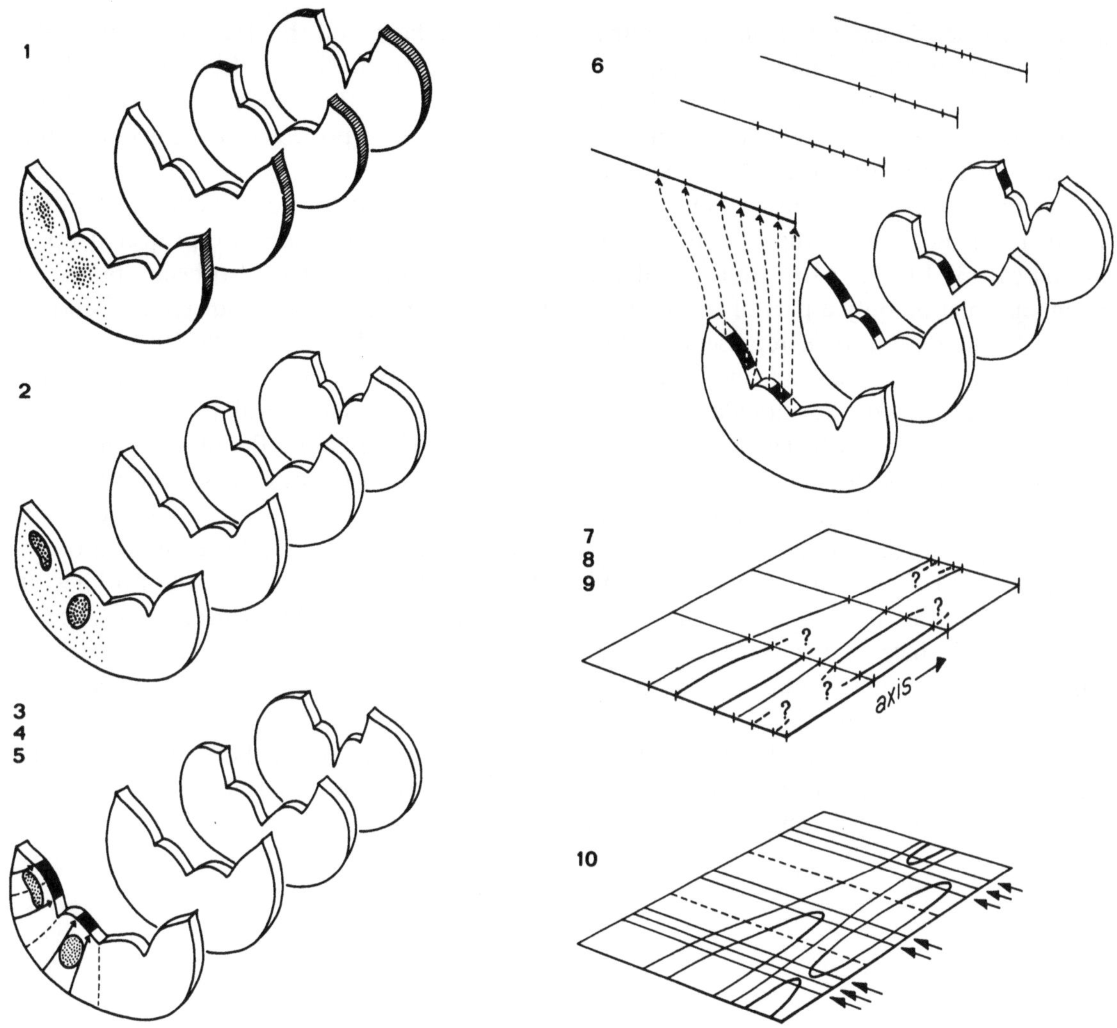

Fig. 2-2. The steps involved in the preparation of a topological reconstruction of the brain stem. *Numbers* correspond to those employed in the text. (After Nieuwenhuys 1974)

with the points of reference were transferred to a reconstruction of the brain stem, using a system of co-ordinates in which the central vertical axis, corresponding to the midline, is crossed by horizontal lines, corresponding to the sections. These lines were plotted in their rostrocaudal sequence with their zero points coincident with the central vertical axis. (8) Best fitting curves were drawn through the sets of points belonging to one and the same structure. (9–10) In some places the number of reference points available appeared to be too small for the drawing of sufficiently exact curves. Where this was the case, additional sections were analyzed and the resultant lines with points of reference were interpolated at their appropriate positions. Three structures, namely the corpus cerebelli, a part of the auriculae cerebelli and the tela choroidea of the fourth ventricle, have been omitted from the reconstructions. Thus, so far as the extent of the reconstructions is concerned, three different conditions obtain (cf. Fig. 2): (1) The entire ventricular surface is mapped in the most caudal part of the brain stem and in the midbrain. Here the reconstruction extends from the ventromedian sulcus medianus inferior to the dorsomedian sulcus medianus superior. (2) In the caudal part of the rhombencephalon the curves which constitute the lateral limits of the reconstruction represent the sites of attachment of the choroid roof, i.e. the taenia choroidea. (3) In the rostral portion of the rhombencephalon only a part of the neuronally differentiated wall is represented. The areas which are omitted, namely the corpus cerebelli and a part of the auriculae cerebelli, are difficult to map with the present procedure (Nieuwenhuys 1974, p. 267–268). However, since they have a homogeneous structure throughout, their omission is of minor importance. In *Hydrolagus* the tectum and the lateral part of the tegmentum have not been mapped since in the transverse series this region is actually sectioned in an approximately horizontal plane.

The topographical reconstructions of the brain stem, which may be very useful for experimental anatomical and physiological work, are based on the drawings used in the topological analysis. In these figures (Figs. 3, 36, 68, 102) the cell masses of the brain stem have been projected upon a horizontal plane.

A topographical reconstruction technique, in which the cell masses are projected upon a horizontal plane, has also been applied to the telencephalon of the four species studied (Figs. 6, 7, 39, 40, 71, 72, 105, 106).

However as the axis of the diencephalon extends in a dorsoventral rather than in a rostrocaudal direction, the topographical reconstructions of the cell masses and ventricular sulci of this region are projected upon a sagittal plane. To reduce complication in the figures, two separate sagittal outlines of the diencephalon and mesencephalon of each species are provided, one presenting the cell masses, which have a periventricular position or lie in the midsagittal plane, the other presenting the nuclei, which have migrated away from the ventricular surface (Figs. 4, 5, 37, 38, 69, 70, 103, 104).

3 Gross Morphology

The chondrichthyan brain is large and well developed. The investigations of Bauchot et al. (1976), Ebbesson and Northcutt (1976), and Northcutt (1977, 1978) have shown that the polygon encompassing the brain-body-weight ratios of a number of chondrichthyans shows a considerable overlap with comparable graphs (Figs. 3-1, 3-2) of birds and mammals. Comparisons between species of elasmobranchs of similar body size suggest that the ratio of brain-bodyweight in galeomorph sharks is two to six times larger than that of squalomorph sharks. A similar variation can be recognized within the superorder Batoidea, for the rajiforms have low brain-body ratios whereas the more advanced myliobatiforms show the highest brain-body ratios known for elasmobranchs. Unfortunately no data are yet available for holocephalian fishes.

The brains of the four species described in this book are shown in Fig. 3-3 from the dorsal aspect. The spinal cord passes imperceptibly into the rhombencephalon, a structure which, as its name implies, surrounds the rhomboid-shaped fourth ventricle. The floor of this ventricular cavity is predominantly made up of the rhombencephalic basal plates, whereas the alar plates form the lateral walls. The boundary between the basal and alar plates throughout most of the rhombencephalon is marked by a ventricular groove, the sulcus limitans of His. Rostrolaterally,

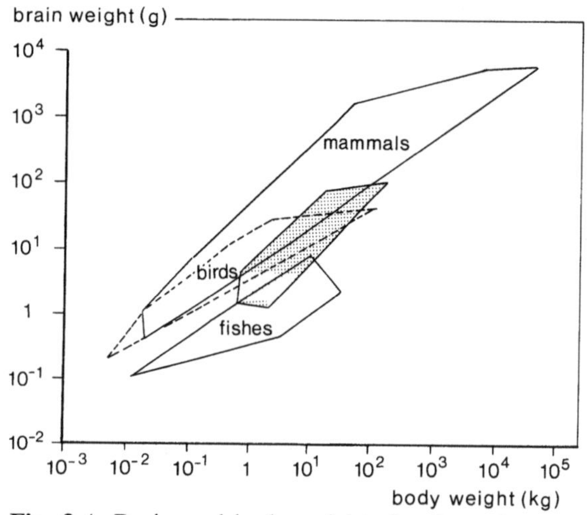

Fig. 3-1. Brain and body weights for four classes of vertebrates expressed as minimum convex polygons. *Stippled polygon* encloses elasmobranch brain-to-body ratios. (Modified from Northcutt 1977)

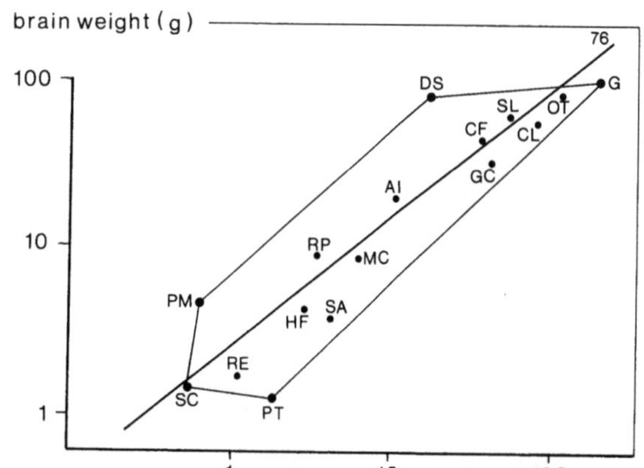

Fig. 3-2. Elasmobranch minimum convex polygon, illustrating positions of various taxa. Interspecific coefficient of allometry is 0.76 with a coefficient of determination of 0.86. *AI Aprionodon isodon; CF Carcharinus falciformis; CL Carcharinus leucas; DS Dasyatis sabina; G Galeocerdo cuvieri; GC Ginglymostoma cirratum; HF Heterodontis francisci; MC Mustelus canis; OT Odontaspis taurus; PM Potamotrygon motoro; PT Platyrhinoidis triseriata; RE Raja eglanteria; RP Rhinobatos productus; SA Squalus acanthias; SC Scyliorhinus canicula; SL Sphyrna lewini.* (Modified from Northcutt 1977)

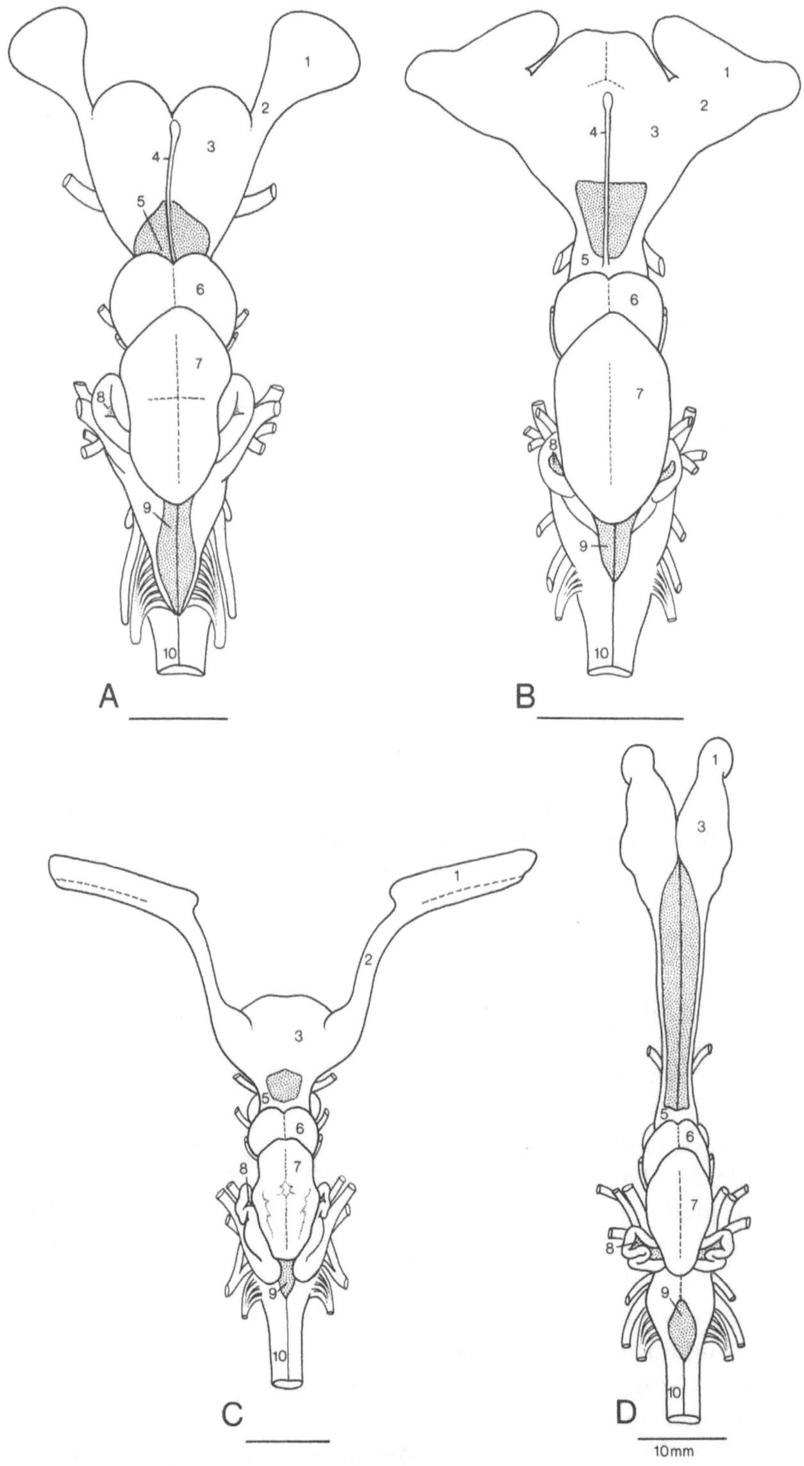

Fig. 3-3 A–D. Dorsal view of the brain of the four species studied: **A** *Squalus acanthias;* **B** *Scyliorhinus canicula;* **C** *Raja clavata;* **D** *Hydrolagus collei.* *1* bulbus olfactorius; *2* pedunculus olfactorii; *3* lobus hemisphericus telencephali; *4* epiphysis; *5* diencephalon; *6* tectum mesencephali; *7* corpus cerebelli; *8* recessus lateralis ventriculi quarti; *9* fossa rhomboidea; *10* medulla spinalis

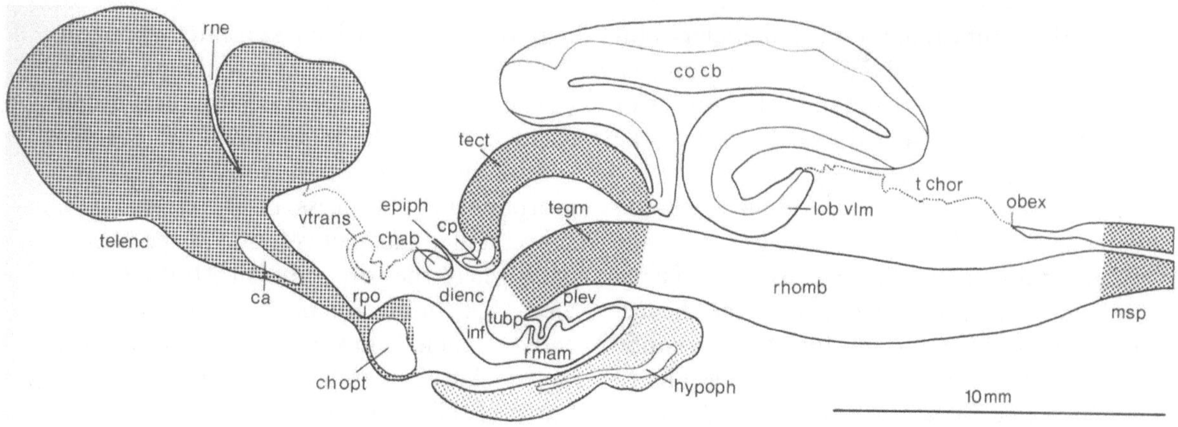

Fig. 3-4. Mid-sagittal section of the brain of *Scyliorhinus canicula* showing the various subdivisions

the fourth ventricle widens on either side to form a recessus lateralis, enclosed by the auriculae cerebelli. The ventricle gradually diminishes in size in the most rostral part of the rhombencephalon and the region which surrounds this tapering part of the ventricle is known as the isthmus rhombencephali. Dorsally, the fourth ventricle is covered by a folded, highly vascular tela choroidea. A feature peculiar to *Hydrolagus* is the fusion over a short distance of the ventricular surfaces of both alar plates (Fig. 3-3).

The cerebellum is strongly developed and consists of a central unpaired corpus cerebelli and the paired auriculae. The evaginated corpus cerebelli, enclosing a large ventricular cavity, expands rostrally over the tectum mesencephali and caudally over the fourth ventricle (Fig. 3-4). The auriculae can be subdivided into a rostromedial upper leaf and a caudolateral lower leaf. The upper leaves of both auricles are connected over the fourth ventricle by a band of nervous tissue known as the "lower lip". Caudolaterally, the auriculae merge into the dorsal parts of the lateral walls of the rhombencephalon, which constitute the prominent areae octavolaterales.

The most rostral part of the brain stem, the mesencephalon, comprises ventrally the tegmentum mesencephali, and dorsally the tectum mesencephali. The tectum is externally demarcated from the tegmentum by a groove, the sulcus tectotegmentalis. The tegmentum is caudally continuous with the rhombencephalic basal plate. The tectum is large and differentiated into bilateral lobes which surround expansions of the mesencephalic ventricle. Caudally the tectum passes into the corpus cerebelli via the relatively thin velum medullare anterius. This velum constitutes the most dorsal part of the isthmus rhombencephali. Rostrally the dorsal border of the mesencephalon is formed by the commissura posterior.

The diencephalon of adult cartilaginous fishes has retained the tube-shape of the early embryo. The thickened lateral walls enclose a slit-like median ventricle, which ventrocaudally opens into lateral expansions within the hypothalamus. The diencephalon, as in other anamniotes, is wedge-shaped in lateral view with the apex directed dorsally (Fig. 3-4). The short dorsal border displays the distinct habenular ganglia, which are connected dorsocaudally by the commissura habenulae. Rostral to this commissure, the diencephalon is roofed by a membranous structure, the tela diencephali, which passes over rostrally into a deep transverse fold, the velum transversum. This velum transversum is attached to the lateral walls of the brain in front of the habenulae, at a point which defines the dorsal border of the tel-di-encephalic boundary (Johnston 1911). The diencephalic roof again thins out caudal to the habenular commissure to form a tube-like evagination, the epiphysis. Further caudally, the posterior commissure forms the dorsal border of the di-mes-encephalic boundary.

The rostral boundary of the diencephalon extends from the lateral attachment of the velum transversum through to the chiasma opticum. Rostral to the chiasm, the diencephalic base thins out and forms the preoptic recess and thickens again further rostrally to hold the anterior commissure. In *Hydrolagus*, the preoptic region is extremely elongated due to the far rostral position of the telencephalon. The caudal boundary of the diencephalon passes from the posterior commissure to the deep transverse external fold of the brain floor known as the plica encephali ventralis. Rostral to this fold there is a marked intraventricular protrusion of the brain wall, the tuberculum posterius. This structure splits off a separate recess (the infundibulum) from the main ventricular system, which tapers caudally into the cavity of the saccus vasculosus. The recesses of the hypothalamic ventricle are the recessus mamillaris, a dorsal ventricular outpouching appearing caudally in the tuberculum posterius, and the bilateral recesses surrounded by a massive nervous wall, that form the lobi inferiores hypothalami. In *Squalus* and *Scyliorhinus* this hypothalamic recess is very large, but in *Raja* it is almost completely obliterated. The floor plate of the diencephalon is thin with the exception of its rostral part, which contains the optic chiasm, the postoptic commissure and the nucleus medius hypothalami. Part of the diencephalic floor contributes to the neural part of the hypophysis. The epithelial part of the hypophysis, the adenohypophysis, lies against the ventrocaudal surface of the diencephalon.

Two external sulci can be useful for orientation. The first is the sulcus tectodiencephalicus, which marks the boundary between the tectum and the diencephalon and which is caudally continuous with the sulcus tectotegmentalis. The second is the sulcus thalamohypothalamicus, which separates externally the thalamus from the hypothalamus.

There are considerable structural differences between the telencephala of different species although evagination is the essential morphogenetic process in this region. The simplest arrangement is found in the *Holocephali* where the forebrain is rostrally evaginated. The bulbi are inserted at the rostral ends of each hemisphere but lack separate ventricular cavities (Fig. 3-5). The studies of Sterzi (1912) have shown that in sharks the telencephalic side plates bulge out laterally and that the olfactory bulbs are formed later through an outgrowth of the lateral walls of the expanded portions. During the formation of these bulbs another outpouching appears, more rostral and lateral, and thus the lateral lobes or hemispheres are formed. Late in development the olfactory bulbs separate more and more from the hemispheres and the interconnections, the crura olfactoria, become stalks of various lengths (Fig. 3-3).

A remarkable feature of the shark brain is the connection of the right and left medial hemispheric walls (Fig. 3-4). Johnston (1911) supposed that this connection resulted from a secondary fusion of these walls but the studies of Bäckström (1924) have shown that the union results largely from a migration of nervous elements into the initially membranous structures which connect the hemispheres. The degree of fusion varies from species to species. There is little fusion in primitive forms such as *Chlamydoselachus* and *Hexanchus*, more in *Squalus* (Fig. 1), while in *Mustelus* and *Scyliorhinus* (Fig. 3-4) the medial hemisphere walls are completely united.

The early development of the forebrain of the Batoidea closely resembles that of sharks, but the two groups diverge later in development with regard to the extent of the ventricular cavities. Whereas the lateral ventricles remain wide in sharks, in rays they become gradually reduced to narrow slit-like spaces (Fig. 3-5). According to Bäckström (1924) this diminution of the ventricles is due to an

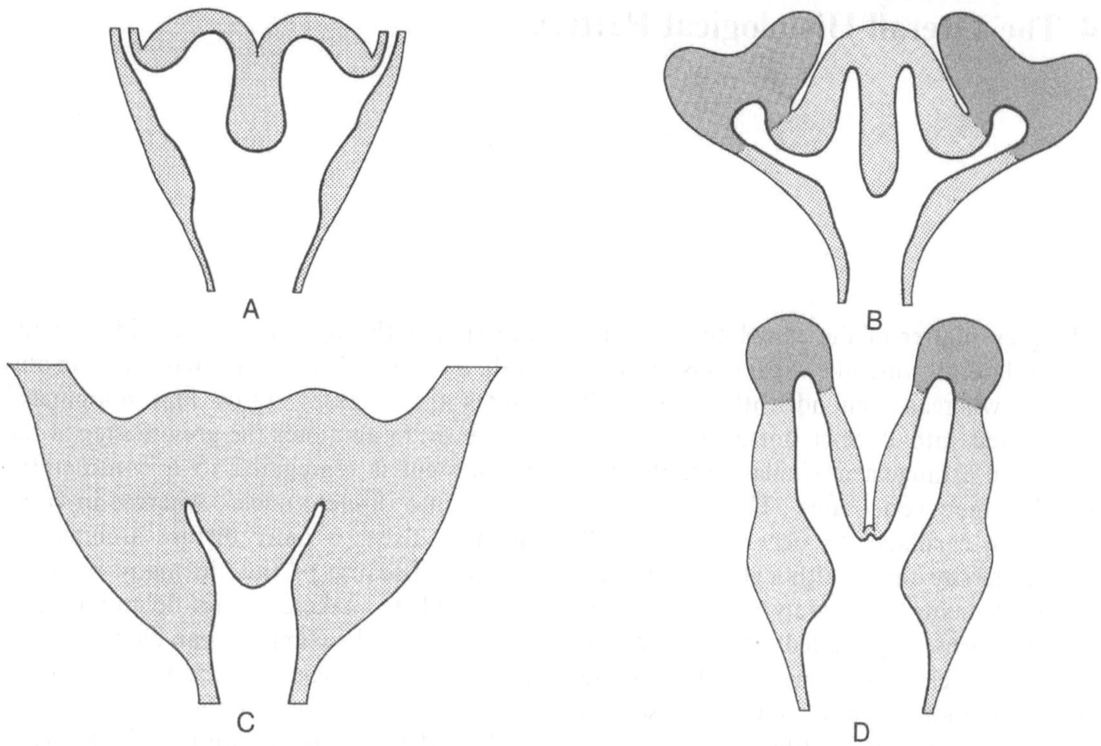

Fig. 3-5. Diagrammatic horizontal sections through the forebrain of the cartilaginous fish *Squalus* (**A**), *Scyliorhinus* (**B**), *Raja* (**C**), and *Hydrolagus* (**D**). *Heavily shaded parts* olfactory bulbs; *lightly shaded parts* cerebral hemispheres

increase in thickness of the hemispheric walls. Finally, the bulbar ventricles and the lateral part of the lateral ventricles close completely and in full-grown rays the lateral ventricles are merely represented by two short, horn-shaped rudiments. The medial walls of the batoid hemispheres are completely united in the median plane (Fig. 66), as in certain sharks.

In all cartilaginous fishes the olfactory bulbs are well developed structures, but they vary considerably in shape (Fig. 3-3). In sharks such as *Hexanchus* and *Squalus* the bulbi are ovoid, but in most rays they have a flattened saucer-shaped appearance. Other species like *Scyliorhinus* lie between these extremes. The rostral or rostroventral surface of the bulbs lies near to the olfactory organs and so the olfactory nerve fibres are short; in holocephalians the olfactory bulbs connect directly with the telencephalic hemispheres.

4 The Overall Histological Pattern

The grey matter of the spinal cord consists of two lateral, wing-like expansions in cyclostomes, whereas in chondrichthyans it is differentiated into distinct dorsal and ventral horns, an arrangement similar to that found in all other vertebrates. The spinal grey matter in cartilaginous fishes contains cells of a wide size range, although in adults the larger elements are mainly confined to the ventral horns. Myelinated fibres are absent from the cyclostomes but the white matter of the chondrichthyan cord comprises numerous well-myelinated fibres.

Whereas the neuronal perikarya in the brain stem of many groups of anamniotes (e.g. cyclostomes, lungfishes, amphibians) are chiefly located in a continuous zone of central grey, in cartilaginous fishes these cells are spread through the entire width of the wall, a condition similar to that found in amniotes (i.e. reptiles, birds and mammals). However, in amniotes the grey matter of the brain stem is segregated to a considerable extent into discrete nuclei, whereas in cartilaginous fishes a more diffuse architecture prevails. Because of this, in many places it is difficult to make an exact delimitation of cell masses and a complete parcellation is impossible. Nevertheless, more than 30 nuclei have been delineated in the brain stem of each of the four species studied on the basis of cell arrangement and the usual cytoarchitectonic criteria: (1) the size and shape of the neurons, (2) the disposition of the Nissl substance and (3) the density and the pattern of cell arrangement.

Table 4-1. Survey of cell masses in the brain stem and the average size of their constituting neurons in μm

	Squalus	*Scyliorhinus*	*Raja*	*Hydrolagus*
Somatomotor				
columna motoria spinalis	46	37	40	44
n. nervi abducentis	42	28	27	31
n. nervi trochlearis	42	29	24	28
n. nervi oculomotorii	42	33	26	28
Visceromotor				
n. motorius n. vagi	—	30	33	—
n. motorius n. vagi p. caudodorsalis	—	—	—	22
n. motorius n. vagi p. lateralis	37	—	—	—
n. motorius n. vagi p. medialis	39	—	—	—
n. motorius n. vagi p. rostroventralis	—	—	—	32
n. motorius nervi glossopharyngei	37	33	33	34
n. motorius nervi facialis	35	35	29	31
n. motorius nervi trigemini	49	37	34	41
n. of Edinger-Westphal	18	20	—	18

Table 4-1 (continued)

	Squalus	*Scylio-rhinus*	*Raja*	*Hydro-lagus*
Formatio reticularis				
n. raphes inferior	—	46	52	43/33
n. raphes superior	13	11	15	11
n. reticularis inferior	72	49	52	52
n. reticularis medius	81	64	53	57
n. reticularis superior	73	55	49	53
n. of the flm	70	54	45	48
Viscerosensory				
lobus vagi	11	10	9	11
General somatosensory				
n. mesencephalicus n. trigemini	61	50	50	44
n. radicis nervi trigemini	35	29	27	16
n. sensibilis nervi trigemini	15	19	15	16
Special somatosensory				
n. dorsalis areae octavolateralis	26	23	21	28
n. intermedius areae octavolateralis	32	25	22	26
n. octavus ascendens	28	24	23	30
n. octavus magnocellularis	57	55	48	51
"Relay" centres				
n. funiculi lateralis	29	26	24	27
n. cerebelli	27	25	18	21
n. ruber	26	21	28	23
n. tegmentalis lateralis	17	15	16	19
n. tori semicircularis	11	—	16	—
oliva inferior	16	14	12	13
n. isthmi	15	12	9	13
n. interpeduncularis	10	9	8	8
griseum centrale rhombencephali	11	9	9	—
Nuclei of unknown relationships				
nucleus A	51	—	—	—
nucleus B	25	27	25	20
nucleus C1	—	—	21	16
nucleus C2	18	13	17	19
nucleus D	—	—	—	33
nucleus E	—	—	—	17
nucleus F	16	11	16	14
nucleus G	—	19	—	—
nucleus H	25	22	23	23
nucleus J	—	—	22	—
n. intercollicularis	—	15	—	—

The nerve cells in the brain stem of the species studied here show considerable differences in size, ranging from 8–81 μm (Table 4-1). For convenience we have subdivided the neurons into four size categories which we shall call small, medium-sized, large and very large. Such a classification is, of course, arbitrary but since some correlations exist between size, shape and distribution of Nissl substance, this size classification can also be

Table 4-2. Survey of cell masses in the diencephalon and the average size of their constituting neurons in µm

	Squalus	*Scylio-rhinus*	*Raja*	*Hydro-lagus*
Epithalamus				
ganglion habenulae[a]	8/13	7/10	7/12	8/11
Thalamus				
thalamus dorsalis p. medialis	11/10[b]	8	10	10/8[b]
thalamus dorsalis p. lateralis	12	13	11	9
thalamus ventralis p. medialis	11	8	11	10
thalamus ventralis p. lateralis	10	12	12	10
eminentia thalami	15	17	19	(16)
corpus geniculatum laterale	29	27	24	23
Hypothalamus				
n. preopticus p. magnocellularis	15	12	11	15
n. preopticus p. parvocellularis	11	8	8	9
n. preopticus p. compacta	—	—	—	11
n. suprachiasmaticus	17	17	15	—
n. medius hypothalami	11	12	15	9
n. periventricularis hypothalami	12	9	9	12
n. lateralis tuberis	14	11	—	13
n. lobi lateralis p. dorsalis	11	13	—	18
n. lobi lateralis p. medialis	13	14	17	—
n. lobi lateralis p. ventralis	14	15	—	18
n. lobi lateralis p. lateralis	—	—	—	13
Synencephalic region				
n. opticus basalis	14	—	17	19
n. tori lateralis	—	—	18	—
n. tuberculi posterioris	12	14	13	10
n. sacci vasculosi	16	17	19	16
nucleus K	—	12	11	12
nucleus L	—	—	—	12

[a] Densely packed and more loosely arranged neurons, respectively.
[b] Pars dorsalis and pars ventralis, respectively.

Table 4-3. Survey of cell masses in the telencephalon and the average size of their constituting neurons in µm

	Squalus	*Scylio-rhinus*	*Raja*	*Hydro-lagus*
Telencephalon impar				
n. interstitialis commissurae anterioris	13	11	10	10
n. interstitialis fasciculi basalis telencephali	14	13	14	12
n. entopeduncularis	15	12	13	14
nucleus M	12	10	11	10
nucleus O	—	—	14	—

Table 4-3 (continued)

	Squalus	*Scylio-rhinus*	*Raja*	*Hydro-lagus*
Subpallium				
area superficialis basalis	12	12	15	13
striatum	10	9	—	—
area periventricularis ventrolateralis	17	12	—	—
n. septi medialis	17	—	—	—
n. septi medialis p. dorsalis	15	—	—	—
n. septi medialis p. anterior	—	15	—	15
n. septi medialis p. posterior	—	17	—	19
n. septi p. caudodorsalis	22	14	—	—
n. septi p. caudoventralis	14	10	12	—
n. septi lateralis	17	—	—	—
nucleus N	23	16	15	20
SP1	—	—	15	—
SP2	—	—	10	—
SP3	—	—	17	—
SP4	—	—	14	—
SP5	—	—	—	10
SP6	—	—	—	22
SP7	—	—	—	10
SP8	—	—	—	16
SP9	—	—	—	12
SP10	—	—	—	22
SP11	—	—	—	19
SP12	—	—	—	8
Pallium				
pallium dorsale p. superficialis	19	—	—	—
pallium dorsale p. superficialis anterior	—	14	—	—
pallium dorsale p. superficialis posterior	—	15	—	—
pallium dorsale p. centralis	17	—	16	—
pallium dorsale p. centralis lateralis	—	13	—	—
pallium dorsale p. centralis medialis	—	11	—	—
pallium laterale	16	14	14	17/12[a]
pallium mediale	14	14	—	18
area periventricularis dorsolateralis	21	15	—	—
area periventricularis pallialis	9	8	—	—
nucleus P	16	—	—	—
P1	—	—	21	—
P2	—	—	13	—
P3	—	—	14	—
P4	—	—	13	—
P5	—	—	12	—
P6	—	—	14	—
P7	—	—	14	—
P8	—	—	—	13
P9l	—	—	—	15
P9m	—	—	—	16
P10	—	—	—	15
P11	—	—	—	14
P12	—	—	—	15

[a] Pars anterior and pars posterior, respectively.

employed for a characterization of the neurons.

Small cells (8–20 μm) are round or ellipsoid and their narrow rim of cytoplasm generally contains only a few granules of chromophilic substance. *Medium-sized cells* (20–30 μm) are chiefly ellipsoid or triangular in shape and contain distinct granules or clumps of Nissl substance. *Large cells* (30–43 μm) are bipolar, tripolar or, less frequently, quadrupolar, and contain abundant Nissl substance, the individual particles of which are usually elongated and arranged in strands. A characteristic feature of the cells is that the Nissl substance is not confined to the perikarya but also extends into the initial segments of the dendrites. *Very large cells* (43–81 μm) are generally bipolar or quadrupolar. The distribution and appearance of the Nissl substance is similar to that of the large cells.

Golgi studies on the chondrichthyan brain stem (Houser 1901, Ramón-Moliner and Nauta 1966) and our own silver-impregnated material indicate that cells of medium-sized, large and very large categories have scanty, long, rectilinear dendrites, that have relatively few branches. This means that the great majority of the elements in the brain stem of these fish are "leptodendritic" and "isodendritic", in terms of the general categories of Ramón-Moliner (1967, 1969) and Ramón-Moliner and Nauta (1966). The long dendrites of these cells unite in bundles which invade and subdivide the white matter. This highly characteristic reticular organization is found throughout almost the entire extent of the rhombencephalon and the tegmentum mesencephali. Thus, the dendrites of neurons in these parts of the chondrichthyan brain extend far beyond the cytoarchitectonic boundaries of the various nuclei, a fact which is of great importance for the interpretation of the results of experimental hodological studies.

In the diencephalon most cells retain a periventricular position, possibly indicating a more primitive condition than that observed in the brain stem. The variation in cell size is also less than in the brain stem for all cell masses are composed of cells that fall into the class of small neurons, except for the corpus geniculatum laterale, which consists of medium-sized elements (Table 4-2). However, despite the predominantly periventricular position and the small differences in size of the cells, about 20 cell masses have been delineated in the diencephalon of the four species studied.

In the telencephalon the cells are in most places spread through the entire width of the wall. The variation in cell size is, like that in the diencephalon, much less than in the brain stem (Table 4-3). Though the telencephalic grey matter consists mainly of a continuum of diffusely arranged cells, some 20 more or less individualized nuclei could be delineated within this brain part.

Finally some comments on the non-neuronal elements of the central nervous system should be made. The studies of Horstmann (1954) and Klatzo (1967) have revealed that the elasmobranch brain comprises at least two types of glial elements, tanycytes and astrocytes. The tanycytes, a term introduced by Horstmann, are characterized by long processes that span the entire width of the neural wall (Fig. 2-1). Although the processes of these glial cells show some degree of ramification, an overall radial pattern can be recognized. The cell bodies of the tanycytes lie mostly in the ependymal layer, but migration in the direction of the meningeal surface may occur. The astrocytes do not posses such elongated processes, but have rather numerous, short ramifications. Though some astrocytes lie within the ependymal layer, the majority can be found in the proximity of neurons, blood vessels and the meningeal surface. Both tanycytes and astrocytes are closely applied to blood vessels or form characteristic end feet which entwine around the blood capillaries. However, species differences may exist, because Horstmann (1954), working on *Scyliorhinus, Torpedo* and *Raja,* could identify such end feet, whereas Klatzo

(1967) found only direct contacts between glial cells and blood vessels in *Negaprion*.

In addition to the two glia types mentioned above, Klatzo identified oligodendrocytes and microgliacytes. Oligodendrocytes have an oval or roundish cell body, and slender branching processes. These elements appeared to be very abundant in the various parts of the brain. The microgliacytes show an irregular outline, short thorny expansions, and are found mainly in the caudal part of the brain.

A striking feature observed by Klatzo (1967) in the central and rostrodorsal regions of the telencephalon of *Negaprion* is a clustering of glial cells including astrocytes, oligodendrocytes and microgliacytes around nerve cells. A similar satellitosis has been recognized by the present authors in some telencephalic areas of *Raja clavata* (Fig. 10-2F).

5 The Spinal Cord

5.1 Introductory Notes

The spinal cord of chondrichthyan fishes merges rostrally with the rhombencephalon and ends caudally in the tail fin, surrounded by lymph tissue. It is enclosed in a single sheet of tissue, the meninx primitiva of Sterzi (1900), and is restrained within the vertebral canal by ligaments. It retains the tube shape of the early embryo, being cylindrical (*Raja*) or slightly flattened (sharks), and may be considered to consist of a series of segments, essentially identical in organization, that match the segmental divisions of the body musculature. In *Scyliorhinus* there are between 60 and 70 such segments and this number does not increase as the body grows. On each side a segment provides a dorsal and ventral root which unite as a segmental spinal nerve that passes to the periphery.

5.2 Peripheral Spinal Nerves

The peripheral nerves contain the axons of neurons that convey sensory information from the skin (excluding the lateral line organs), musculature and viscera and those that supply the myotomal muscle fibres. Our limited knowledge of the organization of the peripheral sensory system of elasmobranchs has recently been reviewed (Roberts 1978), and so only a brief outline will be given here. General sensory information is provided by the "free-nerve" terminations of nerve fibres which form a network near to the skin surface. Specialized nerve terminals are found deeper in the skin, in connective myocommata, between vertebrae and amongst fin muscle fibres. The corpuscular endings of Wunderer (1908) provide general mechanoreceptive information and may serve a proprioceptive function; the intermuscular endings, found in the pectoral fins of rays (Poloumordwinoff 1898), are stretch receptors. Muscle spindles and Golgi tendon organs are apparently absent from chondrichthyan musculature.

Although the great majority of the sensory nerves have cell bodies (about 70 μm diameter; Campbell 1946) located in the dorsal root ganglia close to the vertebral column, at least some sensory fibres appear to enter the spinal cord via a few ventral roots (*Squalus,* Grillner et al. 1976). The extent of the areas of the body supplied by the dorsal roots (the dermatomes) has been determined in *Scyliorhinus* (van Rijnberk 1904; Roberts 1969), and *Raja* (Ten Cate 1927).

The motor axons that supply the body musculature leave the spinal cord through the ventral root and pass along the myocomma connectiva to innervate the adjacent two myotomes. The myotomal musculature can be divided broadly into two main types—the "red" tonic and the "white" phasic systems—which have quite different structure, metabolism and pattern of innervation. Full descriptions of these muscle types have been provided by Bone (1978) and Johnston (1981).

In each segment the ventral root is located rostral to the dorsal root and leaves the cord as three rootlets which unite before passing through the vertebral cartilage. The dorsal and ventral roots become enclosed in a common connective sheath and form the mixed spinal nerve, which gives off dorsal, medial and ventral rami. The mixed nerve also sends a small ramus communicans, carrying sympathetic fibres, to the chain of sympathetic ganglia; there is no grey ramus communicans (Young 1933a).

Although the motor and sensory fibres intermix in the peripheral nerves of most chordates, in many elasmobranchs they remain separate, although they are contained within a common nerve (Roberts 1969; Coggeshall et al. 1978); only in some rays (*Raja*) do sensory and motor fibres appear to intermix within the nerve.

The most rostral spinal nerves can be distinguished from all others by their larger size and different organization. The number of these nerves may vary but usually the first two, which lie within the chondrocranium, lack dorsal roots. The hypobranchial nerve trunk that innervates several head muscles is derived from the occipital and first two spinal nerves (Norris and Hughes 1920).

A ventral root in the mid-body region in *Scyliorhinus* contains approximately 300–400 fibres that have an average diameter of about 14 μm, compared with over 500 fibres in the dorsal root which have an average diameter of only 6 μm (Roberts 1969). Coggeshall et al. (1978) found that in the stingray *Dasyatis* a dorsal root contains around 1260 fibres of average diameter 4.2 μm, whereas a ventral root has 680 fibres of average diameter of 9.2 μm. The study just mentioned and that of Leonard et al. (1978a) revealed two interesting features of spinal nerve organization in *Dasyatis:* firstly, the number of fibres increases as the animals grow; and secondly, unmyelinated fibres are virtually absent from these nerves. This fish lack, therefore, any equivalent of the important C-fibre system of the mammal.

5.3 General Organization of the Spinal Cord

Transverse sections of the spinal cord of the four species studied in this book are provided in Figs. 5-1 to 5-4. The spinal neurons retain their primary position around the border of the central canal and form a column of grey matter that is surrounded by a white matter composed of nerve fibres.

The white matter lacks any truly giant axons although large axons are found in the ventral funiculi and in the deepest part of the lateral funiculi. Counts of axons of a segment in the mid-body region of the cord of *Scyliorhinus* show that there are approximately 50,000 myelinated axons in the white matter; degeneration studies indicate that the majority of these axons originate within the spinal cord.

As in all gnathostomes, the grey matter is arranged as dorsal and ventral horns but the dorsal horns, which are separated by small dorsal funiculi fuse centrally as a mass, called by Keenan (1928) the *corpus commune posterius,* just dorsal to the central canal. Keenan (1928) thought that almost the entire dorsal horn of these fishes constitutes a substantia gelatinosa. Comparisons show that there are large differences between cartilaginous fish species in the extent and development of this region.

All the fibres contained within the dorsal root of chondrichthyans originate in adults from extramedullary spinal ganglion cells. Their axons bifurcate on entry and run rostrally and caudally as the intracornual bundles, but the extent of the distribution is unknown. Fine fibres from these bundles can be seen to penetrate into the dorsal horn. In some species (*Pristiurus*) dorsal root fibres have been seen making contact with motoneurons, thus providing a pathway for monosynaptic activation of the latter (von Lenhossék 1894). In *Dasyatis* Leonard et al. (1978b) give evidence from intracellular recordings of monosynaptic activation of motoneurons from dorsal root stimulation, but in *Scylio-*

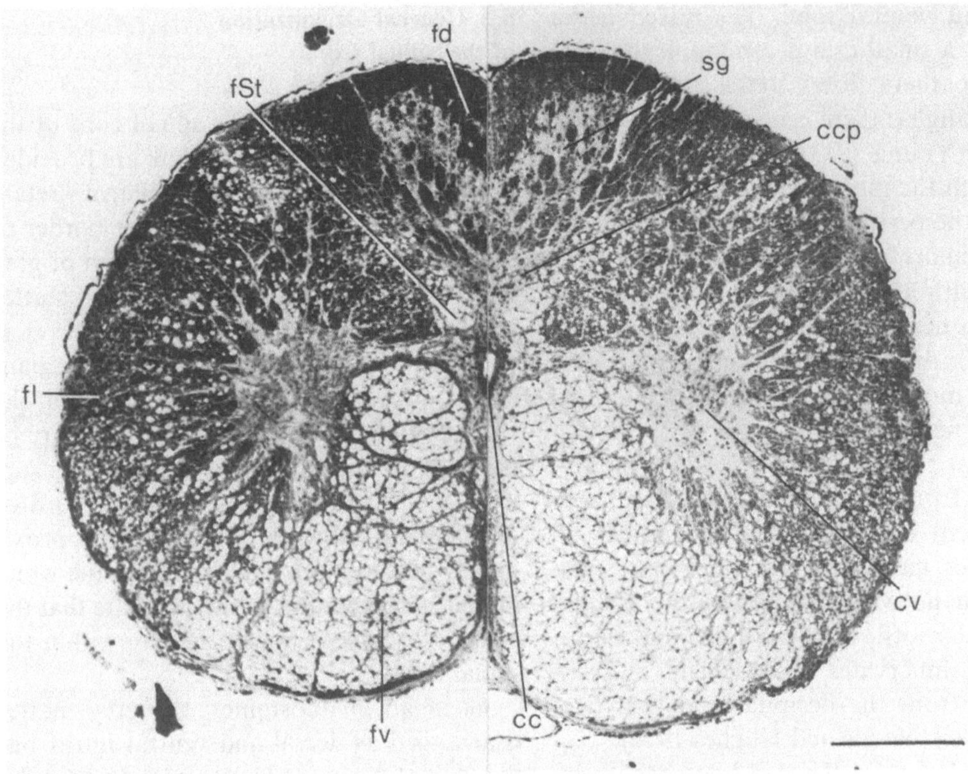

5-1

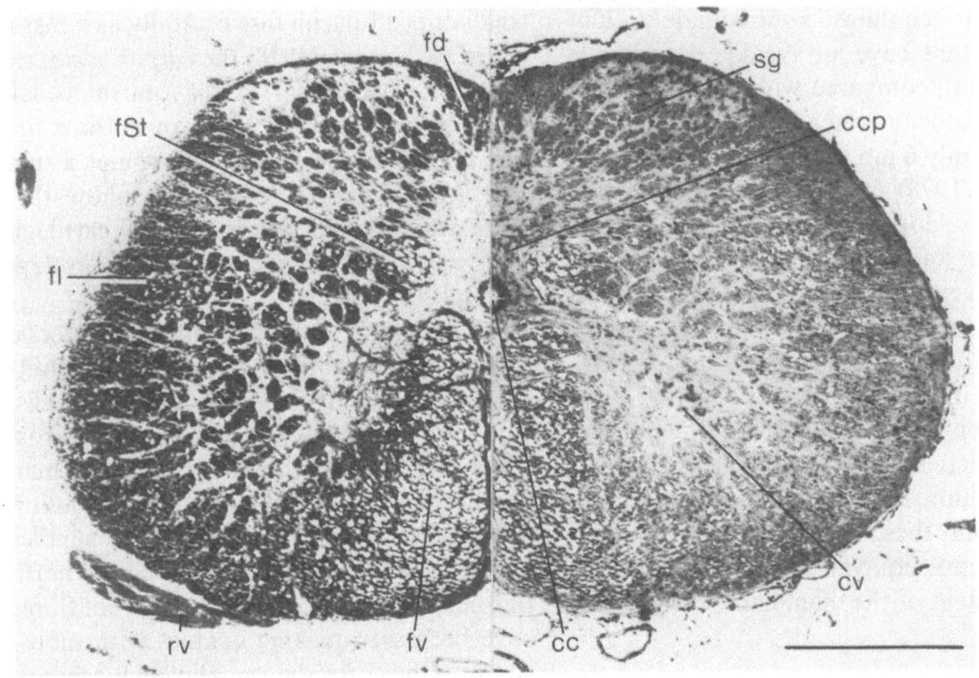

5-2

Fig. 5-1. Photomicrograph of a transverse section through the "cervical" cord of *Squalus acanthias*. In Fig. 5-1 to 5-4 the *left side* is stained according to Klüver and Barrera, whereas the *right half* is stained with cresyl violet. *Bar scale* = 500 μm

Fig. 5-2. Photomicrograph of a transverse section through the "cervical" cord of *Scyliorhinus canicula*

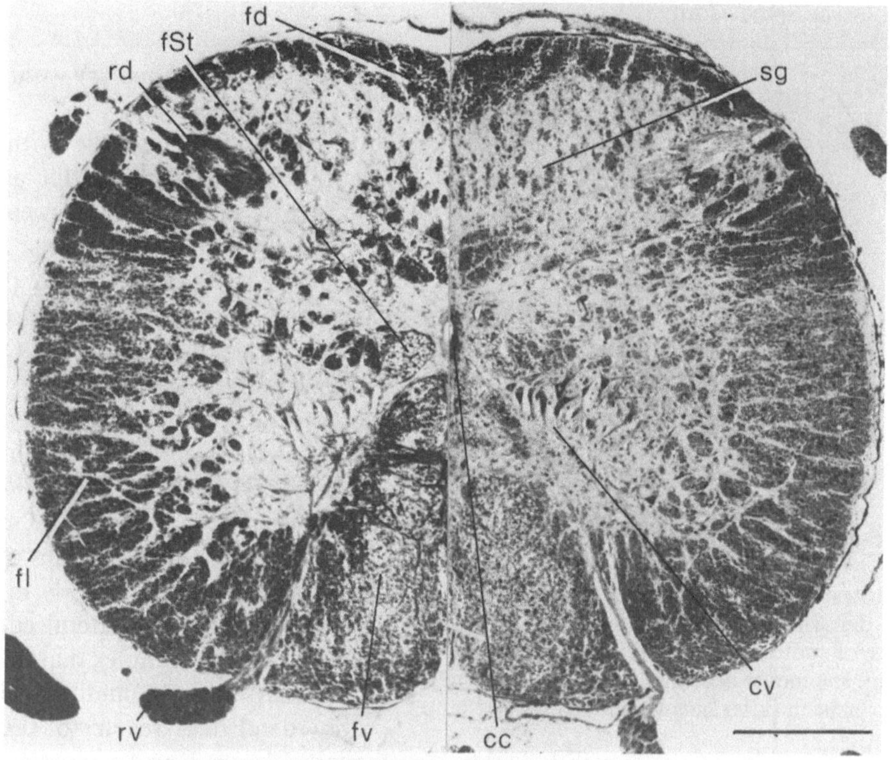

5-3

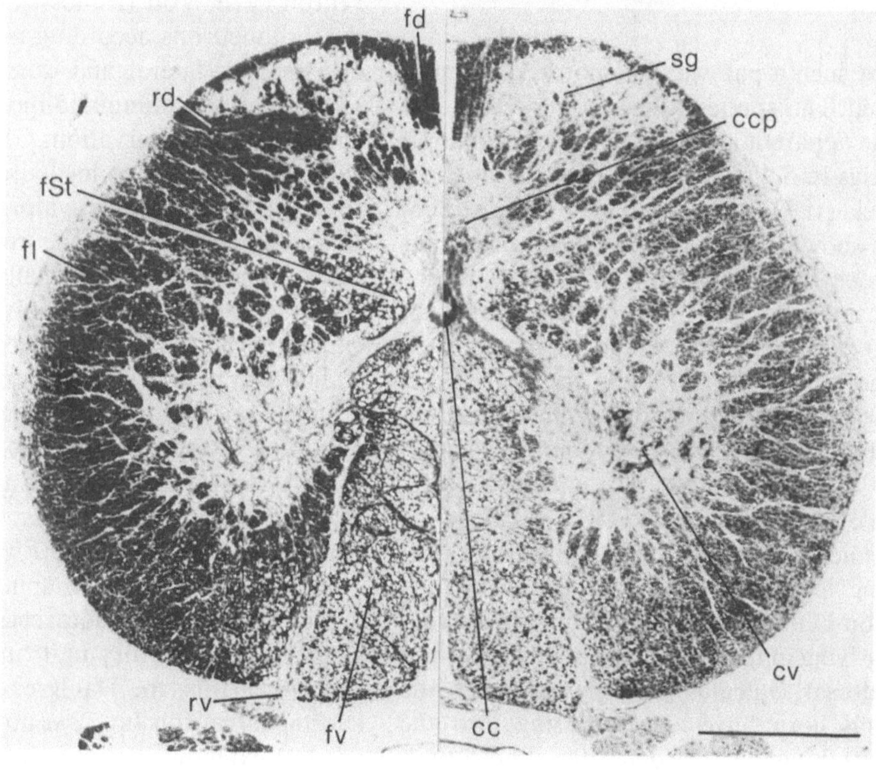

5-4

Fig. 5-3. Photomicrograph of a transverse section through the "cervical" cord of *Raja clavata*

Fig. 5-4. Photomicrograph of a transverse section through the "cervical" cord of *Hydrolagus collei*

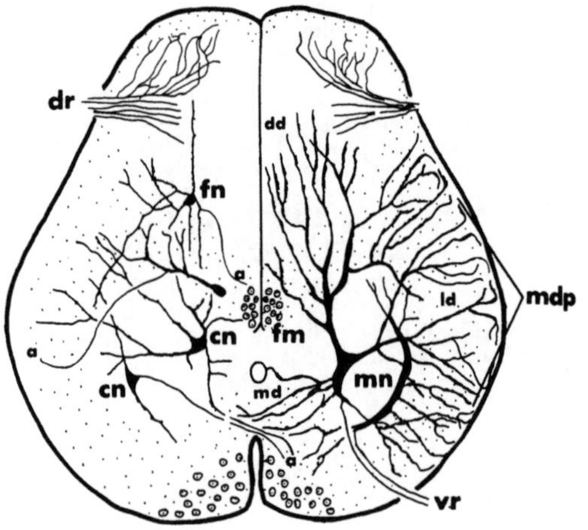

Fig. 5-5. Neurons in the spinal cord of a young ray. *a* axon; *cn* commissural neuron; *dd* dorsal dendrites of motor neuron; *dr* dorsal root; *ld* lateral dendrites of motor neuron; *md* medial dendrites of motor neuron; *mdp* marginal dendritic plexus; *mn* motor neurons; *vr* ventral root; *fn* funicular neuron (after Nieuwenhuys 1964)

rhinus such a pathway has not been demonstrated. This species difference may be related to the apparent absence of Poloumordwinoff endings in *Scyliorhinus*. Studies by von Lenhossék (1892, 1894) and by Retzius (1895) have shown that there are many cord interneurons that have axons that cross to the opposite side (Fig. 5-5, commissural neurons). These neurons, which are mostly sited in the ventral horn, vary in size and shape. Their axons cross the raphe immediately beneath the central canal (the commissura anterior) or more ventrally (commissura accessoria), and then bifurcate.

Funicular interneurons (with axons projecting ipsilaterally) have been documented by von Lenhossék (1894), who reported that those lying in the dorsal horn send axons into the dorsal funiculus, and those sited in the ventral horn have axons passing into the ventral funiculus. The dendrites of both funicular and commissural cells are long and some extend to the marginal plexus (see below). Our knowledge of interneurons that have only short axons (Golgi's type 2) is very sparse and von Lenhossék (1894, *Raja*) and Retzius (1895, *Squalus*) were even doubtful if they are present.

The motoneurons lie within the ventral horn and are large multipolar cells with extensive dendrites that interweave and interact with the white matter axons. Although most synaptic input must be on these dendrites, synapses have also been reported on the perikarya of ventral horn cells (McClung et al. 1978, *Ginglymostoma*). The motoneurons constitute a continuous column throughout the cord. Retrograde labelling with cobalt salts and HRP in *Scyliorhinus* has shown that there are approximately 700 motoneurons in each segment. The dendrites of the motoneurons spread predominantly in the transverse plane, reaching the lateral edge of the cord and forming a complex marginal plexus. Processes crossing the midline, which are probably dendrites, reach to the contralateral ventral horn. Von Lenhossék (1894) divided the motoneurons according to their position into medial, lateral and dorsal groups and on the basis of stimulation experiments and embryological observations Glees (1940) concluded that the dorsomedially situated cells represent preganglionic sympathetic neurons. In the lamprey the motoneurons that supply red muscle fibres are smaller than those innervating white fibres (Teräväinen and Rovainen 1971), but it is not known if a similar size difference exists in chondrichthyans.

In the tail region of rays (Rajidae) the innervation of the small electric organs is provided by motoneurons which are much larger than adjacent motoneurons that innervate conventional musculature (Ewart 1893).

In the tail region the spinal cord of all chondrichthyans includes cells that form a caudal neurosecretory unit (urophysis). One element of this, the Dahlgren cell, is a truly enormous neuron (in *Raja* up to 300 μm diameter) extending several short processes to the vascular meninx (Dahlgren 1914; Speidel 1922). This cell has a highly developed cytoplasm, which is presumably related to its neurosecretory function. The urophysis is thought to serve a role in osmoregulation.

5.4 Fibres Ascending to the Brain in the Funiculus Dorsalis

In mammals the dorsal funiculus contains numerous branches of dorsal root primary afferent fibres many of which extend to the hindbrain to synapse in the dorsal column nuclei and then, relaying on in the medial lemniscus, to the thalamus and cerebral cortex. We have already emphasized that the dorsal funiculus in elasmobranchs is small but the important question of whether any of its fibres extend to the obex remains unanswered. Dorsal column nuclei cannot be distinguished on cytoarchitectonic grounds in the four species used in the present study, although such nuclei have been described by Wallenberg (1907) in the nursehound *Scyliorhinus stellaris* and by Ebbesson and Hodde (1981) in the nurse shark *Ginglymostoma cirratum* (Fig. 5–6).

5.5 Fibres Ascending to the Brain in the Funiculus Lateralis

On the basis of normal material, several authors have recognized ascending systems in the lateral funiculi that pass to the rhombencephalon, mesencephalon and cerebellum. These fibres are collectively described as the lemniscus spinalis (Fig. 5–6). Hayle (1973a, b) charted the degeneration that resulted from spinal cord hemisections in *Scyliorhinus*. He found ascending spinal pathways reaching to the predominantly ipsilateral rhombencephalon and mesencephalon, and ending as mossy fibres in the cerebellar corpus. Ebbesson and Hodde (1981), in a degeneration study in *Ginglymostoma,* found that the spinal lemniscus comprises three fasciculi in the brain stem (Fig. 5–6): one that reaches the cerebellum (receiving a small input from the dorsal funiculus as well); a second that extends as far as the thalamus; and a third that ends in the rhombencephalon, the latter being by far the largest. Most of the spinorhombencephalic fibres terminate

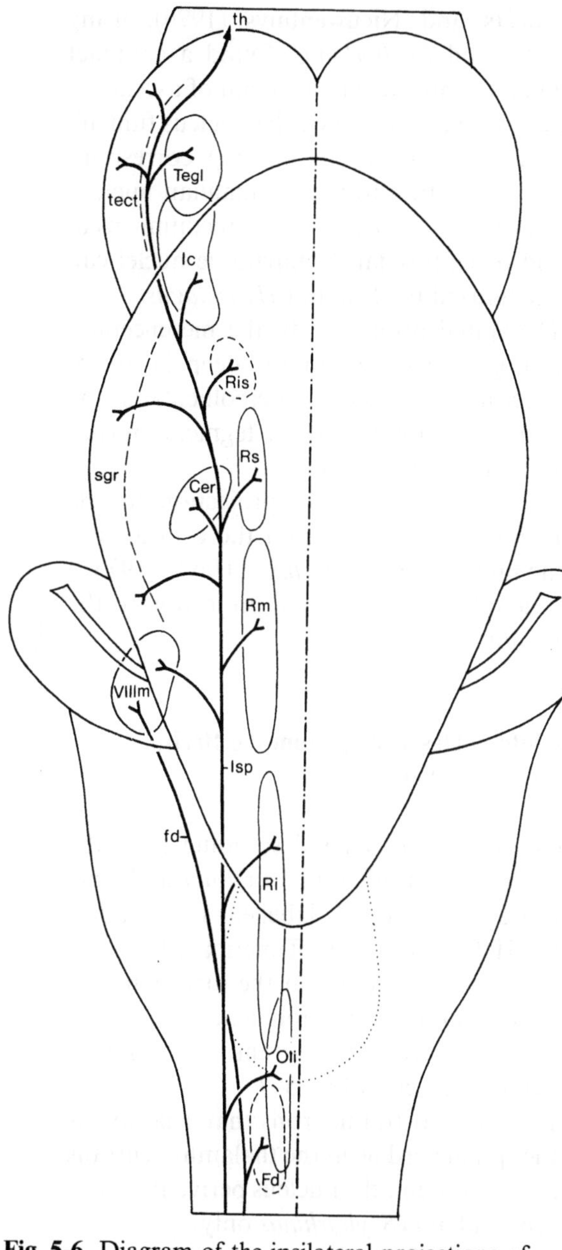

Fig. 5-6. Diagram of the ipsilateral projections of pathways ascending from the spinal cord to the brain, as viewed from the dorsal surface of the brain

in the reticular formation, but some project to the dorsal motor nucleus of the vagus. The projection to the cerebellum terminates as mossy fibres amongst the granule cells of the corpus cerebelli and also distributes to the cerebellar nuclei.

Smeets and Nieuwenhuys (1976), using *Squalus* and *Scyliorhinus,* found a compact cell mass, embedded in a stream of ascending fibres, which they termed the nucleus funiculi lateralis, because of its striking correspondence in position to the mammalian nucleus of that name. Its neurons are medium-sized, round or ellipsoidal. Comparable nuclei can be recognized in *Raja* and *Hydrolagus.*

The spinal projection to the mesencephalon (Fig. 5–6) passes to the deep layers of the tectum, the nucleus intercollicularis, the periaqueductal grey and the tegmentum (nucleus tegmentalis lateralis).

In *Ginglymostoma* the projection to the thalamus terminates within the central thalamic nucleus; in *Scyliorhinus* Hayle (1973a) was unable to find a spinal projection to the thalamus.

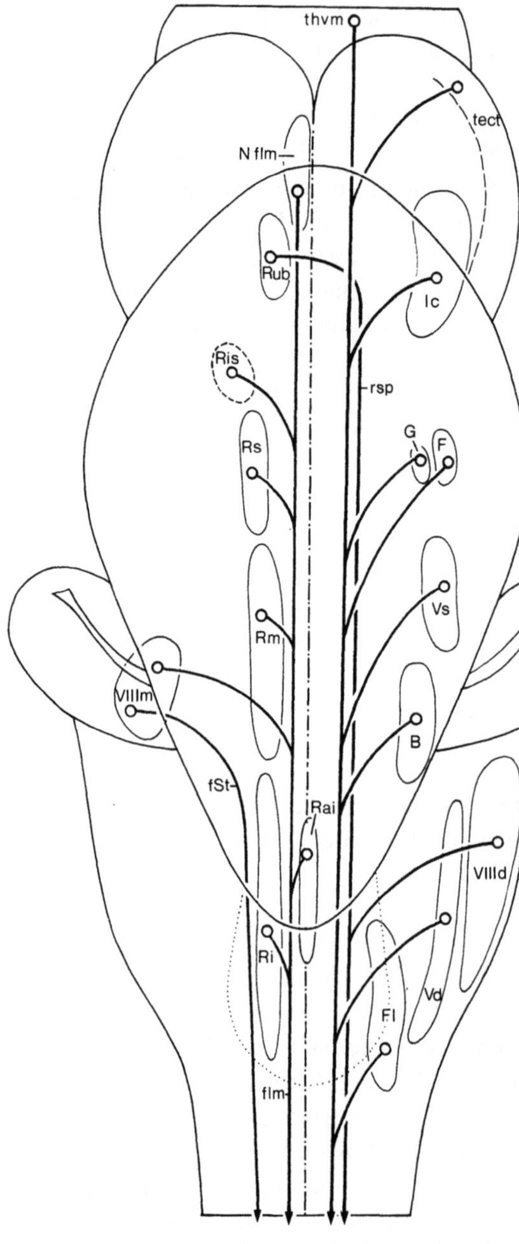

Fig. 5-7. Diagram of the projections of pathways descending from the brain to the spinal cord, as viewed from the dorsal surface of the brain. Only the main ipsilateral pathways are illustrated although most projections are, in fact, bilateral

5.6 Fibres Descending from the Brain to the Spinal Cord

Long axonal pathways descending from the brain have been identified in *Raja* and *Scyliorhinus* in material labelled retrogradely with HRP (Smeets and Timerick 1981). Labelled cells were found in the diencephalon, mesencephalon and rhombencephalon, as can be seen from a diagram of the major projections given in Fig. 5–7.

The most rostral neurons that send axons to the spinal cord lie in the thalamus ventralis pars medialis and the nucleus periventricularis hypothalami (*Scyliorhinus* only).

Projections from the mesencephalon arise from the nucleus of the fasciculus longitudinalis medialis, the tectum mesencephali, the nucleus intercollicularis, the tectotegmental junction zone and from diffusely arranged tegmental neurons. A contralateral rubrospinal pathway could be identified in *Raja,* but not in *Scyliorhinus.*

The major descending pathways from the rhombencephalon originate from reticular and octaval neurons. The nuclei B, F and G, the nucleus tractus descendens nervi trigemini, the nucleus funiculi lateralis and the nucleus tractus solitarii also send fibres into the cord.

For the most part the route followed by descending fibres in the spinal cord and the sites of their termination are unknown. An electrophysiological study in *Scyliorhinus* (Paul and Roberts 1978) demonstrated that the inferior reticular nucleus projects mainly to the spinal cord and that about half of the neurons in the nucleus reticularis medius project to the same target. A rubrospinal system could not, however, be established. A more recent study (Roberts and Williamson 1982) also in *Scyliorhinus* has shown that fast conducting axons terminate on ipsilateral motoneurons and interneurons. Stimulation of these axons, which are probably contained within the fasciculus longitudinalis medialis and Stieda's fasciculus, evokes synchronous excitation of motoneurons.

5.7 Regeneration of the Spinal Cord

There are now several studies indicating that the spinal cord of fishes, particularly of teleosts, shows considerable powers of regeneration following spinal transection (reviewed by Healey 1957). This regeneration is known to be associated with substantial synaptic reorganization (Bernstein and Gelderd 1973). There have been fewer reports of regeneration in elasmobranchs but Marón (1963) observed complete repair in the cord of *Scyliorhinus* embryo. Gelderd (1979) has shown that, although regeneration appears to be much more limited in the adult (*Ginglymostoma*), there is evidence of synaptic readjustment. Leonard et al. (1978a) suggested that the good regenerative capacity of the nervous system of fishes may be related to the fact that in these animals neuronal cell number continues to increase throughout adult life.

5.8 Organization of the Spinal Cord in the Embryo

The study by Harris and Whiting (1954), following on that made by Wintrebert (1920), established that the contractions of the myotomal musculature of the early embryo of *Scyliorhinus* are myogenic and are not initiated by neural tissue, which is undifferentiated at this stage of development. Subsequently, the frequency of muscular contraction changes, as the motoneurons innervate the muscle fibres, and myotomal contraction then becomes initiated neurogenically. Later still, these movements may be modified by sensory stimulation. Harris and Whiting associated these events in the pattern of development with changes in the neural organization of the cord; motoneurons and interneurons interconnected with each other before becoming associated with the large dorsal sensory neurons. The latter neurons are a special feature of the embryo and disappear during the course of development; they have peripheral processes which provide sensory endings to the skin. These neurons are usually named after Rohon (1884) and Beard (1896) who first described them for a number of species, including elasmobranchs. The organization of these cells throughout the vertebrates has been reviewed by Nieuwenhuys (1964).

5.9 Functional Correlations

The output from the spinal cord is involved in body movement maintenance and control; in co-ordinating the viscera, through the sympathetic system; in osmoregulation via the urophysis (Bern 1969); and in the firing of the electric organ in the Rajidae (Bennett 1971).

Locomotor body movements are the basis of most of the behaviour of elasmobranch fishes and in most species movement is brought about by body undulation (sharks and *Torpedo*); in most batoids and in chimaeroids locomotion involves the specialized movements of the enlarged pectoral fins.

The importance of the spinal cord in determining the fundamental locomotor pattern is evident from the fact that a dogfish in which the spinal cord has been totally severed close to the brain can perform continuous locomotor movements that are essentially of normal form. These properties appear not to be restricted to any specific part of the cord; in the lamprey motor rhythms can be produced by as few as four segments of isolated spinal cord (Cohen and Wällen 1980).

Although spinal preparations of rays do not show spontaneous locomotor movements, motor rhythms can be elicited by stimulation of the mesencephalic locomotor region (Droge and Leonard 1979).

Studies on the spinal cord of *Scyliorhinus* and *Dasyatis* (Leonard et al. 1978b; Roberts and Williamson to be published), have revealed details of the cord organization, but it remains to be seen how these features result in locomotor output. Recordings from motor nerves of spinal dogfish, in which all movements have been prevented by curare paralysis of the musculature, have shown that a rhythmical motor output continues to be produced which is organized within and between segments in a manner that is appropriate for locomotion (Roberts 1981a). Microelectrode studies within the cord reveal that only a small fraction of the motoneurons are rhythmically active, the inactive neurons presumably supplying the white muscle fibres which are not used in spinal swimming.

Interneurons can be segregated into three groups on the basis of their patterns of spontaneous activity: inactive cells; continuously active cells and neurons showing clear rhythmical patterns. It is possible that the rhythm of these interneurons is produced by specific pacemaker neurons but it is more probable that it results as an emergent property of a network of neurons (Williamson and Roberts 1980).

Although motor rhythms persist in the spinal cord in the absence of locomotor-generated sensory signals, it is clear that in the swimming fish these signals play a role in the coordination of movement. The sensory inflow enhances the excitability of the spinal neurons, influences the stability of the motor patterns and regulates relationships within and between segments.

6 The Brain Stem

6.1 Introductory Notes

The brain stem (truncus cerebri) comprises the rhombencephalon and the mesencephalon; caudally it grades into the spinal cord while rostrally it borders the diencephalon. In all vertebrates the brain stem harbours the centres of origin and termination of all cranial nerves, except for cranial nerve I. In cartilaginous fish the brain stem contains a fairly well developed reticular formation and a number of relay centres and their associated ascending and descending connections. The cerebellum, although ontogenetically and phylogenetically a derivative of the rostral part of the rhombencephalon, is not included in the brain stem and will be considered separately in Chap. 7. However, the cerebellum and the brain stem are strongly interconnected and many of the cerebellar afferent systems originate, and almost all cerebellar efferent fibres terminate, within the brain stem.

The tectum mesencephali, the dorsal portion of the mesencephalon, will also be dealt with separately (Chap. 8).

6.2 Nerve Components and Longitudinal Zones

The fundamental studies of Gaskell (1886, 1889) revealed that the fibres which compose the spinal nerves can be split into four categories: somatic afferent; visceral afferent; visceral efferent and somatic efferent fibres. Peripherally these fibres are connected with specific end organs while centrally they converge on local areas or zones. The four categories are also found in the cranial nerves but these contain several further fibre classes not represented in spinal nerves, which are related to certain specializations that do not occur at the spinal level. The nerve components which are present only in cranial nerves are generally considered to belong to "special" systems, whereas the ones which occur at both cranial and spinal levels are designated as belonging to "general" systems. With the aid of this terminology the nerve components of the cranial nerves can be indicated as follows:

1. general somatic afferent fibres, which transmit impulses from the skin, skeletal muscles, joints and ligaments;
2. special somatic afferent fibres, which are associated with the receptor cells in the special sense organs (retina, internal ear, lateral-line organs);
3. general visceral afferent fibres, which convey impulses from receptors in visceral organs and blood vessels;
4. special visceral afferent fibres, which supply visceral sense organs (i.e. the organs of taste and smell);
5. general visceral efferent fibres, which supply the smooth musculature of internal organs, the cardiac muscle and glands;
6. special visceral efferent fibres, which innervate muscles derived from the mesenchym of the visceral branchial arches; and
7. general somatic efferent fibres, which supply striated muscles derived from the myotomes.

Whereas most spinal nerves usually contain fibres of all four "general" categories, the cranial nerves vary widely in the types they carry. Some have only one type of fibre, while others carry two or more.

The afferent centres are located in the brain stem of all vertebrates in the dorsally and laterally situated alar plate whereas the efferent centres lie in the medially and ventrally located basal plate. The boundary between these two plates in cartilaginous fishes is marked by a distinct ventricular groove, the sulcus limitans of His. The "motor" basal and "sensory" alar plates containing the cranial nerve nuclei can be subdivided into separate longitudinal zones which, as was discovered by Herrick, Johnston and others, are related specifically to nerve fibres of one of the main categories mentioned previously. The basal plate contains a medial somatomotor and a lateral visceromotor zone, whereas the alar plate contains a ventral viscerosensory and a dorsal somatosensory zone. The boundaries between these zones are generally marked by sulci, such as the sulcus intermedius ventralis and the sulcus intermedius dorsalis. Although these four functional zones, or columns, were originally considered as entities which contained only primary efferent and afferent centres it was later observed that certain grisea of secondary or higher order also fit into the longitudinal zonal pattern (e.g. Herrick 1913; Bartelmez 1915). Rostral to the rhombencephalon, the four functional zones are less readily recognized but at least one, the somatomotor zone, extends throughout the midbrain.

The various zones, their relations to particular nerves or nerve components and their constituent centres will now be briefly reviewed (see Figs. 2, 35, 67, 101).

In the four species studied there is a distinct sulcus limitans that extends from the spinal level to the pretrigeminal region and divides the rhombencephalon into two basic morphological entities, the basal and alar plate.

The *basal plate* can be separated into a medial area ventralis and a lateral area intermedioventralis which appear as longitudinal columns that project into the ventricle on either side of a distinct ventricular groove, the sulcus intermedius ventralis. In *Squalus* this sulcus extends into the caudal part of the mesencephalon but it is discontinuous with any groove in the tegmentum mesencephali in the other three species studied. The area ventralis contains a large fibre bundle, the fasciculus longitudinalis medialis, together with a number of cell masses and its functional designation as a somatomotor zone is appropriate in so far as it contains the rostral end of the spinal motor column, the nucleus nervi abducentis and almost the entire rhombencephalic median and medial reticular formation. The latter is believed to project to somatomotor centres in the spinal cord via the fasciculus longitudinalis medialis. Moreover the area ventralis may be considered to be continuous with the medial parts of the tegmentum isthmi and the tegmentum mesencephali and these regions contain the somatomotor nuclei of III and IV and the nucleus fasciculus longitudinalis medialis, a nucleus which probably belongs to the somatomotor coordinating apparatus. The nucleus ruber, which probably also forms part of the somatomotor mechanism, is likewise situated within the medial part of the tegmentum mesencephali. However, it should be emphasized that the area ventralis and its rostral continuation contains a number of centres which cannot be considered to be somatomotor in function. The inferior olive, for example, is believed to project to the cerebellum while the efferent connections of the interpeduncular nucleus and of the griseum centrale are entirely unknown. Finally, we cannot exclude the possibility that elements of the medial reticular formation serve as links in ascending projections.

The area intermedioventralis contains the motor nuclei of V, VII, IX and X and together these centres constitute a visceromotor column, of which the mesencephalic nucleus of Edinger-Westphal may be considered to be the rostral extreme. However, the nucleus funiculi lateralis, the nucleus of VI and particularly the nucleus reticularis superior encroach upon the area intermedioventralis from the medial side. This area also contains nuclei A (*Squalus*) and B (*Squalus* and *Scyliorhinus*) and some nuclei in the isthmus region, the connections of which are not fully known. Consequently, the term visceromotor zone seems to be inapplicable to the whole of the area intermedioventralis.

In the caudal half of the rhombencephalon the sulcus intermedius dorsalis subdivides the *alar plate* into an area intermediodorsalis and an area dorsalis which occur as elongated intraventricular protrusions. The area intermediodorsalis is predominantly the common centre of termination for general visceral afferent and special visceral afferent fibres, which enter the brain by way of nerves VII, IX and X. However, the general somatosensory nucleus of V is found in the superficial part of the intermediodorsal zone and its most rostral part contains the nucleus octavus magnocellularis. According to Hugosson (1955, 1957, *Torpedo ocellata*) this nucleus develops from the same cell column that gives rise to the viscerosensory nuclei. The nucleus octavus magnocellularis receives direct octavus nerve afferent fibres and thus represents a special somatosensory centre. However, since it projects directly to somatic efferent centres in the spinal cord it may also be considered to be part of the brain stem motor apparatus. As the area dorsalis is entirely occupied by two large lateral-line centres it may be designated as a special somatosensory zone.

The zonal pattern is less clear in the subcerebellar and isthmus regions of the rhombencephalon of the chondrichthyans studied. In *Squalus* and *Hydrolagus* sulcus e may perhaps be considered to indicate the boundary between the intermediodorsal area and the dorsal area and, if so, the medial part of nucleus octavus ascendens and nucleus V would then fall within the intermediodorsal area, whereas the nucleus cerebelli and the nucleus isthmi would be contained within the area dorsalis. However, the sulcal pattern is more complex in *Scyliorhinus* and *Raja* for the single sulcus e of *Squalus* and *Hydrolagus* seems to be represented by two sulci, e1 and e2. These relations imply that a distinct four-column pattern cannot be recognized in the rostral part of the rhombencephalon of *Scyliorhinus* and *Raja*.

The general arrangement of the cell masses in the mesencephalon of the cartilaginous fishes studied does not exhibit a clear morphological zonal pattern, although functionally this part of the neuraxis may be divided into primarily motor and sensory zones. The motor zone, which includes the nuclei of III and IV, the nucleus of Edinger-Westphal, the nucleus of the fasciculus longitudinalis medialis and the nucleus ruber, occupies the medial part of the tegmentum mesencephali. The sensory zone comprises the lateral part of the tegmentum and the tectum. In the four species studied the boundary between these two zones is marked on the ventricular side by sulcus d, which some authors (Ariëns Kappers et al. 1936; Kuhlenbeck and Niimi 1969) have apparently interpreted as being part of the sulcus limitans. However, the topological reconstructions show clearly that these two sulci are quite distinct.

The sensory centres of the midbrain belong to several functional categories. The nucleus mesencephalicus of V, which occupies the most ventral part of the tectum, is a primary general somatosensory centre. The tectum is the chief end-station of the optic nerve (Ebbesson and Ramsey 1968; Graeber and Ebbesson 1972; Northcutt 1979a; Smeets 1981a) but also receives afferent input from the spinal cord (Hayle 1973a; Smeets 1982) and lateral line area (Boord and Northcutt 1979; Smeets 1982). The tectum can be interpreted therefore as a somatosensory correlation area, a designation compatible with the fact that it sends ascending efferents to the thalamus (Smeets 1981b). However, the tectum also receives direct projections from several other cell masses such as the pallium dorsale pars centralis lateralis of the telencephalon (Ebbesson 1972; Smeets 1982) and the functional significance of this connection is entirely unknown. Furthermore, the tectum sends descending efferents to the rhombencephalon (Smeets 1981 b) which, via the reticular formation, impinge upon somatomotor centres in the spinal cord. Because of these connections it is equally possible to consider the tectum as a higher-order somatomotor centre.

It is noteworthy that of the centres situated in the lateral part of the tegmentum mesencephali the nucleus tegmentalis lateralis receives fibres from the spinal lemniscus (Hayle 1973 a) and tectum (Smeets 1981 b), whereas the dorsolateral tegmental area is the end station of a primordial lateral lemniscus (Wallenberg 1907; Boord and Northcutt 1979, 1982). A nucleus intercollicularis, first described in *Raja eglanteria* (Northcutt 1978), could be delineated here only in *Scyliorhinus*. Although the pathways of the efferents of these nuclei are unknown—except for a spinal projection (Smeets and Timerick 1981)—it may be suggested that together these nuclei constitute a zone of somatosensory relay centres.

6.3 Centres and Fibre Tracts

6.3.1 The Afferent Components of the Nervus Trigeminus and Their Centres of Termination

The sensory fibres of the trigeminal nerve are carried by all three peripheral branches (maxillary, ophthalmic and mandibular). In *Scyliorhinus* this nerve contains around 11,000 fibres (Roberts and Witkovsky 1975). In the brain stem they can be divided into three pathways: (a) the tractus descendens nervi trigemini; (b) a number of fibres which pass medially and terminate at the level of entrance of the nerve; and (c) the radix mesencephalicus nervi trigemini (Fig. 6-1).

A considerable portion of the entering fibres of V constitutes a superficially situated tractus descendens which gradually shifts during its caudal course from a lateral to a dorsolateral position and finally enters the spinal cord. Wallenberg (1907), using the Marchi method, could trace the descending trigeminal tract to the caudal border of the rhombencephalon. On the basis of normal material Ariëns Kappers et al. (1936) stated that somatosensory branches of VII, IX and X separate off from the rest of the centripetal fibres of these nerves and join the spinal trigeminal tract.

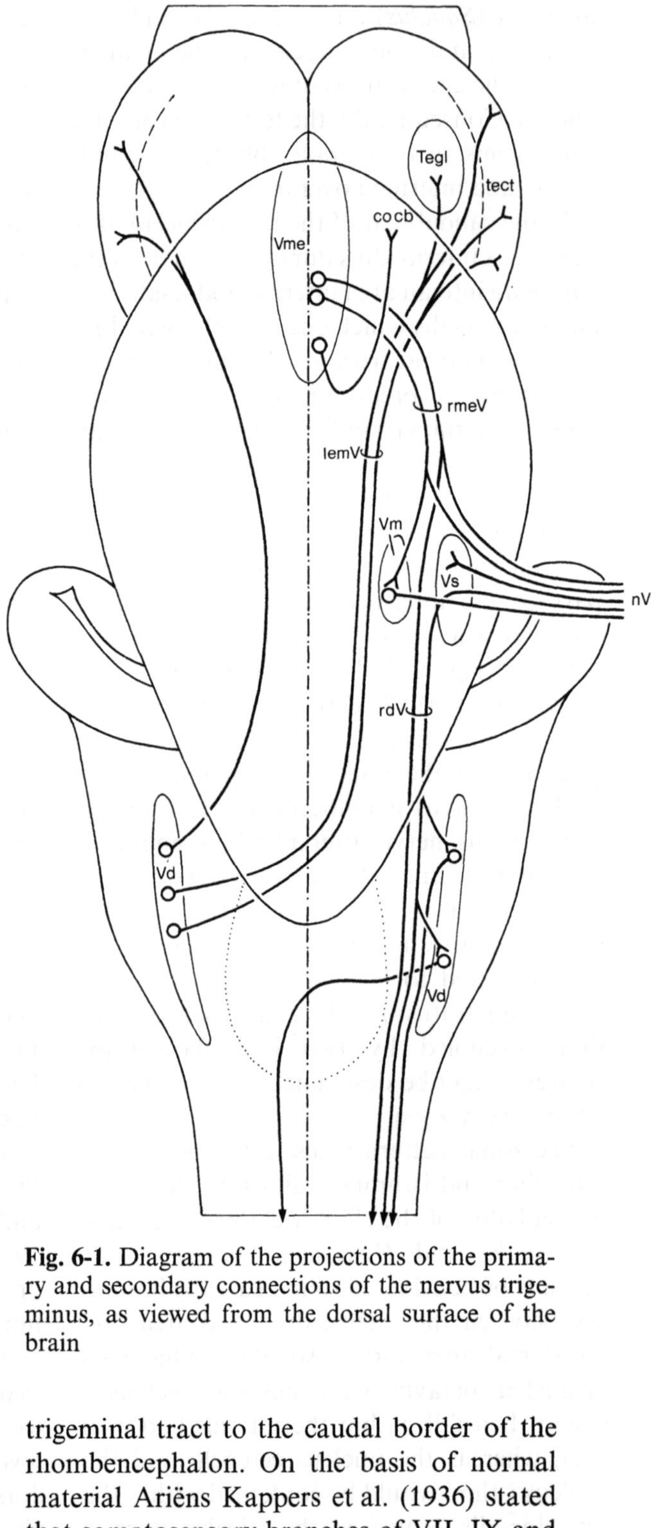

Fig. 6-1. Diagram of the projections of the primary and secondary connections of the nervus trigeminus, as viewed from the dorsal surface of the brain

According to Smeets and Nieuwenhuys (1976) the nucleus of termination of the tractus descendens of V consists of medium-sized cells that are distributed between the fibres throughout the length of the tract. The cells are diffusely scattered in the most rostral and caudal parts of the tractus descendens, but the neurons in the intermediate part are sufficiently compacted to form a nucleus. A corresponding nucleus has been observed by Wallenberg (1907), Papez (1929) and Ariëns Kappers et al. (1936). The nucleus tractus descendens nervi trigemini of *Hydrolagus* differs from that of the other species examined in consisting of rather small cells which extend into the "cervical" cord.

The efferent connections of the nucleus tractus descendens nervi trigemini comprise (a) "reflex" fibres and (b) fibres ascending to higher levels of the brain. According to Wallenberg (1907) and Ariëns Kappers et al. (1936) the "reflex" fibres pass from the nucleus of the descending root to the ventral horns of the spinal cord and to the efferent nuclei of the rhombencephalon. Recently Smeets and Timerick (1981) found labelled neurons in the nucleus of the descending trigeminal tract after HRP injections in the spinal cord of *Scyliorhinus* and *Raja* and showed that the spinal projection of the descending trigeminal tract reaches at least as far as the twentieth spinal segment. Fibres ascending from the nucleus descendens of V to higher levels of the brain have been described by Wallenberg (1907) in *Scyliorhinus stellaris*, on the basis of the Marchi technique. After a lesion of the rostral part of this nucleus fibres could be traced to the contralateral tectum mesencephali, nucleus tegmentalis lateralis ("lateraler Teil des Mittelhirnhaubenkernes") and, for a very small part, to the caudal part of the thalamus region. In a recent HRP study (Smeets 1982) in *Scyliorhinus* and *Raja* labelled neurons were found in the nucleus tractus descendens nervi trigemini following tectal HRP injections. This tectal projection is bilateral.

Although most primary afferent fibres enter the tractus descendens, an appreciable number terminates at the level of entrance of nerve V, where there is a cell mass consisting of small cells, situated lateral to the motor nucleus of V. In *Squalus*, this cell mass occupies a periventricular position and was termed nucleus C by Smeets and Nieuwenhuys (1976). It probably corresponds to the diffuse cell mass, which van Hoevell (1916, *Cetorhinus*) labelled in his Figs. 3 and 4 as the sensory nucleus of V and which we have now found in *Scyliorhinus*, *Raja* and *Hydrolagus*. However, in *Raja* this nucleus could not be delineated from the surrounding grey matter.

The fibres which constitute the tractus mesencephalicus nervi trigemini originate from a nucleus of the same name which consists of a continuous dorsomedian strip of very conspicuous cells that extends throughout the length of the tectum. These very large neurons deviate considerably from the general description of the elements of that category as presented in Chap. 4 with regard to shape and distribution of the Nissl substance. They are spherical or ovoid in shape and have fine granules of Nissl substance distributed throughout the cytoplasm, with large particles located at the periphery. Many of the cells make contact with the ventricular cerebrospinal fluid and also have a close relationship with the tectal blood supply (Klatzo 1967; Witkovsky and Roberts 1975; MacDonnell 1980a); the cell bodies lie immediately adjacent to large blood vessel walls or have capillaries entwined about them. In the anterior and posterior parts of the nucleus some neurons lie away from the main cluster, but they are also always closely associated with large blood vessels. Other large cells of the brain stem of elasmobranchs do not display a comparable association with blood vessels.

In the cell clusters of this nucleus gap junctions are formed where these neurons contact each other directly. Axons from deep tectal cells and collaterals from commissural fibres give synapses on the perikarya (Witkovsky and Roberts 1975, 1976).

The tractus mesencephalicus nervi trigemini, formed by the major processes of the mesencephalic trigeminal neurons, has been studied by Johnston (1905), Weinberg (1928), and, in more recent years, by Witkovsky and Roberts (1975) and Roberts and Witkovsky (1975). The stout axons of the mesencephalic trigeminal neurons join the mesencephalic trigeminal tract which courses laterally then posteriorly and ventrally along the outer margin of the periventricular zone. This bundle divides into two at the level of the tectocerebellar border, in the velum medullare anterius. The more medial bundle (containing approximately 15% of the axons) enters the cerebellum via the ventral granular eminences of the anterior lobe, whereas the more lateral bundle turns to the edge of the ventricle and then courses posteriorly to enter the rhombencephalon (Fig. 6-1). Experimental physiological confirmation of the cerebellar projection was obtained by recording from neurons in the posterior part of the mesencephalic trigeminal nucleus while the cerebellum was stimulated (Roberts and Witkovsky 1975). The exact cerebellar termination site of these fibres has not been determined, though most of them end in the anterior lobe in close proximity to the layer of Purkinje cell bodies. At the anterior end of the rhombencephalon, the rhombencephalic component of the radix mesencephalica is seen as a compact bundle of fibres, situated close to the lateral edge of the ventricle. As this bundle passes caudally it shifts gradually to a position ventral to the lateral groove of the ventricle. At the level of the motor nucleus of nerve V, mesencephalic V axons run medially and ventrally, crossing the peripherally extending dendrites of trigeminal motoneurons. Roberts and Witkovsky (1975) presented electrophysiological evidence for a monosynaptic mesencephalic V-motor V connection and showed that many mesencephalic V axons leave the brain in the trigeminal nerve. Some mesencephalic V fibres, degenerating as a result of large tectal lesions, were found by Smeets (unpublished) to leave the brain with the trigeminal nerve, while others appeared to terminate at the level of the motor V nucleus, while still others entered the descending trigeminal tract and could be traced to "cervical" cord levels. Conversely, a recent experimental study with the HRP technique (Smeets and Timerick 1981) revealed labelled cells in the ipsilateral nucleus mesencephalicus nervi trigemini after HRP was injected into the spinal cord of *Scyliorhinus*. This study also showed that this spinal projection probably extends only to the "cervical" cord level, since more caudal injections in *Raja* did not label mesencephalic trigeminal neurons.

The organization of trigeminal afferents in elasmobranchs, and of their centres of termination, shows great similarities to the trigeminal system of other classes of vertebrates. Thus primary trigeminal afferents to the nucleus tractus descendens nervi trigemini and to the "cervical" cord level have been experimentally confirmed in several other groups of anamniotes: lampreys (Northcutt 1979b), teleosts (Luiten 1975), and anurans (Joseph et al. 1968; Matesz and Székely 1978). However, a distinct projection to the sensory nucleus of the trigeminal nerve has only been identified in anurans.

Mesencephalic trigeminal connections have been experimentally studied, so far, in frogs (Rubinson 1970; Matesz and Székely 1978), reptiles (Goldstein and Jacobs 1969; Ebbesson 1981; Dacey 1982), birds (Azzena et al. 1970; Bortolami et al. 1972), and mammals (Panneton and Martin 1979; Somana et al. 1980; Matesz 1981). A projection of the mesencephalic trigeminal nucleus to the motor V nucleus has been recognized in frog (Rubinson 1970; Matesz and Székely 1978), reptiles (Ebbesson 1981; Dacey 1982), and mammals (Panneton and Martin 1979; Matesz 1981) and mesencephalic trigeminal projections to the "cervical" cord level have been reported in frog (Rubinson 1970; Matesz and Székely 1978), reptiles (Ebbesson

1981; Dacey 1982) and rat (Matesz 1981). A connection between the mesencephalic trigeminal nucleus and the cerebellum has been experimentally confirmed so far only in birds (Azzena et al. 1970; Bortolami et al. 1972) and the cat (Brodal and Saugstad 1965; Somana et al. 1980). It is noteworthy that the neurons which project to the cerebellum in the cat are also restricted to the caudal part of the mesencephalic trigeminal nucleus. According to Somana et al. (1980), the mesencephalic trigeminal neurons of the cat project only to the vermal region.

A comparative anatomical study of the mesencephalic part of the fifth nerve has revealed the constancy of its organization throughout the jawed vertebrates (Weinberg 1928). In mammals the neurons of the mesencephalic trigeminal nucleus form part of the neural pathway which initiates the "jaw-jerk" reflex (Harrison and Corbin 1942), a rapid jaw-closure produced by contraction of the jaw adductor musculature. In mammals, birds and reptiles mesencephalic trigeminal fibres arise in part from muscle spindles in jaw-closing muscles (Szentagothai 1948; Manni et al. 1965) as well as from tactile receptors around the teeth, but in chondrichthyans only tactile units appear to be involved for muscle spindles are absent from the jaw musculature. The neurons of the mesencephalic trigeminal nucleus of chondrichthyan fish also differ from those of amniotes in other ways: they are larger, have more processes, show more extensive extranuclear interconnections, and are closely related to blood vessels (Klatzo 1967; Witkovsky and Roberts 1975, 1976; MacDonnell 1980a, b). Furthermore, the mesencephalic trigeminal nucleus of amniotes is nowhere contiguous with the ventricular system, whereas in chondrichthyans there are a considerable number of CSF-contacting neurons in the nucleus (MacDonnell 1980b; Kemali and Miralto 1979).

6.3.2 Visceral Afferent Centres and Tracts

The visceral sensory fibres of the VIIth, IXth and Xth nerves terminate in a longitudinal zone situated in the caudal rhombencephalon, just dorsal to the sulcus limitans. This zone constitutes a longitudinal protrusion of the lateral wall of the fourth ventricle, known as the viscerosensory column or lobus vagi (Houser 1901). Many of the primary viscerosensory fibres end at their level of entrance, but a certain portion descends in the lobe and constitutes the fasciculus solitarius (or fasciculus communis). At caudal rhombencephalic levels the number of axons in the fasciculus solitarius declines as they terminate on cells of the surrounding nucleus fasciculi solitarii. This nucleus could not be distinguished as a separate cell mass in the species studied though in some sections an accumulation of cells around the solitary tract could be recognized. Only a relatively small solitary tract remains at the level of the obex. Slightly more caudally, at the commissura infima (Halleri), the fascicles of both sides fuse and exchange fibres which then ascend in the contralateral fascicle (Ariëns Kappers et al. 1936).

The lobus vagi consists almost entirely of small cells that vary in density throughout the nucleus. As in other anamniotes the viscerosensory area in chondrichthyans cannot be divided into separate nuclei that are specific for taste or general visceral sensibility.

The efferent connections of the viscerosensory area are still obscure. It is known in teleosts (Finger 1978a; Morita et al. 1980) and amphibia (Herrick 1948; Larsell 1923; Barnard 1936) that the efferents of the viscerosensory area ascend to a centre situated in the most rostral part of the rhombencephalon, which is termed nucleus visceralis secundarius because of its specific input. We were unable to identify this ascending viscerosensory pathway in our material, but a cell mass (nucleus G) was identified in the isthmus region, which, because of its position, may well represent the nucleus visceralis secundarius. Nucleus G will be discussed in more detail later (Sect. 6.3.10).

It is noteworthy that Wallenberg (1907) in *Scyliorhinus* has traced fibres to the "cervical" cord after lesions of the viscerosensory area. These descending fibres may be equivalent to the tractus gustatorius secundarius descendens of teleosts.

6.3.3 The Octavolateral Area and Its Connections

In preceding sections we have described the terminal centres of the viscerosensory and general somatosensory components of the cranial nerves. The special somatosensory labyrinthine and lateral-line systems dominate the remaining parts of the grey matter of the rhombencephalic alar plate.

The Sense Organs of the Lateral Line System. The lateral line sensory system is well developed in cartilaginous fishes. In accordance with the terminology of Dijkgraaf (1963), the component sense organs can be divided into ordinary and ampullary categories. The former includes the canal and pit organs of the head and body, while the ampullary group, restricted in cartilaginous fishes to the head, is represented by the ampullae of Lorenzini.

The ordinary lateral-line component comprises the neuromasts that are found in a series of superficially located canals, and the pit organs (or "free neuromasts") that are spread over the body surface between pairs of modified scales. Lateral-line neuromasts are sensitive to external water displacements, whereas the pit organs are activated by both mechano- and chemo-stimulation. The biological significance of these sensitivities still remains unclear (see Roberts 1978, 1981b).

The ampullae of Lorenzini consist of jelly-filled canals that end blindly in subcutaneous tissue. Each ampulla is made up of a number of alveoli, the walls of which contain the sensory epithelium, innervated by a small fascicle of nerve fibres. Although ampullae of Lorenzini respond to thermal, mechanical, chemical and electrical stimuli there is now good evidence that they are electroreceptors (Kalmijn 1971, 1974, 1978), highly sensitive to very weak electric fields that originate from external animate or inanimate sources. Recently the electrosensory nature of the Lorenzinian ampullae of *Hydrolagus* has also been established (Fields and Lange 1980).

The Peripheral Lateral-Line Nerves. Adjacent to the brain the lateral-line nerves are associated closely with cranial nerves V, VII and X and have classically been considered to be branches of these nerves. However, they nei-

Fig. 6-2. A Topographic relationships of the various rami of the anterior lateral-line nerve to the Vth, VIIth and VIIIth cranial nerves in *Mustelus canis* from a dorsolateral view of the brain stem. **B** Buccal and otic rami of the anterior lateral-line nerve are removed to show the course of the dorsal and ventral roots of the superficial ophthalmic and external mandibular branches. *Circles* indicate positions of the ganglia. *ALL* anterior lateral-line lobe; *b* buccal ramus of anterior lateral-line nerve; *Gb* ganglion of the buccal ramus; *Gg* geniculate ganglion; *Gme* ganglion of the external mandibular ramus of the anterior lateral-line nerve; *God* ganglion of dorsal root of otic ramus of the anterior lateral-line nerve; *Gos* ganglion of superficial ophthalmic ramus of anterior lateral-line nerve; *Gov* ganglion of the ventral root of the otic ramus of the anterior lateral-line nerve; *h* hyomandibular trunk; *NLLp* posterior lateral-line nerve; *o* otic ramus of the anterior lateral-line nerve; *os* superficial ophthalmic ramus of anterior lateral-line nerve; *Rmed* dorsal root of external mandibular ramus of anterior lateral-line nerve; *Rmev* ventral ramus of external mandibular ramus of anterior lateral-line nerve; *Rosd* dorsal root of superficial ophthalmic ramus of anterior lateral-line nerve; *Rosv* ventral root of superficial ophthalmic ramus of anterior lateral-line nerve; *Rov* ventral root of otic ramus of anterior lateral-line nerve; *Vm* mandibular ramus of V; *Vmx* maxillary ramus of V; *Vop* deep ophthalmic ramus of V; *Vos* superficial ophthalmic ramus of V; *RVIIvs* visceral sensory root of VII; *VIIvs* visceral sensory ramus of VII; *VIII* statoacoustic nerve; *IX* glossopharyngeal nerve; *X* vagal nerve. (After McCready and Boord 1976)

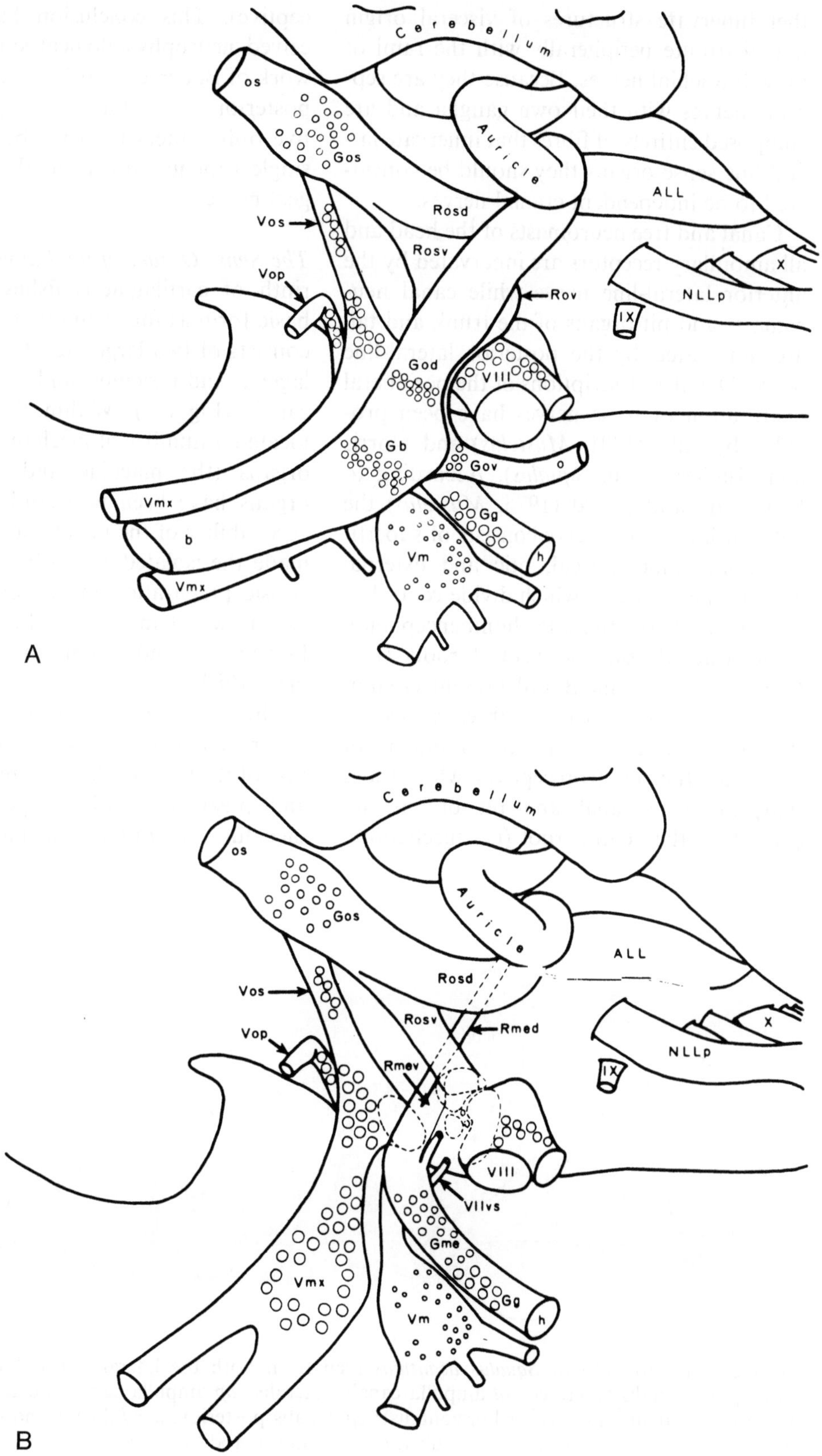
Cerebellum
os
Gos
Auricle
ALL
Rosd
Vos
Rosv
X
Vop
Rov
NLLp
IX
God
VIII
Gb
Gov
o
Vmx
Gg
b
Vm
h
Vmx
A
Cerebellum
os
Gos
Auricle
ALL
Rosd
Vos
Rosv
Rmed
X
Vop
NLLp
IX
Rmev
VIII
VIIvs
Gme
Vmx
Gg
Vm
h
B

ther innervate structures of visceral origin nor distribute peripherally with the rami of these branchial nerves. Because they are separate nerves with their own ganglia and are composed entirely of fibres that innervate lateral-line sense organs they should be considered to be independent cranial nerves.

Canal and free neuromasts of the head and all ampullary receptors are innervated by the anterior lateral-line nerve while canal neuromasts and pit organs of the trunk and tail are innervated by the posterior lateral-line nerve. Detailed descriptions of the peripheral distribution of these nerves have been provided by Allis (1901, *Mustelus*) and Norris and Hughes (1920, *Squalus*). According to McCready and Boord (1976, *Mustelus*) the anterior lateral-line nerve consists of superficial ophthalmic, buccal, otic and external mandibular branches which divide central to the ganglion to enter the rhombencephalon as separate dorsal and ventral roots (Fig. 6-2). McCready and Boord present circumstantial evidence that those fibres carried by the dorsal root innervate the ampullae of Lorenzini (i.e. electroreceptive) while those that innervate canal and pit organs are carried by the ventral root (i.e. mechanoreceptive). This conclusion has recently received neurophysiological support from the work of Bodznick and Northcutt (1980). The posterior lateral-line nerve (mechanoreceptive only) enters the rhombencephalon as a single root just dorsal to the glossopharyngeal nerve.

The Sense Organs of the Labyrinth. The labyrinth of cartilaginous fishes has the same basic form as found in other vertebrates and consists of two large sacs, the sacculus (with lagena) and utriculus, and three semicircular canals (Fig. 6-3). Within the labyrinth are located a number of mechanoreceptive sense organs (the maculae and cristae). These organs have been well studied because the accessibility of the nerve and end organs has made the isolated elasmobranch labyrinth a classic preparation in vestibular physiology (see Lowenstein 1974, O'Leary et al. 1976; Lowenstein and Compton 1978; Montgomery 1980).

The three semicircular canals form a transducer system that detects angular acceleration of the head while the detection of gravity and linear acceleration is performed by the sacculus, utriculus and lagena. Vibrations are

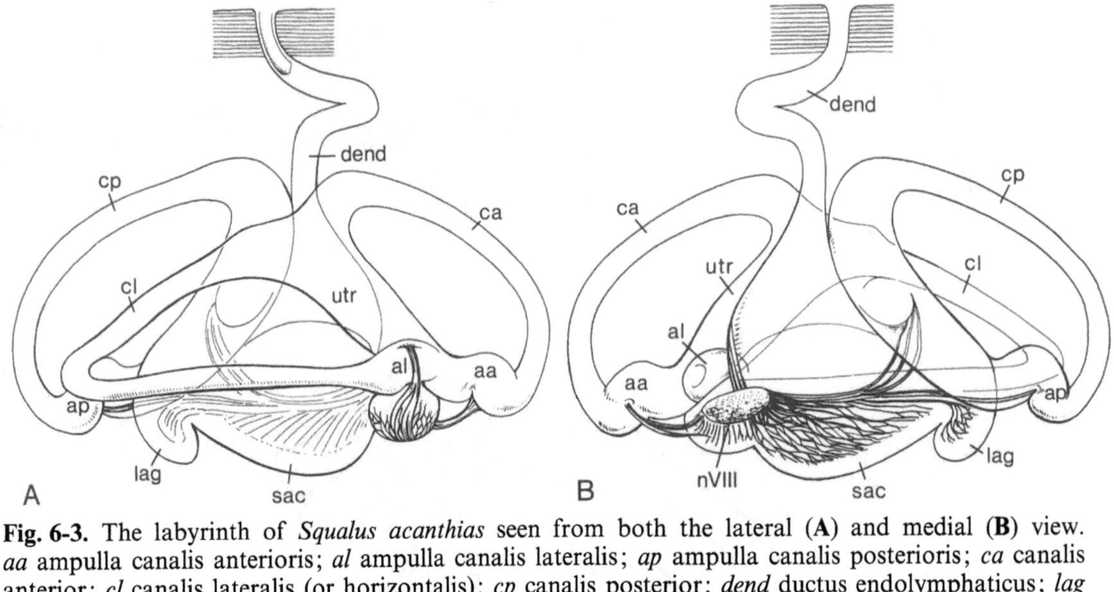

Fig. 6-3. The labyrinth of *Squalus acanthias* seen from both the lateral (**A**) and medial (**B**) view. *aa* ampulla canalis anterioris; *al* ampulla canalis lateralis; *ap* ampulla canalis posterioris; *ca* canalis anterior; *cl* canalis lateralis (or horizontalis); *cp* canalis posterior; *dend* ductus endolymphaticus; *lag* lagena; *nVIII* nervus octavus; *sac* sacculus; *utr* utriculus. (Modified from Retzius 1881)

detected by sensory areas in the utriculus, sacculus and, particularly, the macula neglecta.

When the head is stationary most of the semicircular canal nerve fibres discharge very steadily. This activity increases during ipsilateral acceleration in the canal's plane and declines during contralateral acceleration (Lowenstein 1974; Montgomery 1980). Afferent fibres from the sense organs in the utriculus, sacculus and lagena measure change in static position by changing their firing rate according to the position of the head (Lowenstein and Roberts 1949). Afferent fibres from parts of the maculae sacculi, utriculi and neglecta respond to vibration. Recordings of microphonic potentials from the labyrinth and of compound potentials and single unit activity from nerve VIII all suggest that the maculae sacculi and neglecta play important roles in sound detection (Lowenstein and Roberts 1951; Fay et al. 1974; Corwin 1981a).

The Octavus Nerve. The various end organs of the labyrinth are innervated by the two branches of the octavus nerve. Fibres to the crista of the posterior vertical canal and the maculae of the sacculus, lagena and neglecta run in the posterior ramus of nerve VIII while the anterior ramus is composed of fibres supplying the cristae of the anterior vertical and horizontal canals, and the macula utriculi (Retzius 1881). Early in development, when these rami are separate, the posterior ramus is seen to enter the rhombencephalon at a level ventral to the entry of the anterior ramus. In the adult, however, these rami intermingle and cannot be separated (Boord and Roberts 1980).

Central Projections of the Octavolateralis System. The termination zones of the VIIIth and the lateral-line nerves occupy a large portion of the lateral rhombencephalic wall of cartilaginous fishes. According to Schepman (1918), Ariëns Kappers (1920) and Ariëns Kappers et al. (1936) this region contains three nuclei: the nucleus dorsalis, the nucleus medialis and the nucleus ventralis. We now know that the first two nuclei constitute the end stations of the lateral-line nerves, whereas the nucleus ventralis receives octavus fibres. In recent years several authors have recognized subdivisions within the ventral zone and have given these regions a variety of names. In the present work we shall retain the name nucleus dorsalis (with the addition areae octavolateralis); the nucleus medialis we shall call the nucleus intermedius areae octavolateralis, following Smeets and Nieuwenhuys (1976). The large neurons of the nucleus ventralis constitute a cell mass designated here as the nucleus octavus magnocellularis. Four additional nuclei can be recognized in the four species studied for this work. One, which extends along the ascending branches of VIIIth nerve fibres, was called the nucleus vestibularis superior by Smeets and Nieuwenhuys (1976) and anterior octavus nucleus by Northcutt (1980); it will now be called the nucleus octavus ascendens. Another nucleus extends along the descending branches of the octavus nerve and was called the ventral vestibular nucleus by Montgomery (1977) and Montgomery and Roberts (1979). We shall follow Northcutt's terminology (1980) and call this the nucleus octavus descendens. Another nucleus can be recognized just rostral to the entry of the ventral branch of the anterior lateral-line nerve (Nucleus C2) and in *Raja* and *Hydrolagus*, but not in *Squalus* or *Scyliorhinus*, an additional neuronal cluster lies at the level of entry of the posterior lateral line nerve (Nucleus Cl, Fig. 6-4). We believe these two nuclei are equivalent to cell plate "X" of Northcutt (1978).

Nuclei of the Lateral-Line Nerves. Both lateral-line nuclei are covered dorsally by a highly characteristic layer, the crista cerebellaris (cerebellar crest), which closely resembles the molecular layer of the cerebellum with which it is contiguous rostrally. The cerebellar crest comprises parallel-running unmyelinated

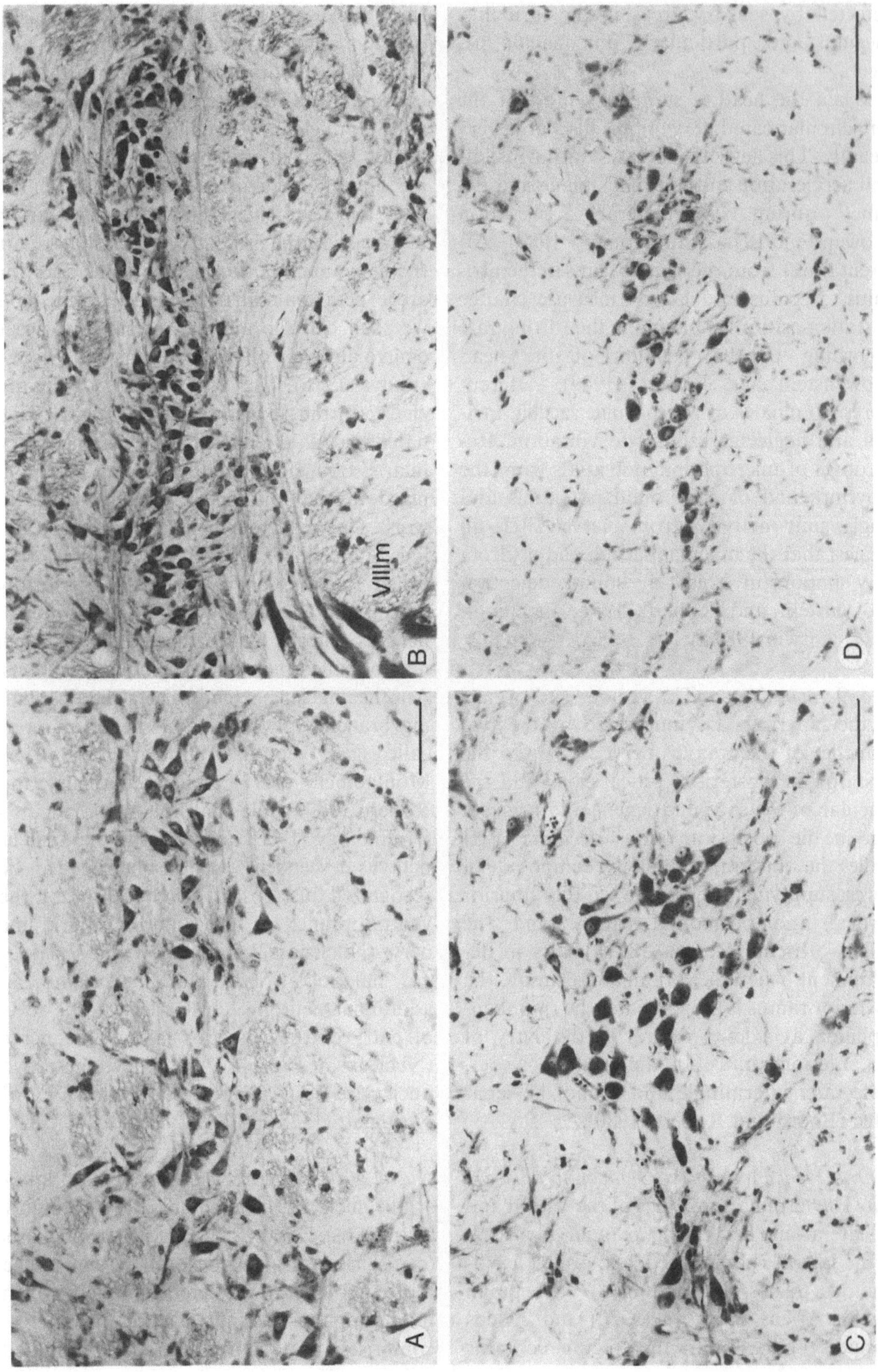
A
B
C
D
VIIIm

axons which arise from granule cells of the auricles (Boord 1977; Paul and Roberts 1977a; Montgomery 1981).

The crista cerebellaris lateral, caudal and caudomedial to the nucleus dorsalis, lies immediately against the ependyma of the fourth ventricle. This subependymal part of the cerebellar crest has been included in the topological charts.

The nucleus intermedius is covered by the lateral part of the cerebellar crest and, since by far its largest diameter is in the direction perpendicular to the ventricular surface, the size of its projection in the topological charts (Figs. 2, 36, 67, 101) is rather small. At the caudal limit of nucleus intermedius in *Platyrhinoidis* Northcutt (1980) recognizes a caudal nucleus, as found in Osteichthyes (McCormick 1981).

The nucleus dorsalis is very large and occupies, as its name implies, the most dorsal part of the lateral wall of the rhombencephalon. In most sharks (Larsell 1967) it shifts during development from a dorsolateral to a dorsomedial position in relation to the nucleus intermedius. However, in *Hexanchus,* a primitive species, the nucleus dorsalis retains a dorsolateral position (Ariëns Kappers 1906). In *Raja*, where the nucleus dorsalis lies dorsal to the nucleus intermedius, the two nuclei are covered by individual cerebellar crests separated by a deep external groove (Figs. 75, 76). In *Hydrolagus*, the dorsal nuclei fuse across the midline (Figs. 110, 111) but no fibres have been seen crossing to the contralateral side.

The dorsal and intermediate nuclei have the same general cellular organization and contain medium-sized cells which are multipolar and resemble cerebellar Purkinje cells in form (Fig. 6-5A). They have an elongated cell body, which grades into a slender dendritic trunk. Some of the dendrites interweave with afferent lateral-line fibres while other spiny dendrites penetrate the cerebellar crest. Unlike those of cerebellar Purkinje cells, however, the dendrites lying within the cerebellar crest are not flattened in a plane orthogonal to the parallel fibres. Stimulation of the overlying parallel fibres has an excitatory action on the principal neurons of the intermediate nucleus, followed by a later inhibition, presumably mediated through stellate-cell action (Montgomery 1981). Parallel fibres pass between the dendrites and form crossing-over synapses on the spines (Paul et al. 1977).

Interneurons of the stellate type (Fig. 6-5B) are relatively abundant in the cerebellar crest. They can be divided into two groups on the basis of their dendritic arborizations. In one type, found mostly at the top and bottom of the crest, there is usually a single primary dendrite from which all others arise. In the other, situated in the middle of this layer, primary dendrites arise from all over the cell body and ramify dorsally and ventrally, running parallel with the dendrites of the large Purkinje-like cells. It has not been possible to trace with certainty the projection of the axons of these cells, but they pass amongst the dendrites of the large Purkinje-like cells and presumably make synaptic connections with them (Paul et al. 1977).

The experimental studies of Boord and Campbell (1977) on *Mustelus* and of Koester and Boord (1978) on *Raja* revealed that the dorsal root of the anterior lateral line nerve enters the nucleus dorsalis and bifurcates to form ascending and descending bundles. The fibres of the ventral anterior lateral line nerve were shown to bifurcate into descending and ascending branches within the nucleus intermedius (Koester 1981a). The fibres of the posterior lateral-line nerve ascend within the territory of the nucleus intermedius but they run lateral to the fibres from the anterior lat-

Fig. 6-4. High power photomicrographs of transverse sections showing accumulations of cells in the octavolateralis area of *Raja* and *Hydrolagus:* **A** nucleus C1, *Raja*; **B** nucleus C2, *Raja*; **C** nucleus C1, *Hydrolagus*; **D** nucleus C2, *Hydrolagus. Bar scale* 100 μm

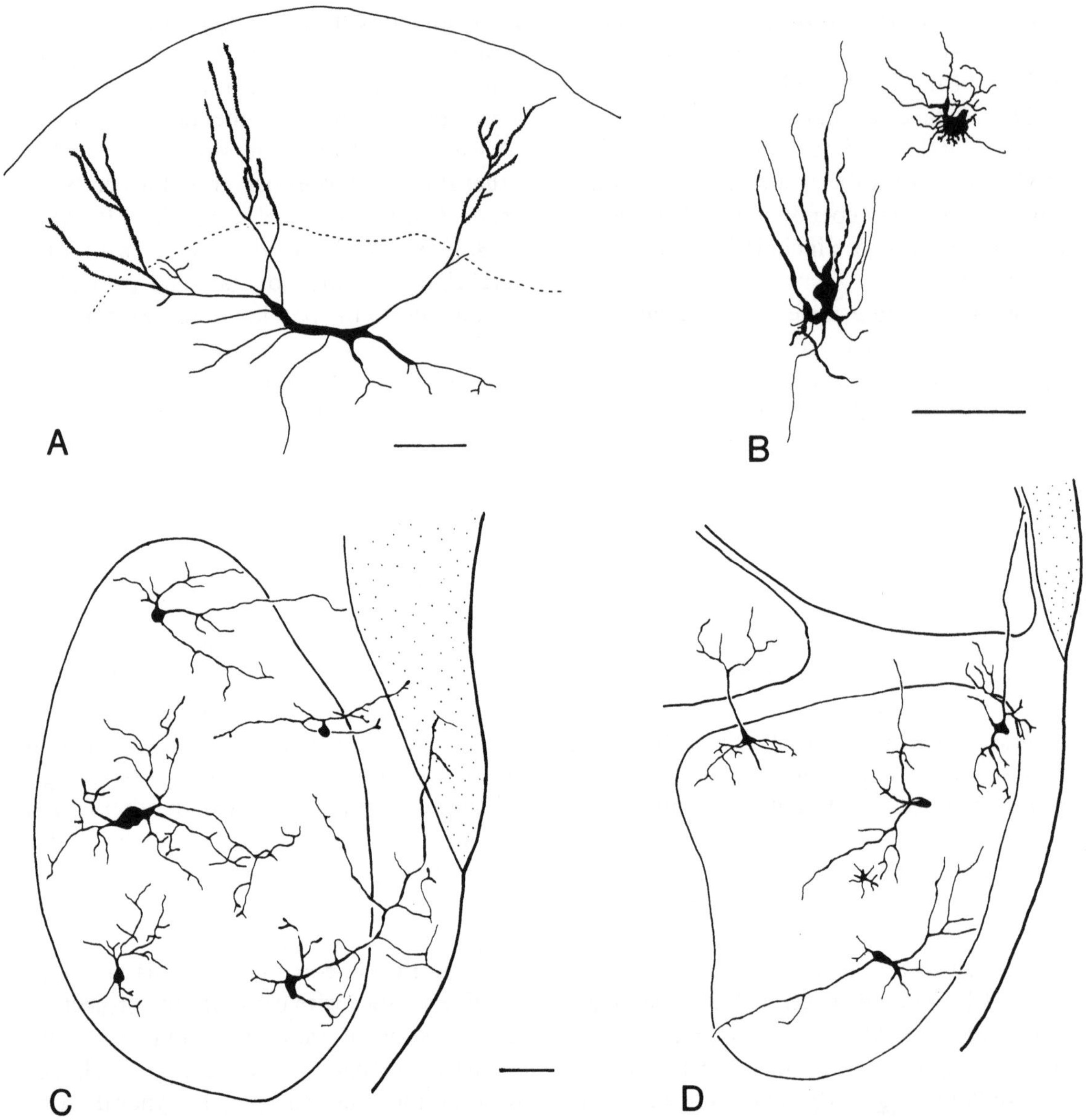

Fig. 6-5. A, B Camera lucida drawings of neurons of the nucleus intermedius areae octavolateralis of *Scyliorhinus*. **A** principal neuron; **B** stellate interneurons. Golgi impregnation, transverse plane. **C, D** Camera lucida drawings of neurons of the nucleus octavus ascendens (**C**) and nucleus octavus descendens (**D**) of *Scyliorhinus*. Golgi impregnations, transverse plane. (After Montgomery and Roberts 1979). *Bar scale* = 100 μm

eral-line nerve. Koester (1981 b) reports that some mechanoreceptive lateral line fibres terminate on nucleus octavus magnocellularis. The fibres from both lateral-line nerves terminate about the cells that lie among the ascending and descending tracts; there is no evidence of degeneration in the crista cerebellaris. Using electrophysiological methods Paul and Roberts (1977 b) confirmed the projection of the dorsal root of the anterior lateral line nerve to the dorsal nucleus and of the posterior nerve to the intermediate nucleus. They suggested that the incoming lateral line fibres make direct connections with the

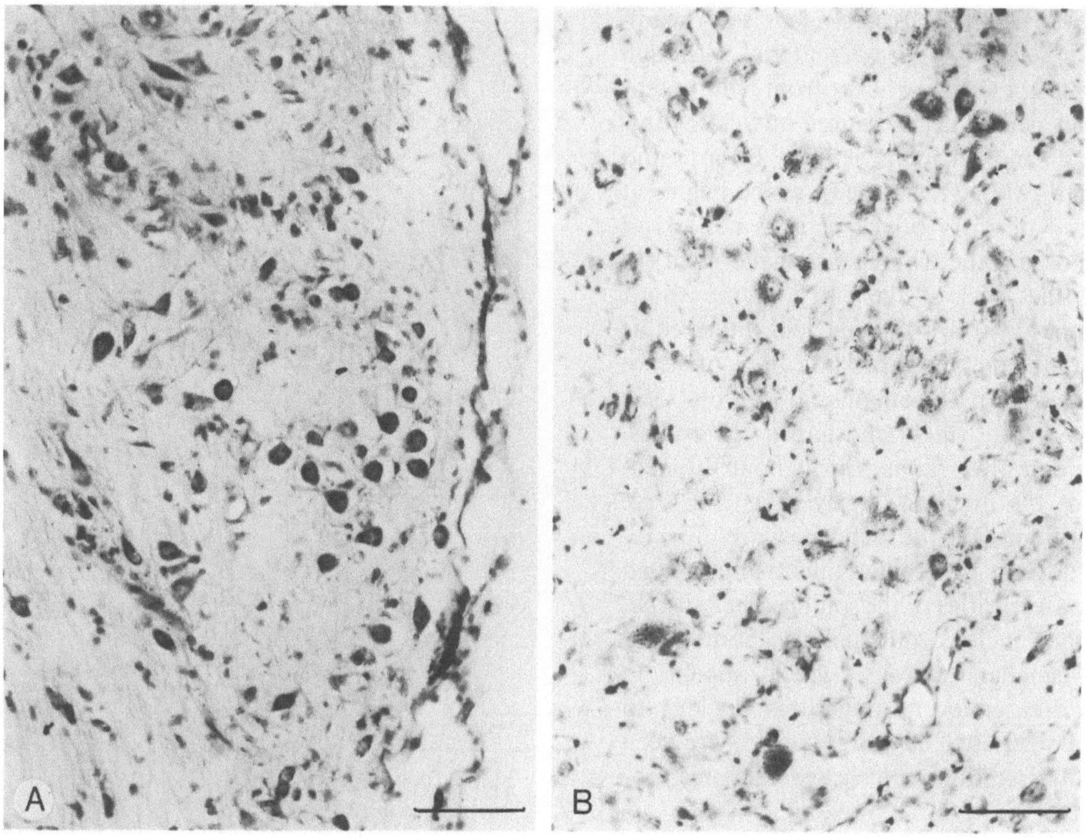

Fig. 6-6. High power photomicrographs of a transverse section showing the neurons of the nucleus octavus ascendens of *Raja* (**A**) and *Hydrolagus* (**B**). Nissl staining. *Bar scale* = 100 μm

ventral dendrites of the large Purkinje-like neurons and, possibly, with the subcrestal parts of the ascending dendrites.

As yet it is not possible to interpret the functional significance of the neuronal circuitry revealed by studies on the octavolateralis centres. There has been no systematic study of secondary neuron responses to natural mechanoreceptive lateral line stimulation, but studies on the responses of neurons to electric fields show that a low threshold and directional sensitivity are retained in these centres (Andrianov et al. 1974).

Koester and Boord (1978) and Koester (1981 b) have shown that some lateral-line afferent fibres pass through the lateral-line nuclei and project to the granule cells of the auricles. Mechanoreceptive fibres from the anterior and posterior lateral line nerves project to the more lateral portions of the lower leaf of the auricles, whereas the electroreceptive fibres take a more medial course.

Nuclei of the Octavus Nerve. The *nucleus octavus ascendens* is situated dorsolateral to the most rostral part of the sulcus limitans. Its posterior margin is adjacent to the entrance of nerve VIII and it extends rostrally as far as the posterior part of the cerebellar nucleus. In *Raja*, it consists of two parts, a more loosely arranged dorsal portion, which lies just beneath the cerebellar crest, and a more compact ventral portion (Fig. 6-6). According to Montgomery and Roberts (1979, *Scyliorhinus*) the neurons of this nucleus vary in form, although their irregularly shaped cell bodies are predominantly multipolar (Fig. 6-5 C).

The *nucleus octavus descendens* is not readily distinguished in normal material, although its limits are clearly evident from field potential and neural degeneration studies (Montgomery and Roberts 1979; Boord and Roberts 1980). This nucleus, which begins at the level of entry of the octavus nerve and descends in the rhombencephalic wall to about the level of the entry of the vagus nerve, lies ventral to the nucleus intermedius and lateral to the descending tract of nerve V (Fig. 6-5D). It contains large multipolar neurons, with irregularly shaped cell bodies, whose dendrites ramify throughout the neuropil of the nucleus, largely in a transverse plane. Some of the dendrites are not restricted to the descending nucleus but extend into the territories of the spinocerebellar tract and the nucleus intermedius. Some very much smaller neurons, which appear in transverse section through this nucleus to have a restricted dendritic spread, are seen in sagittal section to have several long processes.

In Osteichthyes, the most caudal part of this projection has been separated as a distinct cell mass, the nucleus caudalis (McCormick 1981). Northcutt (1980) and Koester (1981b) have recognized a similar division in elasmobranchs (*Platyrhinoidis, Raja*).

The *nucleus octavus magnocellularis* is situated at the level of entrance of the eighth nerve close to the descending trigeminal tract. It consists of very large bipolar and tripolar cells, although the neurons are smaller in the rostral and caudal parts of the nucleus. Several authors, among them Ariëns Kappers (1947), Gerlach (1947) and Larsell (1967), have considered this nucleus to be homologous to the nucleus of Deiters of amniotes. Nucleus D (*Hydrolagus*) may be considered to be a rostral continuation of the magnocellular nucleus.

The relations of the afferent fibres of nerve VIII to these nuclei have been recently investigated experimentally for *Scyliorhinus* (Montgomery 1978; Montgomery and Roberts 1979; Boord and Roberts 1980), for

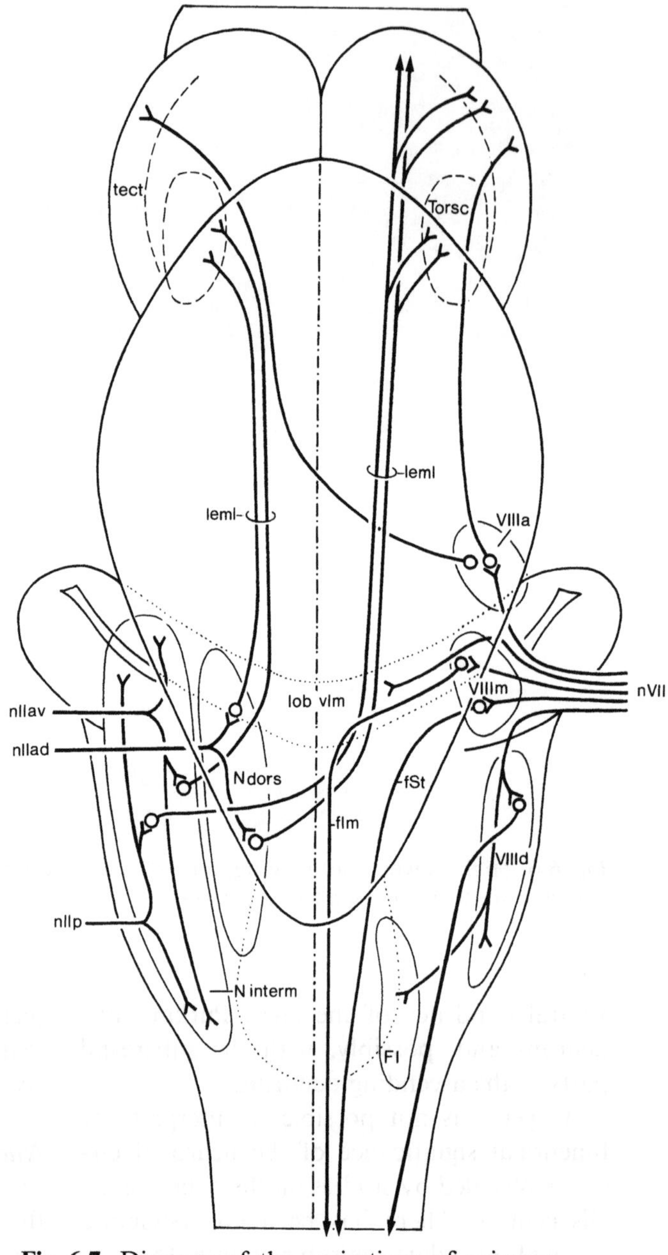

Fig. 6-7. Diagram of the projections of primary and secondary connections of the nervus lineae lateralis and nervus octavus, as viewed from the dorsal surface of the brain

Platyrhinoidis (Northcutt 1980) and for *Raja* (Koester 1981b). On entering the rhombencephalon, the fibre bundles divide into medial, ascending and descending branches (Fig. 6-7). Fibres of the medial pathway terminate about the soma and lateral dendrites of the neurons of the magnocellular nucleus.

Fibres in the descending branch terminate in the descending octavus nucleus while fibres of the ascending tract project to the ascending nucleus and to the granule cells of the medial part of the lobus vestibulolateralis of the cerebellum (lower lip, see p. 69). Boord and Roberts (1980) and Northcutt (1980) have also reported that some primary fibres emanate from fascicles of the descending pathway and form a large field of degenerating axons and terminals within the ventromedial part of the rhombencephalon. Some of these fibres seem to terminate in the region of the lateral funicular nucleus.

Electrophysiological recordings from neurons in the ascending and descending octavus nuclei have been made in response to horizontal head rotation in *Scyliorhinus* by Montgomery (1980). This study has shown that sinusoidal head rotation produces a smooth sinusoidal modulation of discharge frequency in the primary afferent fibres which is rather faithfully represented in the neuronal response of both the octavus nuclei and the vestibulocerebellum. Information from the contralateral horizontal canal is also reflected in the firing patterns of the octavus neurons.

It is unlikely that the projection of the octavus nerve to these brainstem nuclei relates directly to the different sense organs of the labyrinth because units responding to horizontal canal stimulation could be recorded in the ascending and descending nuclei, while field potentials produced throughout the region in response to separate stimulation of the anterior and posterior rami were essentially similar in form and distribution (Montgomery 1980).

The important question of whether there is a separate auditory nucleus in elasmobranchs has been recently examined by Corwin and Northcutt (1982), using 2-deoxyglucose autoradiography. With this procedure they showed that cell plate "X" (Nucleus C1 and C2 of the present study) and the ascending octavus nucleus become active in skates (*Platyrhinoidis*) exposed to auditory stimulation. Evoked potential recordings confirmed acoustic activity in these sites.

Mauthner Neurons in Cartilaginous Fish. A pair of large neurons (the Mauthner cells) that are evidently associated with nerve VIII and project to the spinal cord are found in many groups of fishes. These neurons are believed to mediate the first stage of the "startle response" evoked by sudden visual and vibratory stimuli (Eaton et al. 1981) and have a number of specialized features that distinguish them from other large nerve cells in the hindbrain. These include their location at the level of the root of nerve VIII; the presence of two principal dendrites (lateral and ventral); the large axon which decussates and then passes along the spinal cord; and, quite strikingly, the axon hillock surrounded by a peculiar neuropil forming the axon cap (Faber and Korn 1978). Although these criteria are unambiguous there has been considerable uncertainty whether cells with some of these features can be recognized in chondrichthyan fishes. Solger (1899) reported that Mauthner cells were present in *Chimaera* but Kuhlenbeck and Niimi (1969) were unable to confirm this. A pair of large neurons were observed by Larsell (1967) in *Squalus* embryo and these were shown by Bone (1977) to have some features typical of Mauthner cells; in particular they have large dendrites and crossing axons. These axons, however, are not larger than adjacent axons in the ventral fasciculi. Cells of this form are absent from adult *Squalus* and from both embryo and adult *Scyliorhinus* and there have been no reports of Mauthner cells in other elasmobranchs. Bone considered that the squaloid condition shows only the vestige of a neural apparatus that had been elaborated in some ancestral group.

In the present study a pair of Mauthner cells can be distinguished in *Hydrolagus* (Fig. 6-8). Their axons cross immediately caudal to the cell body and course within the fasciculus longitudinalis medialis to the spinal cord (Fig. 6-9D, fM).

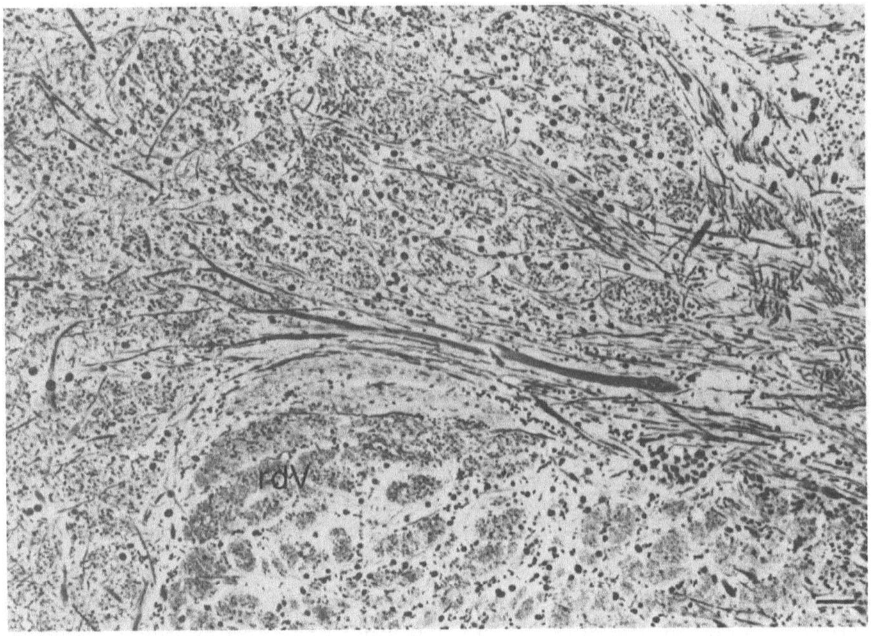

Fig. 6-8. Photomicrograph of a transverse section showing the Mauthner cell of *Hydrolagus collei*. Bodian staining. *Bar scale* = 100 µm

Secondary Connections of the Lateral-Line and Octavus Nuclei. Secondary connections of the lateral-line centres have been studied experimentally by Boord and Northcutt (1982) who charted in *Raja* the passage of proline injected into dorsal and intermediate nuclei and followed the degeneration produced by lesions to the anterior and posterior lateral-line areas. They found that efferent axons from each nucleus give rise to commissural components and to an ipsilateral and contralateral lemniscus. The commissural fibres project to the equivalent nucleus on the contralateral side. The lemnisci, ascending to the mesencephalon, terminate within the nucleus of the lateral-line lemniscus and a dorsolateral tegmental area. Although the lemnisci from both lateral-line nuclei run in parallel, the tract from the dorsal nucleus lies lateral to that from the intermediate nucleus. A substantial number of fibres from each nucleus terminates bilaterally within the central zone of the tectum (Fig. 6-7).

The secondary projections of the statoacoustic centres are presently less well known. Ariëns Kappers (1906) observed that some of the large cells of nucleus octavus magnocellularis give rise to a small but conspicuous bundle of large axons, the fasciculus medianus of Stieda (1873), which extends far caudal into the spinal cord (Figs. 6-7, 6-9). This bundle contains about 50 fibres in *Hydrolagus* and *Raja* and about 25 fibres in the sharks studied. This pathway, which is evident even in normal material, has recently been experimentally established by backfilling magnocellularis neurons with HRP placed in the spinal cord (Smeets and Timerick 1981). This procedure also demonstrated that some of the most rostrally located neurons in the descending octavus nucleus project to the spinal cord.

Metabolic mapping in the brain of *Platyrhinoidis* with ^{14}-C-2-deoxyglucose autoradiography and sound stimulation has indicated that acoustically related activity can be detected in a limited region of the intermediate nucleus (Nucleus C1 and C2, see p. 41), the ascending octaval nucleus, the nucleus of the lateral lemniscus and the ventromedial divi-

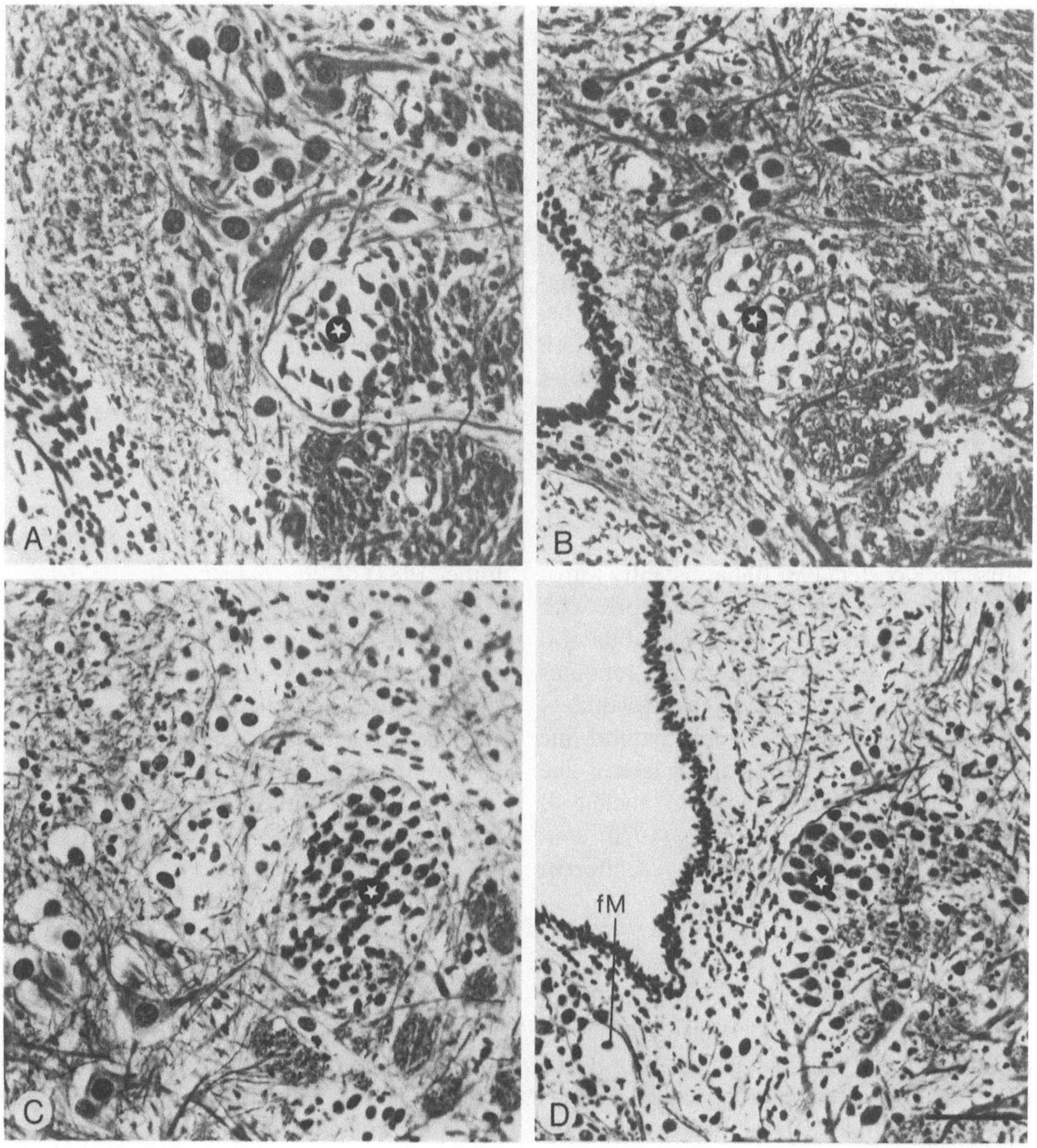

Fig. 6-9. Photomicrographs of transverse sections showing the fasciculus medianus of Stieda (asterisk) of *Squalus* (**A**), *Scyliorhinus* (**B**), *Raja* (**C**) and *Hydrolagus* (**D**). Bodian staining. *Bar scale* = 100 μm

sion of the lateral mesencephalic nucleus. Microelectrode recordings showed that evoked potentials could be obtained from these sites and from the reticular formation and parts of the auricles in response to acoustic stimulation (Corwin and Northcutt 1982).

Efferent Supply to the Lateral-Line and Labyrinthine Sense Organs. An important aspect of the innervation of many labyrinthine and lateral-line receptors is the efferent supply which originates in the brain and terminates on the peripheral sense organs. An efferent innervation has been reported for all of the various types of end organs in elasmobranchs, except for the pit organs and the receptors of the ampullae of Lorenzini. In most cases the efferent supply has an inhibitory impact on the hair cells (see Roberts 1978).

The location of the cell bodies has proved elusive although Paul and Roberts (1977d) reported that the efferent neurons from the anterior lateral line nerve resided in the dorsal nucleus. On several grounds this suggestion now seems unlikely, particularly because recent studies on other fishes with HRP have shown the cell bodies of efferent fibres for both lateral-line and labyrinthine receptors to cluster close to the motor columns of the rhombencephalon (Szabo and Libouban 1979; Claas and Münz 1980; Strutz et al. 1980; McCormick 1981).

In *Raja* Koester (1981b) has recently retrogradely labelled these efferent neurons with HRP and shown them to lie in a bilateral column, dorsolateral to the medial reticular formation, that extends from the entrance of the ventral root of the anterior lateral-line nerve to the entry of the posterior lateral-line nerve. This position is medial to nucleus B which Smeets and Nieuwenhuys (1976) considered to be a possible nucleus of efferent neurons.

Functional Commentary. A substantial portion of the wall of the rhombencephalon of cartilaginous fishes is occupied by the termination sites of the octavus and lateral-line nerves. This region, commonly called the statoacoustic centre, has been regarded as a homogenous zone. Indeed the existence of this common centre has been quoted as evidence for the view that lateral-line and labyrinthine senses are phylogenetically related (van Bergeijk 1966). However it is now clear that, for the most part, the lateral-line and octavus nerves project to separate, although adjacent, portions of the rhombencephalon.

The lateral-line input passes to dorsal and intermediate nuclei whereas the signals from the labyrinth pass more ventrally. Regions of overlap appear to be in nucleus octavus magnocellularis and along the medial edge of the intermediate nucleus, adjoining nuclei C1 and C2. Recent studies by Corwin and Northcutt (1982) indicate that this latter region is a site of auditory input. Apart from the separation of mechanoreceptive from electroreceptive inputs of lateral-line receptors in the intermediate and dorsal nuclei (McCready and Boord 1976; Bodznick and Northcutt 1980) we cannot yet recognize divisions of the other nuclear groups which relate to specific end organs nor do we know whether these projections map, in a somatotopic sense, the distribution of the various sense organs. Recent studies on the teleost *Astronotus* do indicate, however, that the end organs of the labyrinth project to distinct laminae in the octavus nuclei (Meredith and Butler 1982).

The striking similarity of the primary centres of lateral-line mechanoreceptors and electroreceptors in the lateral-line areas strongly suggests that they operate similarly and provide a complementary service to the animal. Examination with the light and electron microscope has shown that the cerebellar crest which forms the dorsal surface of the lateral-line areas consists of numerous small unmyelinated axons, lying parallel to each other; "stellate" interneurons; and the spiny dendrites of the principal neurons whose cell bodies lie in the dorsal and intermediate nuclei. This arrangement strikingly resembles the molecular layer of the corpus cerebelli (Paul et al. 1977). The significance of this circuitry for lateral-line function remains unclear for the present. But it is perhaps important that "cerebellar-like" circuits have been found in other non-cerebellar parts of brains, which suggests that this neuronal arrangement is suitable for a particular type of neural analysis. Several theories have been proposed that relate neuronal architecture to cerebellar function and these emphasize a role in "motor memory" storage for the vast number of synapses present in the cerebellar cortex. It is possible that the cerebellar crest acts similarly in the storage of lateral-line information and provides a spatial "map" of the world surrounding a fish, derived from past experiences, to which incoming electro- and mechanoreceptive data may be related (see Roberts 1981 b).

The deeper-lying octavus nuclear neurons are not associated with the cerebellar crest but have a cerebellar connection because they receive input from, and are apparently inhibited by, the Purkinje cells of the auricular lower lip (Montgomery 1980). This portion of the auricle receives direct projections from nerve VIII and so appears to constitute a "vestibulo-cerebellum" (Boord and Roberts 1980). The appearance of a specialized vestibulocerebellar circuit so early in vertebrate history indicates that this region has a fundamental role in coordination. In many respects fishes differ in their movements from terrestrial vertebrates but they have the same requirement to maintain a stable visual field during locomotion. Vestibulo-cerebellar interaction is required for eye movement control in mammals and this is perhaps the function of the lower lip in cartilaginous fishes.

There have been numerous studies on brain centre activity in electrosensory teleost fishes in response to electric fields but surprisingly few data are available on elasmobranch electroreception (Bullock 1981). Knowledge of the connections of the rhombencephalic centres indicates that lateral-line information is predominantly passed rostrally to mesencephalic and possibly higher centres, where evoked activity has been recorded (Platt et al. 1974; Bullock 1979).

Information on the consequences of natural stimulation of the labyrinth is now very complete with respect to the horizontal semicircular canals, with data being available for one species (*Scyliorhinus*) for the behaviour of primary afferent, vestibular nuclear and cerebellar (lower lip) neurons (Montgomery 1980). There have been no studies on octavus nuclear responses to other vestibular inputs, but a recent study (Timerick 1982) has indicated the role of the semicircular canals in the correction of pitch and roll during locomotion and the relationship of this activity to fast pathways descending to the spinal cord. Studies by Corwin and others (Bullock and Corwin 1979; Corwin 1981 a, b; Corwin et al. 1982) have shown that mesencephalic and rhombencephalic regions become active during sound stimulation but we have as yet little information on auditory processing in cartilaginous fishes. Nevertheless the information now available on the organization of these regions provides a good basis for further neurophysiological exploration, in conjunction with natural stimulation.

6.3.4 Projections to the Brain Stem from Higher Levels

Inputs from higher centres of the brain to the brain stem are carried along the basal optic tract, the fasciculus retroflexus, fibres from the ventral part of the diencephalon and telencephalotegmental fibres. A direct telencephalotectal connection has also been demonstrated but this will be discussed separately in Chap. 8.

The *basal optic tract* consists of a group of retinal axons that leaves the lateral optic tract and terminates in a nucleus situated in the most rostral part of the mesencephalic tegmentum. This nucleus, composed of small cells, is known as the accessory or basal optic nucleus. On the basis of normal material we could delineate a presumed basal optic nucleus in *Squalus, Raja* and *Hydrolagus,* but not in *Scyliorhinus*.

An accessory optic tract has also been recognized in *Raja* (Witkovsky et al. 1980; Smeets 1981 a) and in *Squalus* (Northcutt 1979 a) where it runs caudally into the tegmentum, splits into ventral and dorsal fascicles and terminates in a circumscribed ventral optic nucleus and in a vaguely defined tegmental area. In other elasmobranchs this nucleus is less evident. In *Galeocerdo* (Ebbesson and Ramsey 1968), *Negaprion* (Graeber and Ebbesson 1972) and *Scyliorhinus* (Smeets 1981 a) degenerating fibres were observed to leave the lateral optic tract ventrally and to course caudalward to the midbrain tegmentum, but their termination has not been observed except for *Galeocerdo*. However the visualization of an accessory optic tract and nucleus may depend on the experimental technique being used. Thus Ebbesson (1968), Sharma (1972) and Vanegas and Ebbesson (1973), using silver impregnation methods in teleosts, were unable to observe an accessory optic tract, whereas this tract has been found in all teleosts that have been examined with the autoradiographic technique (Voneida and Sligar 1976; Northcutt and Butler 1976; Repérant et al. 1976). Recently Northcutt (1978), using the autoradiographic tracing technique, has found a distinct basal optic system in the galeomorph *Mustelus*, a result which emphasizes the necessity for more autoradiographic studies.

There are at present no experimental data on the efferent connections of the basal optic nucleus in chondrichthyan fish but it would be very interesting to know whether the presumed basal optic nucleus in these fishes makes efferent connections with the oculomotor nucleus and cerebellum, as has been demonstrated for some teleosts (Finger and Karten 1978).

The *tractus habenulo-interpeduncularis*, or *fasciculus retroflexus*, originates from the ganglion habenulae which receives projections from several telencephalic centres by way of the stria medullaris. The nucleus interpeduncularis is an elongated cell mass, consisting of small granular cells, situated in or near the raphe. In the four species investigated this nucleus could be subdivided into a superficially situated pars ventralis and a slightly more dorsal pars dorsalis. The topological charts show that the pars ventralis is located in the isthmus region, whereas the pars dorsalis extends from the most rostral part of the isthmus into the tegmentum mesencephali. Gerlach (1947) also subdivided the nucleus interpeduncularis into a dorsal and a ventral part but noted that the dorsal part lies caudal to the ventral part. It is possible that the pars dorsalis of the nucleus interpeduncularis of Gerlach's (1947) description corresponds to our nucleus raphes superior. The nucleus interpeduncularis receives fibres from the ganglion habenulae by way of the fasciculus retroflexus. The efferent connections of the interpeduncular nucleus have not been experimentally investigated.

Sagittal Klüver-Barrera and Bodian sections reveal that the ventral part of the diencephalon and the tegmental regions of the brain stem are connected by numerous fibres which have been interpreted in part by previous authors as providing descending pathways. Ariëns Kappers et al. (1936) and Gerlach (1947), for example, described a *tractus lobobulbaris* arising from the lobus inferior hypothalami and a *tractus mamillopeduncularis* taking origin from the area around the recessus posterior or recessus mammilaris. The fibres of both tracts pass reportedly to the rhombencephalon where they probably interact with visceral efferent cell masses, particularly motor trigeminal and facial nuclei. Ariëns Kappers et al. (1936) also mentioned the presence of a tractus lobocerebellaris which passes directly from the hypothalamus to the cerebellum. However, none of these fibre systems has been experimentally confirmed.

Some fragmentary data are available on the presence of telencephalotegmental fibres. After a large telencephalic lesion, Ebbesson (1972, *Ginglymostoma*) found a bilateral projection to the tegmentum isthmi, and probably a contralateral projection to the oliva inferior. In normal material we could trace some fibres of the basal forebrain bundle to the ventrolateral part of the tegmentum mesencephali.

6.3.5 Formatio Reticularis

The reticular formation in the brain stem of amniotes is usually subdivided in the current literature into three longitudinal zones: median, medial and lateral (Taber 1961: cat; Petrovicky 1966: pigeon; Cruce and Nieuwenhuys 1974: turtle; ten Donkelaar and Nieuw-

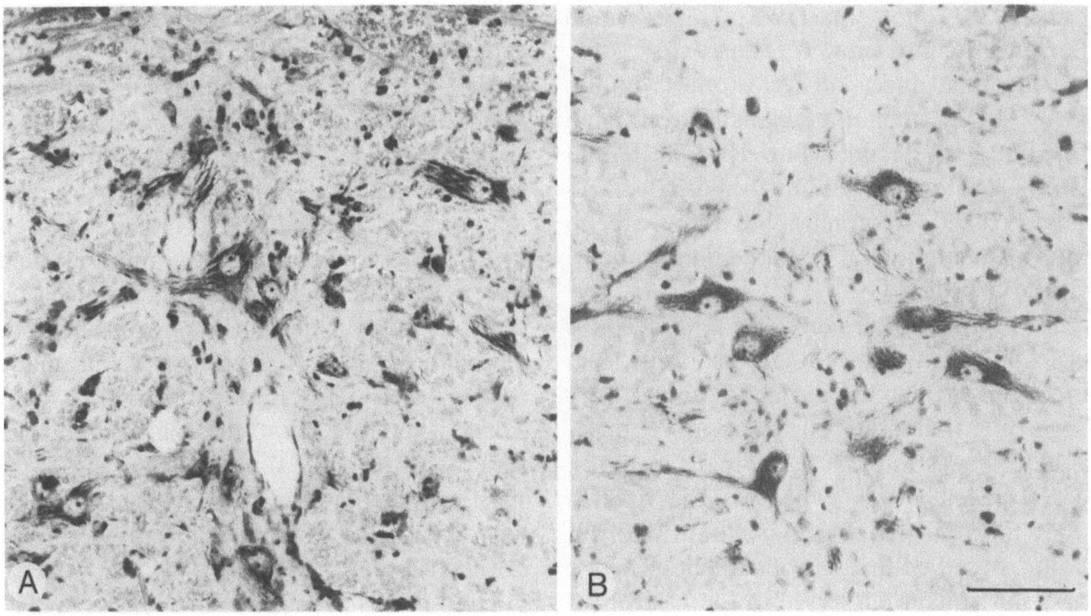

Fig. 6-10. High power photomicrograph of a transverse section showing the neurons of the caudal (**A**) and rostral (**B**) part of the nucleus raphe inferior of *Hydrolagus collei*. Nissl staining. *Bar scale* = 100 μm

enhuys 1979: several reptiles). This subdivision will be followed in the present account as it has recently been shown to apply to anamniotes (Opdam et al. 1976; Nieuwenhuys and Opdam 1976: frogs; Opdam and Nieuwenhuys 1976: axolotl; Kremers and Nieuwenhuys 1979: coelacanth; Thors and Nieuwenhuys 1979: lungfish) including chondrichthyans (Smeets and Nieuwenhuys 1976: shark). The median zone is confined to the rhombencephalon and consists of cells situated in or near the raphe. The medial zone constitutes by far the largest part of the chondrichthyan reticular formation and extends throughout the brain stem. The lateral zone cannot be sharply delimited and lies mostly in the rhombencephalon in the vicinity of the sulcus limitans. Each of these zones will now be considered, but most emphasis will be given to the medial zone.

The *median reticular zone* is composed of the nucleus raphes superior and nucleus raphes inferior. The nucleus raphes superior contains small, ellipsoidal elements, situated in or immediately adjacent to the rhombencephalic raphe at the level of the rostral pole of the motor trigeminal nucleus. A nucleus raphes inferior was first described in chondrichthyans by van Hoëvell (1911), who found a large-celled raphe nucleus in the caudal part of the rhombencephalon of *Raja*. This observation has been confirmed by Northcutt (1978, *Raja*) but Smeets and Nieuwenhuys (1976) could not recognize a distinct inferior raphe nucleus in *Squalus* or *Scyliorhinus*. The present analysis has confirmed the presence of this nucleus in *Raja*, *Hydrolagus* and *Scyliorhinus*, which consists of very large elements situated in or immediately adjacent to the rhombencephalic raphe. The inferior raphe nucleus of *Hydrolagus* can be subdivided into a rostral part of very large neurons and a caudal part characterized by large neurons (Fig. 6-10). In *Squalus* the inferior raphe nucleus, at least the magnocellular part, appears to be absent.

Descriptions of the ultrastructure of the portion of the nucleus raphes inferior thought to be involved in the control of electric discharge have been provided by Nakajima (1970, *Raja*), Fox (1977, *Torpedo*) and Fox and Richardson (1979, *Torpedo*).

The raphe nuclei of cyclostomes (Baumgarten 1972), actinopterygians (Parent et al. 1978) and amphibians (Bartels 1971; Parent 1973) are known to contain 5-hydroxy-tryptamine (5-HT, serotonin), but there are no equivalent studies on the raphe nuclei of cartilaginous fishes.

The *medial reticular zone* extends for almost the entire length of the brain stem and consists of rather loosely arranged, very large cells that occupy a position ventrolateral to the fasciculus longitudinalis medialis. Van Hoëvell (1911), who was the first to study this reticular zone in a chondrichthyan (*Raja*), observed that its constituent elements were not equally dispersed in the rostrocaudal direction. He noted one accumulation in the tegmentum mesencephali and three in the rhombencephalon which he called the nucleus reticularis superior, medius and inferior, respectively. Corresponding aggregations of large reticular elements have been described by Beccari (1930), Saito (1930) and Gerlach (1947) for various sharks, albeit under different names (Smeets and Nieuwenhuys 1976).

In the present study the reticular formation was analyzed as previously (Smeets and Nieuwenhuys 1976). All conspicuous neurons situated in the rhombencephalic basal plate and the tegmentum mesencephali which did not belong to cranial nerve nuclei were plotted on the topological maps. Only somata with distinct nucleoli were charted so as to avoid the inclusion of any cell more than once. In the topological as well as in the topographical charts only every fifth cell is represented (for *Squalus* every tenth cell), for clarity.

In *Squalus* the rhombencephalic part of the medial reticular zone comprises three distinct accumulations of cells, whereas a fourth accumulation is present in the rostral part of the tegmentum mesencephali. The rhombencephalic cell groups appear to be closely comparable to the inferior, medius and superior reticular nuclei of van Hoëvell (1911) and so have been given the same names. The mesencephalic reticular cell group was designated as the nucleus of the fasciculus longitudinalis medialis. In *Scyliorhinus* and *Raja* the division into three rhombencephalic groups is also very distinct. However, in *Hydrolagus* the rhombencephalic medial reticular formation is represented by a more continuous field of cells, though slight but distinctive local reductions in cell-density permit a tripartitioning of this field corresponding to that made in the other three species. It may be concluded therefore, that the medial reticular zone comprises all four nuclei in the four species studied. All of these nuclei consist of very large cells, which are generally tripolar or quadrupolar in shape.

The nucleus reticularis inferior is situated medial to the rostral part of the cell column that contains the motor nuclei of VII, IX and X. The reticular cells are arranged, garland-like around the fasciculus longitudinalis medialis, within the ventral column. However, in *Hydrolagus* a group of reticular elements lies more laterally, between the motor facial nucleus and the ventricular surface (Figs. 101, 102, 110).

The nucleus reticularis medius is roughly coextensive with the motor nucleus of V and contains the largest neurons of the brain (except for the Mauthner neuron of *Hydrolagus*).

The nucleus reticularis superior is represented by a rather narrow elongated accumulation of cells that lies in front of the motor trigeminal nucleus.

The nucleus of the fasciculus longitudinalis medialis constitutes the most rostral part of the medial reticular zone and consists of a group of rather diffusely arranged cells, for the most part situated anterior to the oculomotor nucleus. In *Raja,* however, these neurons lie at the same level as the nucleus oculomotorius. This designation, as the nucleus of the fasciculus longitudinalis medialis, should be considered to be tentative until experimentally established, although in various other groups of vertebrates a cell mass occupying a corresponding position, does send axons into the fasciculus longitudinalis medialis. It is, however, important to note that many neurons in this nucleus were labelled after a spinal cord HRP injection (Smeets and Timerick 1981).

Counts of cells in the four nuclei of the medial reticular zone of the sharks *Squalus* and *Scyliorhinus* were obtained in a previous study (Smeets and Nieuwenhuys 1976). Counts performed as part of the present work are given in Table 6-1 and these differ considerably from the previous figures. It appears therefore that the rhombencephalic reticular nuclei do not contain constant numbers of neurons, a variation which might depend on different sizes of the animals used in the two studies. Leonard et al. (1978a) found in the stingray, *Dasyatis,* that dorsal root ganglion cells, dorsal root axons and "motor" cells in the ventral horn increase in number as the animals increase in size and that this accumulation continues much further into adult life than is the case for other vertebrates (cf. Cowan 1973).

It is worth mentioning that, in contrast, the mesencephalic reticular nucleus is constant with regard to the number of its constituent neurons within a species (Table 6-1). This nucleus contains about twice as many neurons in *Squalus* and *Hydrolagus* as in *Scyliorhinus* and *Raja,* a difference which may well be related to the more developed visual system.

The tegmentum of the isthmus region contains numerous scattered, medium-sized cells most of which are intercalated between fibre bundles. Although these cells correspond in position to the nucleus reticularis isthmi of other non-mammalian species (Cruce and Nieuwenhuys 1974: *Testudo hermanni;* Opdam et al. 1976: *Rana esculenta*), they are very diffusely arranged in all four chondrichthyan species. However, the cell mass indicated by H may be equivalent to the nucleus reticularis isthmi. In *Hydrolagus* this nucleus can be subdivided into a superficial and a more periventricular part (H′, H″).

The *lateral reticular zone.* In the mammalian rhombencephalon the area situated between the medial reticular formation and the nucleus of the tractus descendens nervi trigemini is occupied by a zone of small cells, which has been termed the nucleus reticularis parvocellularis or the lateral part of the reticular formation. This part of the reticular formation is probably represented in cartilaginous fishes by the small and medium-sized cells that lie ventral and ventrolateral to the sulcus limitans. Although this region contains at some levels a more or less circumscript group of cells, the delimitation of a distinct lateral reticular zone on a cytoarchitectonic basis seems to be impossible.

In *Squalus,* as well as in *Scyliorhinus,* the area ventral to the sulcus limitans at the level of entrance of the eighth nerve contains a distinct cell mass, termed nucleus B. This nu-

Table 6-1. Average number of cells in the nuclei of the medial and median reticular zone

	Squalus		*Scyliorhinus*		*Raja*	*Hydrolagus*
Series no.	7,829	7,375[a]	7,929	7,338[a]	7,831	7,840
Rai	–	–	130	–	476	257/193[b]
Ri	833	274	280	509	184	283
Rm	502	272	264	323	244	431
Rs	220	103	108	100	148	138
Nflm	225	193	143	129	121	212

[a] Smeets and Nieuwenhuys (1976)

[b] Nucleus raphes inferior pars rostralis and pars caudalis, respectively

cleus, which consists of medium-sized cells, may well represent a more condensed part of the lateral reticular zone, or give rise to efferent fibres to labyrinthine and/or lateral-line receptors. Another accumulation of medium-sized, bi- and tripolar neurons could be identified rostral to the trigeminal motor nucleus in *Raja,* which has been termed nucleus J. As this nucleus projects to the spinal cord it may also represent a more condensed part of the lateral reticular formation (Smeets and Timerick 1981, Fig. 6K).

Connections of the Reticular Formation. The connections of the nucleus raphes superior are currently unknown for no experimental studies have been done on this nucleus in cartilaginous fishes. The HRP-studies of Leonard and Willis (1977: *Dasyatis*) and of Smeets and Timerick (1981: *Scyliorhinus* and *Raja*) have revealed, however, that the nucleus raphes inferior projects extensively to the spinal cord (Fig. 5-7). Northcutt and Boord (1981, *Raja*) and Smeets (1981b, *Scyliorhinus* and *Raja*) have shown that an afferent projection to the nucleus raphes inferior originates from the tectum mesencephali (Fig. 6-11).

The afferent connections of the rhombencephalic medial reticular formation arise from neurons in the spinal cord, cerebellum, tectum and telencephalon (Figs. 6-11). Spinoreticular projections have been revealed by the experimental neuroanatomical studies of Ebbesson (1972: *Ginglymostoma*), Hayle (1973a: *Scyliorhinus*) and Smeets (unpublished, *Raja*) who found that the long peripherally extending dendrites of the large and very large neurons of the medial reticular formation made synaptic contact with fibres that ascend from the spinal cord. This projection has been confirmed by the electrophysiological study of Paul and Roberts (1978: *Scyliorhinus*). The cells of the medial reticular formation also contact the direct and crossed fibres that originate from the tectum mesencephali (Northcutt and Boord 1981: *Raja;* Smeets 1981b: *Scyliorhinus* and *Raja*) and

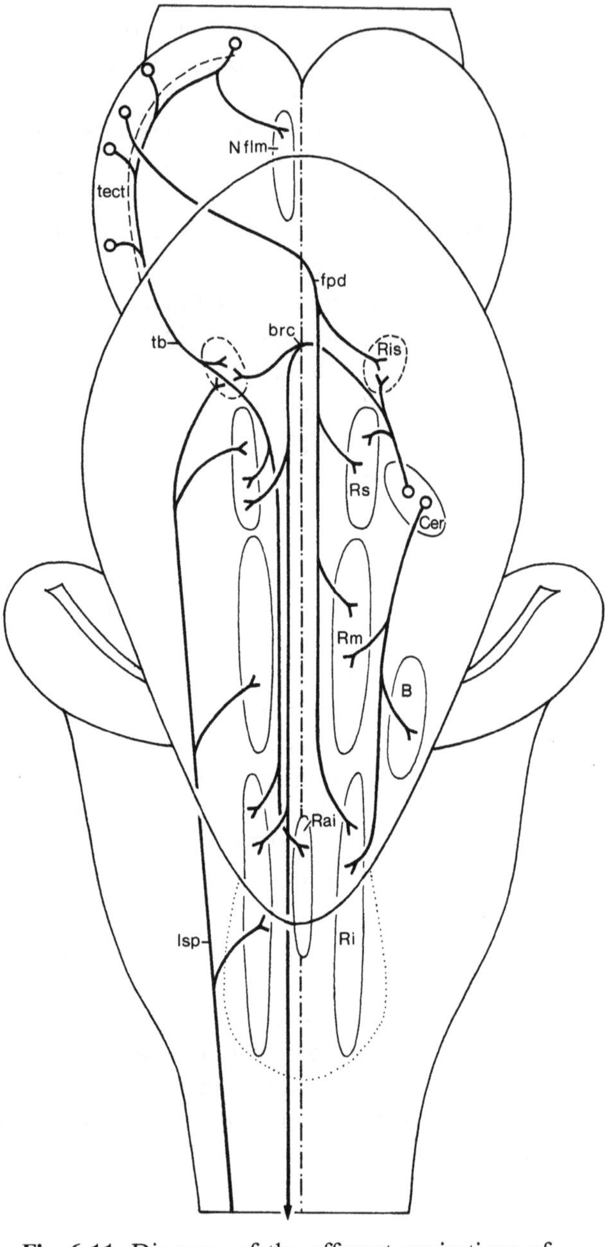

Fig. 6-11. Diagram of the afferent projections of the reticular formation, as viewed from the dorsal surface of the brain

nucleus cerebelli (Ebbesson and Campbell 1973: *Ginglymostoma*), a cerebelloreticular pathway that has been confirmed electrophysiologically by Paul and Roberts (1978) in *Scyliorhinus*. Finally, evidence for a direct telencephaloreticular projection has been obtained by Ebbesson (1972) for *Ginglymostoma* where degenerating fibres could be traced to all levels of the brain stem after large telencephalic lesions. His Fig. 9D indicates that a large telencephalic input attains to at least the nucleus reticularis isthmi.

Data on the efferent connections of the medial reticular zone are sparse. A preliminary HRP study by Leonard and Willis (1977, *Dasyatis*) revealed that the nucleus reticularis medius in particular projects to the spinal cord, a result which has been confirmed by the electrophysiological study of Paul and Roberts (1978). A more detailed HRP study by Smeets and Timerick (1981, *Scyliorhinus* and *Raja*) has revealed that projections from the medial reticular formation to the spinal cord arise from all levels of this zone and that the nucleus of the fasciculus longitudinalis medialis, the nuclei reticularis isthmi, superior, medius, and inferior, mostly on the ipsilateral side, also contain many cells, the axons of which descend to the spinal cord (Fig. 5-7). In addition to this well developed reticulospinal projection, a reticulocerebellar projection has been suggested by Paul and Roberts (1975, *Scyliorhinus*) based on electrophysiological recordings and recently Smeets (1982, *Scyliorhinus* and *Raja*) found labelling in some medium-sized reticular cells after a tectal HRP injection.

One afferent connection of the lateral reticular formation was demonstrated by Ebbesson and Campbell (1973) in an experimental study of the cerebellar efferent connections in *Ginglymostoma*. These authors described an ipsilateral descending cerebellobulbar pathway, which runs just medial to the spinal tract and nucleus of the trigeminus as far caudal as the first spinal segment. This tract issues fibres along its course in a medial direction to an area which corresponds to the lateral reticular zone of the present description, including also our nucleus B. A recent HRP-study (Smeets and Timerick 1981) has revealed that nucleus B projects bilaterally to the spinal cord and, therefore, can be considered to be a cerebellospinal relay centre.

6.3.6 Fasciculus Longitudinalis Medialis

The fasciculus longitudinalis medialis (flm) is one of the most ancient and, in respect to position, one of the most constant pathways in the vertebrate neuraxis. It extends from the rostral part of the mesencephalon throughout the brain stem and continues caudally in the ventral funiculus of the spinal cord. In the rhombencephalon the fasciculus longitudinalis medialis occupies a paramedian position but in the midbrain it is usually situated somewhat more laterally. It is a composite bundle which in some groups of vertebrates (bony fishes, crossopterygians, most groups of reptiles, certain "lower" mammals) attains a very large size.

The fasciculus longitudinalis medialis is very large in *Squalus, Scyliorhinus* and in *Hydrolagus*. In these species the rhombencephalic part of the bundle protrudes into the fourth ventricle and forms a longitudinal ridge which is laterally bounded by the sulcus intermedius ventralis. However, the bundle is less well developed in *Raja*.

We will now give a brief survey of the components of the fasciculus longitudinalis medialis that have been described so far.

1. Fibres originating from the nucleus of the fasciculus longitudinalis medialis descend to spinal cord levels. Although a preliminary study by Leonard and Willis (1977) did not reveal a direct spinal pathway, a recent HRP study by Smeets and Timerick (1981) has shown that fibres originating from this nucleus do descend to the spinal cord, extending at least as far caudal as the twentieth spinal segment.

2. Ariëns Kappers et al. (1936) suggested, on the basis of normal material, that fibres leaving the tectum along with the fasciculus tectotegmentalis dorsalis, descend in the ipsilateral and contralateral fasciculus longitudinalis medialis. The existence of these fibres was not confirmed, however, by the experimental degeneration study of Smeets (1981b). From this study it appeared that fibres descending from the tectum occupy a

position ventrolateral to the fasciculus longitudinalis medialis.

3. According to Sterzi (1909), Ariëns Kappers et al. (1936) and Ariëns Kappers (1947) secondary fibres from the dorsal and intermediate nuclei of the octavolateral area enter the fasciculus longitudinalis medialis. However, the experimental study of Boord and Northcutt (1979, 1982) provided no evidence for such a projection.

4. Larsell reported (1967) that the nucleus octavus magnocellularis and the descending VIII root contribute fibres to the fasciculus longitudinalis medialis. Smeets and Timerick (1981) have shown with HRP that some fibres from the magnocellular octavus nucleus do enter the fasciculus longitudinalis medialis (and also Stieda's fasciculus). No evidence for a direct input from nerve VIII was obtained, however, from the experimental studies by Boord and Roberts (1980, *Scyliorhinus*) or by Northcutt (1980, *Platyrhinoidis*).

5. The fasciculus longitudinalis medialis has been thought to contain a tractus cerebellomotorius cruciatus et rectus, which arises from the corpus cerebelli and the auriculae and reaches the motor and reticular nuclei of the lower brain stem, largely by way of the ipsilateral and contralateral fasciculus longitudinalis medialis (Ariëns Kappers 1906). Some of these fibres reportedly pass rostrally and reach the eye muscle nuclei (Ariëns Kappers et al. 1936). Although, in our normal material, a presumed tractus cerebellomotorius could be identified, Ebbesson and Campbell (1973) were unable to confirm experimentally the existence of such a pathway in the nurse shark (*Ginglymostoma*).

6. In *Hydrolagus* the axons of the two Mauthner cells cross and descend in the contralateral fasciculus longitudinalis medialis to the spinal cord. During their caudal course these axons shift within the bundle from a dorsomedial to a ventrolateral position.

7. Axons from the median and medial reticular formation make up the greater part of the medial longitudinal fascicle at lower brain stem levels. The recent HRP-study of Smeets and Timerick (1981) has confirmed that the very large neurons of the median and medial reticular formation project to the spinal cord by way of the fasciculus longitudinalis medialis.

It is evident from this summary that the fasciculus longitudinalis medialis is the principal encephalospinal pathway in cartilaginous fishes.

6.3.7 Visceromotor Nuclei

On the basis of comparative inference it might be expected that in chondrichthyans the IIIrd, VIIth, IXth and Xth cranial nerves would contain general visceral efferent or parasympathetic preganglionic fibres. However, practically nothing is known concerning the presence of these fibres and the localization of their parent neurons. A nucleus of Edinger-Westphal, consisting of small, spherical and ellipsoidal cells, situated dorsolateral to the somatomotor nucleus of III, was indicated by Smeets and Nieuwenhuys (1976) in *Squalus* and *Scyliorhinus*. In mammals this nucleus supplies visceral-efferent fibres in the oculomotor nerve that bring about the constriction of the pupil on illumination of the eye (the light reflex). The elasmobranch eye also constricts the pupil when it is illuminated. The iris sphincter is directly light responsive (Young 1933b), as well as being innervated by oculomotor neurons. The iris constricts fastest in diurnal sharks, more slowly in nocturnal sharks and slowest of all in the Rajiformes. These differences may reflect the relative importance of neurally mediated compared with local responses of the iris (Kuchnow 1971). It is noteworthy that although the Edinger-Westphal nucleus can be identified in *Squalus,* and *Scyliorhinus,* it cannot be delineated in *Raja* and *Hydrolagus*.

Addens (1933) observed that in the shark *Cetorhinus* and in the Holocephali the vagal part of the visceromotor column can be divided into separate rostral and caudal parts and he suggested that the rostral portion is

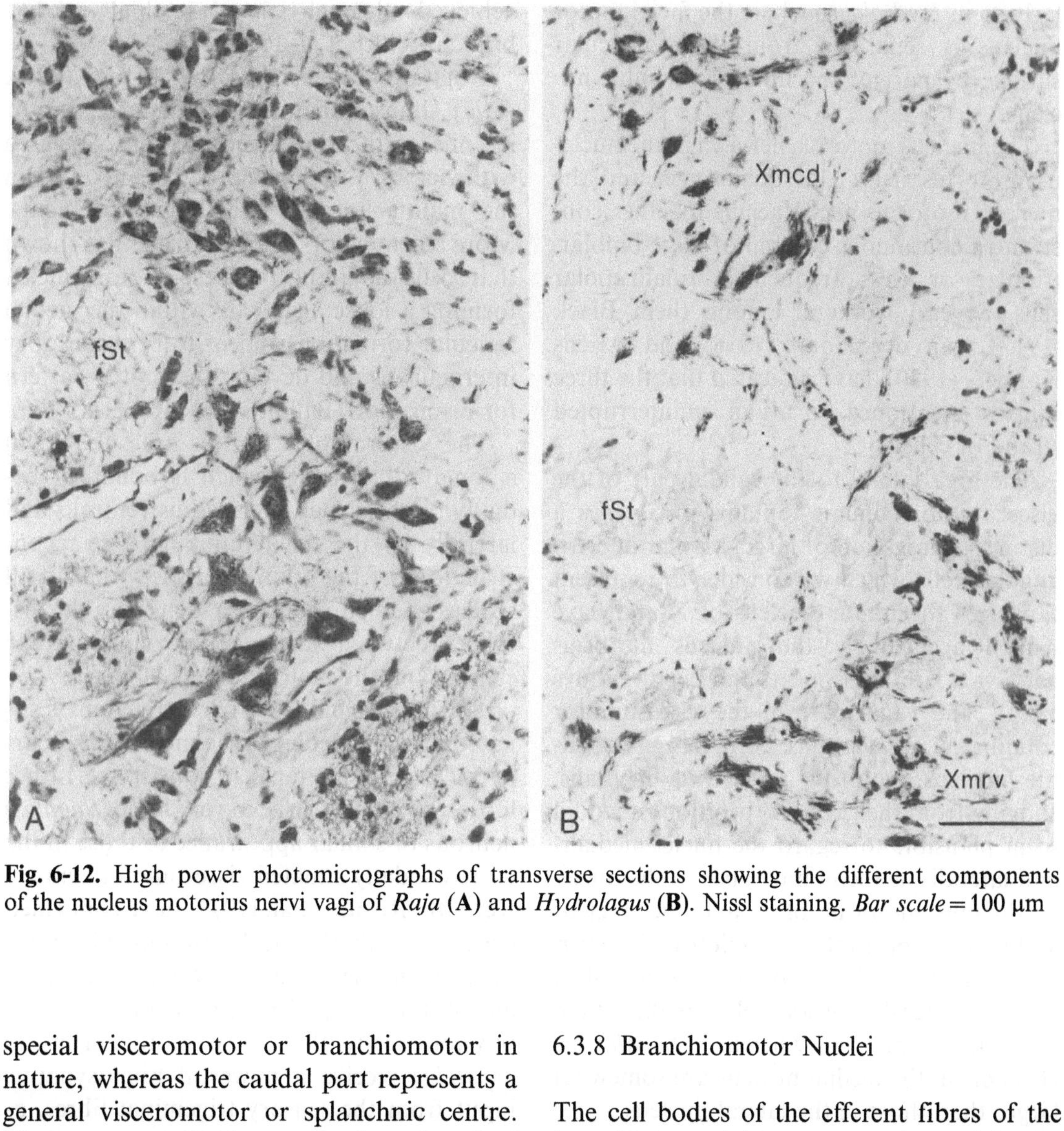

Fig. 6-12. High power photomicrographs of transverse sections showing the different components of the nucleus motorius nervi vagi of *Raja* (**A**) and *Hydrolagus* (**B**). Nissl staining. *Bar scale* = 100 µm

special visceromotor or branchiomotor in nature, whereas the caudal part represents a general visceromotor or splanchnic centre. The present analysis reveals that the vagal nucleus in *Hydrolagus* can be easily divided into a caudodorsal nucleus and a rostromedial nucleus, the latter being continuous with the nuclei of IX and VII (Fig. 6-12). A similar subdivision into a dorsal and a ventral vagal nucleus could be recognized at some levels in *Raja*. The dorsal portion appears to contain somewhat smaller neurons than the ventral portion (Fig. 6-12).

6.3.8 Branchiomotor Nuclei

The cell bodies of the efferent fibres of the branchial cranial nerves are situated in the lateral part of the rhombencephalic tegmentum and include the motor nuclei of X, IX, VII and V, and a cell mass of unknown connections indicated tentatively as nucleus A. The nuclei of X, IX, VII and V constitute a column which shows a wide gap at the level of entrance of the octavus nerve. The motor nucleus of V lies rostral to this gap, while the other nuclei lie caudal to this gap. The motor nuclei of IX and X occupy a periventricular position but the motor nuclei of VII and part of that of V have migrated away from the ventricular surface. This is particularly so in *Hydrolagus* where the facial motor nucleus is separated from the ventricular surface by part of the inferior reticular nucleus.

The nucleus motorius nervi vagi, the nucleus motorius nervi glossopharyngei and the nucleus motorius nervi facialis together constitute a continuous column of large bipolar, tripolar or (less frequently) quadrupolar cells. Several workers, among them Black (1917), van der Horst (1918) and Ariëns Kappers (1920), have indicated that the three centres mentioned form an uninterrupted strand of cells.

The area lateral to the caudal part of the visceromotor column contains in *Squalus* a distinct aggregation of large, bipolar and triangular cells which we consider to represent a part of the motor nucleus of X and have named accordingly the nucleus motorius nervi vagi lateralis (Smeets and Nieuwenhuys 1976). The vagal part of the visceromotor column of *Squalus* has been designated as the nucleus motorius nervi vagi medialis. Judging from their relative positions it would seem plausible to regard the nuclei medialis and lateralis as the homologues of the nucleus dorsalis motorius of X and the nucleus ambiguus respectively of amniotes. However, the cells in the dorsal motor nucleus of X are in general considerably smaller than those in the ambiguus whereas in *Squalus* the elements in the medial nucleus are somewhat larger than those in the lateral nucleus.

In mammals these nuclei are involved in the regulation of the heart beat, gut motility and gut secretions. In elasmobranchs (*Scyliorhinus* and *Raja*) recent studies by Young (1980) have indicated that the nervus vagus plays little role in the control of gut movement although it may function in secretory regulation.

The elasmobranch heart is supplied by separate vagal rami, the branchial cardiac and visceral cardiac branches. It seems that the visceral branch may have a primarily sensory function and that the cardiac inhibition that results from vagal stimulation is mostly achieved through the branchial cardiac branch (Short et al. 1977).

Ablation experiments by Hyde (1904, *Raja*) showed that the sensory and motor regions of the rhombencephalon associated with nerves VII, IX and X are involved in the maintenance of breathing movements. More recent work on teleost fishes has shown that cells adjacent to these motor nuclei, forming a loose aggregate within the medial reticular formation, appear to be respiratory interneurons and develop the motor pattern for respiration (Ballintijn and Roberts 1976).

The visceromotor centres are situated in a longitudinal zone, which is bounded medially by the sulcus intermedius ventralis and laterally by the sulcus limitans. The reconstructions of the brain stem show that in all species studied there is a considerable distance between the caudal visceromotor centres and the motor nucleus of V.

The nucleus motorius nervi trigemini is situated directly medial to the most rostral part of the sulcus limitans. It consists of rather loosely arranged, large to very large (*Squalus*) neurons of similar appearance to those in the more caudally situated visceromotor centres. Recently Rosiles et al. (1977) have confirmed experimentally the localization of this motor nucleus in the stingray, *Dasyatis*. Roberts and Witkovsky (1975) demonstrated electrophysiologically that these motoneurons in *Mustelus* receive a powerful monosynaptic input from the sensory trigeminal fibres, as well as from the mesencephalic trigeminal nucleus.

In a previous paper (Smeets and Nieuwenhuys 1976) we mentioned the presence of a conspicuous nucleus, rostral to the motor nucleus of VII, which is found only in *Squalus* and consists of very large, elongated, spindle-shaped neurons, oriented in the transverse plane and lying parallel to the ventricular surface. As far as we are aware this cell mass has never been described before and since it cannot be readily homologized with any centre in other vertebrates, we term it provisionally nucleus A.

The rhombencephalon of electric rays (torpedoids) in this region has been extensively modified because of the large electric lobes which account for nearly 60% of the total weight of the brain (Roberts and Ryan 1975). These lobes contain approximately 120,000 multipolar neurons, the axons of which pass on each side along four electric nerves to supply the electric organs. Considerable debate has taken place over the homology of these nerves and lobes, but it is probable that in *Torpedo* the electromotoneurons are derived from the motor nuclei of nerves VII, IX and X (Ewart 1890; Hugosson 1957). The electromotoneurons of *Narke* have been examined electrophysiologically by Saito (1966).

6.3.9 Somatomotor Nuclei

The somatic efferent cell masses include a rostral extension of the spinal motor column as well as the abducens, trochlear and oculomotor nuclei. All of these nuclei lie close to the median plane and although of variable topographical position show a distinct relationship to the fasciculus longitudinalis medialis. The spinal motor column and the abducens nucleus lie lateral to the fasciculus longitudinalis medialis, whereas the trochlear and oculomotor nuclei are situated dorsal and dorsomedial to the bundle. The spinal motor column continues for some distance rostral to the obex; it consists of large to very large tripolar and quadrupolar cells.

As in other vertebrates, the abducens, trochlear and oculomotor nuclei are involved in the movement of the eyeball, through the action of 6 muscles, the 4 recti and 2 obliques (Oliva 1967). The superior, anterior and inferior recti and the inferior oblique are innervated by the oculomotor nerve; the superior oblique is supplied by the trochlear nerve and the posterior rectus by the abducens. In Klüver-Barrera and Bodian sections the roots of the nervus abducens could be traced towards the lateral aspect of the medial longitudinal fasciculus, where an ill-defined group of medium-sized to large neurons may be identified as the nucleus of origin of that nerve. A recent HRP study by Rosiles and Leonard (1980) on the organization of the extraocular motor nuclei in the stingray, *Dasyatis,* has revealed, however, that the motoneurons of the abducens nerve are scattered over a large area. The majority of these cells are located in a relatively wide zone near the base of the rhombencephalon, but they also spread dorsomedially to the medial longitudinal fascicle. The distribution of abducens motoneurons resembles that described for teleost fishes (Sterling 1977; Luiten and Dykstra-de Vlieger 1978; Leonard and Willis 1979).

The *nucleus nervi trochlearis* and the *nucleus nervi oculomotorii* are well developed and constitute together a continuous aggregation of rather loosely arranged, fusiform and triangular cells. It is impossible to draw a sharp boundary between these nuclei, but the study by Rosiles and Leonard (1980) in *Dasyatis* has revealed that these nuclei overlap slightly. However, in *Raja* the position of the trochlear nucleus is ventrolateral to the fasciculus longitudinalis medialis and in this species there appears to be a sharp separation between trochlear and oculomotor nuclei.

Rosiles and Leonard (1980) found that the trochlear nucleus innervates the contralateral superior oblique muscle and that the motoneurons of this nucleus form a relatively compact group of cells most of which are located immediately ventrolateral to the medial longitudinal fascicle as found in *Raja.* Many dendrites of the trochlear motor nucleus penetrate the fascicle. The same authors subdivide the nucleus oculomotorius into dorsal and ventral groups. The cells of the dorsal group, which are mainly located around the fasciculus longitudinalis medialis, show to some degree a segregation with respect to the extrinsic eye muscles they innervate. The rostral part of the dorsal group is composed of neurons which innervate the inferior rectus muscle. Caudally there is some intermingling with neurons which project to the superior and medial recti. There is no segregation between the medial rectus motoneurons and the superior rectus motoneurons, though Rosiles and Leonard (1980) found that the medial rectus motoneurons appear further rostral at the level of the oculomotor nerve exit, while superior rectus motoneurons appear slightly more caudal. The neurons of the dorsal oculomotor group, like those of the trochlear nucleus, appear to be intimately associated with the medial longitudinal fascicle into which their dendrites penetrate. Although not recognized by Ariëns Kappers et al. (1936) or Smeets and Nieuwenhuys (1976), oculomotor cells were identified by Rosiles and Leonard (1980) near the base of the midbrain. The majority of the cells of this ventral group are situated within the confines of the nucleus ruber of the present authors. The neurons of the ventral oculomotor group innervate exclusively the inferior oblique muscle and unlike the cells of the trochlear nucleus and the dorsal oculomotor group they do not appear to spread their dendrites into the medial longitudinal fascicle. A striking finding of the study by Rosiles and Leonard is that the inferior rectus and inferior oblique motoneurons lie ipsilateral to their respective muscles, whereas the motoneurons innervating the superior rectus and medial rectus muscle are contralateral. The contralateral innervation of the medial rectus muscle contrasts with the situation in all other vertebrates.

Data about afferent connections of the extraocular motor nuclei are very scarce. Ebbesson and Campbell (1973) have provided some evidence of an input from the cerebellum, via the brachium conjunctivum, to the trochlear and oculomotor nuclei in *Ginglymostoma*.

Movements of the eyes of vertebrates, among other functions, serve to compensate for changes in the position of the body so that the visual frame is maintained approximately constant. The compensatory movements of the eyes of *Rhinobatos* in response to rotation of the body have been described by Maxwell (1923) who showed that these movements depend on the integrity of the labyrinth. The compensatory movements of the eyes that accompany locomotor movements have been well studied by Harris (1965) in *Squalus*. He distinguished between eye movements that require labrynthine information and those that need inputs from the spinal cord. His results indicate the existence of interactions between eye muscle motor nuclei, vestibular centres and other motor regions, although the site of integration is unknown.

The neural circuits involved in the control of eye movements have been little studied in fishes. There are indications that some oculomotor neurons are electrotonically coupled together, and that they receive vestibular inputs (Korn and Bennett 1975).

6.3.10 Certain Centres in the Isthmus Region and in the Tegmentum Mesencephali

The following cell masses could be recognized in the isthmus region: griseum centrale, nucleus isthmi, and three nuclei that have been non-committally labelled as F, G, and H. The nucleus interpeduncularis, which also lies in this region, has been described in a previous section (6.3.4).

The *griseum centrale rhombencephali* could be recognized in all three elasmobranchs studied, but not in *Hydrolagus*. It is an elongated zone of periventricularly-situated small granular cells, which extends throughout the rostral one-third of the rhombencephalic basal plate. Although the caudal limit is clear, its rostral boundary is indistinct. The connections of this cell mass are entirely unknown.

The *nucleus isthmi* is an accumulation of small spherical cells, situated directly rostral to the nucleus cerebelli (Fig. 6-13). Johnston (1905) suggested that this nucleus receives afferent fibres from the vagal lobe and sends an efferent projection to the hypothalamus. He regarded this nucleus as being homologous to the nucleus visceralis secundarius of

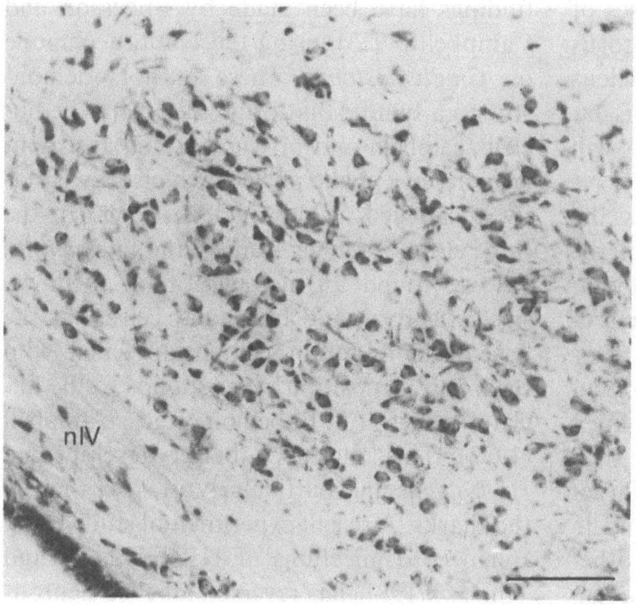

Fig. 6-13. High power photomicrograph of a transverse section showing the neurons constituting the nucleus isthmi of *Hydrolagus collei*. Nissl staining. *Bar scale* = 100 μm

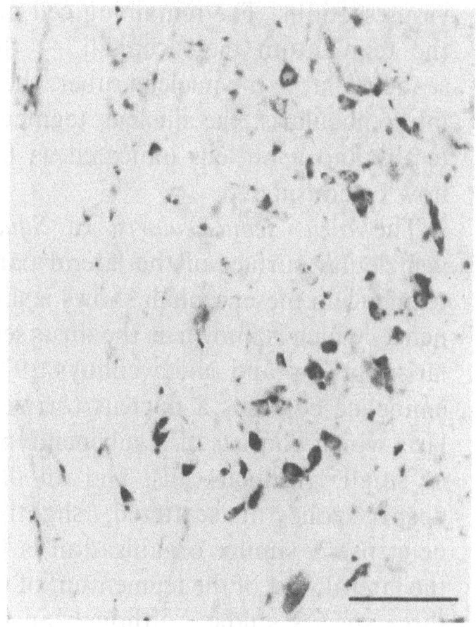

Fig. 6-14. High power photomicrograph of a transverse section showing the nucleus indicated as E of *Hydrolagus collei*. Nissl staining. *Bar scale* = 100 μm

other vertebrates. Ariëns Kappers (1921, 1947) believed that this cell mass is a correlation centre for octavolateral and optic impulses. However recent experimental studies (Smeets 1981 b, 1982) in *Scyliorhinus* and *Raja* do not provide any evidence of a reciprocal connection between the tectum mesencephali and the nucleus isthmi as has been found in teleosts (Ebbesson and Vanegas 1976; Sligar and Voneida 1976; Grover and Sharma 1981; Luiten 1981 c) and amphibians (Rubinson 1968; Wilczynski and Northcutt 1977; Gruberg and Udin 1978; Glasser and Ingle 1978). Nevertheless, we continue to call this nucleus the nucleus isthmi, since there are no other candidates.

Nucleus F consists of relatively densely packed small cells and lies ventrolateral to the nucleus cerebelli. It has been recognized in all four species studied and is large in its rostrocaudal extent, especially in *Raja*. Another cell mass, tentatively labelled *nucleus G*, could be identified only in *Scyliorhinus*. This nucleus lies immediately ventral to the nucleus cerebelli and contains somewhat larger neurons than nucleus F. Nuclei F and G both have a strong projection to the spinal cord (Smeets and Timerick 1981) and do not show any connections with the tectum mesencephali (Smeets 1981 b, 1982). In position, nucleus G resembles the nucleus visceralis secundarius of actinopterygian fishes (Finger 1978 a).

The isthmic basal plate contains a cell mass labelled *nucleus H*, that consists of medium-sized cells that lie ventral to the lateral lemniscus. Similar neurons could be recognized in *Raja*, but because of their diffuse arrangement these have not been designated as a nucleus. As already mentioned (Sect. 6.3.5) this nucleus may represent a nucleus reticularis isthmi.

Of the nuclei of the *tegmentum mesencephali*, the motor nuclei of III and IV, and the nucleus of the fasciculus longitudinalis medialis have already been considered in previous sections. The remaining cell masses of the tegmentum mesencephali — the torus semicircularis, the nucleus ruber, the nucleus intercollicularis, the nucleus tegmentalis lateralis and a nucleus indicated as E — will now be considered.

The *torus semicircularis*. In *Squalus* the ventricular surface of the lateral part of the tegmentum mesencephali shows a slight eminence, which is known as the torus semicircularis (Smeets and Nieuwenhuys 1976). This eminence contains a nucleus tori semicircularis which consists of a subependymal layer of small, granular cells, and an ill-defined deeper zone of scattered, slightly larger neurons. A similar organization is found in the lateral part of the tegmentum of the other three species studied, although an intraventricular protrusion is lacking. In *Raja,* a distinct nucleus tori semicircularis consisting of relatively densely packed neurons can be recognized, but not in *Scyliorhinus* and *Hydrolagus.* Wallenberg (1907, *Carcharias, Scyliorhinus, Torpedo*) described a "Medialer Mittelhirn-Haubenkern", which most probably corresponds to the torus semicircularis of the present description. On the basis of Marchi experiments, Wallenberg suggested that this nucleus receives afferent inputs from the octavolateral area via a primordial lateral lemniscus. However, a recent experimental study of the efferent projections of the octavolateral area (Boord and Northcutt 1982: *Raja*) has revealed that nucleus dorsalis and nucleus intermedius areae octavolateralis project predominantly contralaterally via the lateral lemniscus to a subtectal region, which does not coincide with our nucleus tori semicircularis. These authors also found evidence that there is a segregation within this region between the inputs from the dorsal and intermedial nuclei.

The *nucleus ruber*. In Marchi preparations of cerebellar-lesioned *Scyliorhinus* Wallenberg (1907) traced an entirely crossing bundle ascending to the tegmentum mesencephali, distributing fibres to a cell mass that he designated as the nucleus ruber tegmenti. Similar findings have been made by Ebbesson and Campbell (1973), using the Nauta technique on *Ginglymostoma.* These authors demonstrated a bundle that originates from the nucleus cerebelli and provides a massive input to a mesencephalic nucleus which they tentatively labelled as the nucleus ruber. In the three elasmobranchs studied and in *Hydrolagus* we found an aggregation of medium-sized, fusiform and triangular cells near the base of the midbrain tegmentum which, with regard to its position, corresponds to the nucleus ruber of Ebbesson and Campbell (1973). This nucleus is better developed in *Hydrolagus* and particularly in *Raja* than in the sharks. A recent experimental study using spinal cord injections of HRP (Smeets and Timerick 1981) has revealed labelled neurons in the nucleus ruber of *Raja.* However, the existence of a rubrospinal pathway could not be established for *Scyliorhinus.* Paul and Roberts (1978) were also unable to demonstrate the presence of a rubrospinal tract in *Scyliorhinus,* using electrophysiological techniques, though neurons of the nucleus ruber could be activated from the contralateral cerebellar nucleus. It seems, therefore, that although a nucleus ruber is present in *Scyliorhinus,* it does not provide a rubrospinal projection. On the other hand, such a pathway evidently exists in *Raja.* In this respect, it is interesting to note that: (1) in the snake, *Python reticulatus,* a rubospinal tract could not be demonstrated (ten Donkelaar 1976a, b); (2) a distinct rubrospinal tract has been demonstrated in quadrupedal reptiles (ten Donkelaar 1976a, b; ten Donkelaar et al. 1980). Moreover, a recent HRP study of ten Donkelaar et al. (1981) on the clawed toad *Xenopus laevis* also has revealed the existence of a rubrospinal tract in anuran amphibians. Therefore, it seems possible that a rubrospinal tract forms a basic part of the descending supraspinal pathway in vertebrates with limbs or limb-like structures such as the pectoral fins of the ray.

A rather unexpected connection of the nucleus ruber is the ipsilateral rubrotectal projection described by Smeets (1982). The significance of this projection is unknown, but a similar pathway has been reported in teleosts (Luiten 1981c).

The *nucleus intercollicularis*. In the caudal half of the tegmentum mesencephali of *Scyliorhinus* an aggregation of small cells could be identified which lies between the fasciculi tectotegmentales dorsales and ventrales. We have labelled this cell mass as the nucleus intercollicularis, since its position resembles that of the intercollicular nucleus in *Raja eglanteria* (Northcutt 1978). Such a nucleus could not be delineated from the surrounding grey matter in the other species studied. It has been shown recently that this nucleus receives a bilateral tectal input (Smeets 1981b) and projects to the contralateral spinal cord (Smeets and Timerick 1981). The intercollicular nucleus may be considered to be a relay centre in an indirect tectospinal pathway, supplementing the tectoreticular and tectospinal pathways.

The *nucleus tegmentalis lateralis* is a large aggregation of small, ellipsoid cells, situated in the lateral part of the tegmentum. This nucleus has been previously described by Ariëns Kappers (1921), Gerlach (1947) and Smeets and Nieuwenhuys (1976). Gerlach reported that this nucleus extends throughout the length of the midbrain; however, according to Smeets and Nieuwenhuys (1976) and the present analysis, this nucleus is confined to the rostral part of the tegmentum. A nucleus tegmentalis lateralis has been recognized in all four species studied. In *Hydrolagus* this nucleus has been erroneously considered to be the nucleus dorsalis thalami (Ariëns Kappers and Carpenter 1911) and the nucleus rotundus mesencephali (Kuhlenbeck and Niimi 1969). Our reconstructions show clearly that in its position this nucleus corresponds to the lateral tegmental nucleus of elasmobranchs.

Some experimental data on the connections of the nucleus tegmentalis lateralis are available. Hayle (1973a) found that spinal cord lesions in *Scyliorhinus* resulted in preterminal and terminal degeneration located in a discrete area of the midbrain that corresponds to the nucleus tegmentalis lateralis. A small ipsilateral tectal input to this nucleus has also been described (Smeets 1981b). A projection to the tectum is the only efferent connection known for this nucleus (Smeets 1982). Northcutt (1978) suggested that the nucleus tegmentalis lateralis may be homologous to the substantia nigra of other vertebrates. This suggestion would be strengthened if it could be demonstrated that its cells synthesize dopamine and that the cell mass as a whole is reciprocally connected with the telencephalic basal area.

A cell mass labelled E could be recognized only in *Hydrolagus*. This nucleus lies ventrolateral in the caudal tegmentum and contains small, bi- and tri-polar cells (Fig. 6-14). Our normal material suggests that this nucleus has a close relationship with the tectobulbar pathways.

6.4 Functional Correlations

Work on the mammalian brain stem has established the nature of the reticular formation as a central integrating pool which is involved in the reception and processing of information from all portions of the central nervous system. Physiological studies in the cat have identified fairly localized areas responsible for the control of limbs, trunk and neck, and the movements of the eyes. Differentiated regions have also been recognized that appear to be involved in the control of sleep–wakefulness states, sensorimotor performance and the reception of nociceptive information.

At the present time we have little idea whether the brain stem regions of fishes have essentially the same roles as those in mammals, as there have been very few experimental physiological studies. For example, apart from anecdotal observations, there is little known about sleep–wakefulness states in elasmobranchs or, indeed, whether these terms have any meaning for these animals. It is known that some species show rhythmical patterns of locomotor activity (e.g. *Heterodontus, Cephaloscyllium,* Nelson and Johnson 1970; Finstad and Nelson 1975) but it is uncertain how these patterns relate to sleep and there are no indications of how brain stem actions might affect these patterns.

We do know however that the brain stem plays a significant role in determining the "state" of the spinal cord and is important for the control of locomotion.

Brain Stem and Locomotion. When the brain stem in batoids is separated from the spinal cord by a transection made at the obex, the resulting spinal preparation does not show spinal "shock", typical of mammals, but displays instantly a wide range of reflex movements (Crow 1927, *Raja;* Campbell 1951, *Dasyatis*); spontaneous locomotor movements, however, are absent. In contrast, in the shark-like elasmobranchs (e.g. *Mustelus, Squalus, Scyliorhinus, Squatina*), spontaneous locomotor movements of essentially the normal form appear immediately after spinal section, a result which suggests that the motor "program" for undulatory motion is produced by spinal-cord circuits. In species such as *Scyliorhinus,* where locomotor activity varies diurnally, these centres are presumably restrained by pathways that descend from the brain. The sites of origin of these pathways are unknown but transections made at various levels through the brain stem of *Scyliorhinus* indicate that they are located in the rhombencephalon close to the VIIIth nerve root. The study by Paul and Roberts (1979) on cerebellar function in *Scylirohinus* (cf. p. 77) suggests that the inhibitory impact on spinal-cord centres is modulated by the cerebellum.

Inhibitory and excitatory regions related to locomotion have been described in the mammalian brain stem (Mori et al. 1980), where it has been shown that activation of the lateral reticular formation results in controlled locomotion. This region extends as a continuous zone, rostral to the cuneiform nucleus, in the mesencephalon. Stimulation of this zone, the "mesencephalic locomotor region" (MLR), induces locomotion in the decerebrated cat. Kashin et al. (1974) reported that in teleost fish electrical stimulation of a limited region in the caudal part of the tegmentum evokes co-ordinated fin and body movements. They suggest that this region is equivalent to the mammalian MLR. The position of such a region is less clear in elasmobranchs, however, for although locomotion can be elicited in decerebrated preparations of *Dasyatis* by means of midbrain stimulation (Droge and Leonard 1979), Demski (1977) obtained locomotor responses from several regions of the brain in *Ginglymostoma.*

Brain Stem and Electric Organ Discharge. A modified motor control function for the brain stem is seen in those elasmobranchs that possess electric organs. Weak electric discharges are produced in Rajiformes from the small electric organs that are located in the tail. In torpedoids the pectoral discs contain large electric organs that produce substantial shocks. Both types of electric organ are activated only intermittently. In torpedoids this discharge is important for prey capture (Belbenoit and Bauer 1972) but the significance of the weak electric discharge of *Raja* is uncertain.

As in other electric fishes (teleosts) the coordination of electric discharge from the electric organs is achieved in most cases by a series of nerve cells that are activated in sequence. The cells which initiate the discharge are called "control" cells or "pacemakers" and the following neurons are "relay" cells (Bennett 1968). In *Torpedo* the electromotoneurons are contained in the rhombencephalic electric lobes; in *Raja* they lie in the spinal motor column. In both species the electromotoneurons are activated by a nucleus located in the inferior raphe nucleus. Little experimental work has been done on the properties of this nucleus (Szabo 1955) and it remains uncertain whether the nucleus is the "command" centre for electric discharge in these fishes. One somewhat surprising feature of the organization of this centre, particularly in *Torpedo,* is the apparent lack of gap junctions on these cells (Fox 1977; Nakajima 1970) and presumably, therefore, the absence of electrotonic coupling between the neurons which is such a striking feature of the command centres in other electric fish (Bennett 1968).

7 Cerebellum

7.1 Introductory Notes

It is difficult to delineate the extent of the cerebellum in fishes because the easily recognized corpus cerebelli lies adjacent to other regions which are essentially cerebellar-like in neural organization. These regions, the auricles and lateral-line areas, relate to the octavolateral sensory system and were considered by Larsell (1967) to form the octavolateral lobe. Nicholson et al. (1969), following Johnston (1902), considered this part of the brain to be the phylogenetic base for the vertebrate cerebellum. The octavolateral region certainly arises from the same rhombencephalic zone as the cerebellar anlage (Rüdeberg 1961), but it is important to recognize that cerebellar-like neural circuits are found in other regions of the vertebrate brain and even in invertebrate nervous systems, and that similarities of internal organization do not necessarily imply homology or a common function (Roberts 1981 b).

7.2 Structure

The cerebellum of cartilaginous fishes is much larger and further differentiated than is that of the cyclostomes. In early development it consists of a simple plate-like structure, similar to the cerebellum of the adult lamprey, but its form changes as development proceeds. The cerebellar anlage lengthens and the caudolateral portions protrude to form the auricles, while the corpus cerebelli develops from a dorsal evagination of the rostromedial parts.

The auricles are entirely covered on their dorsal surface by extensions of the tela chor-

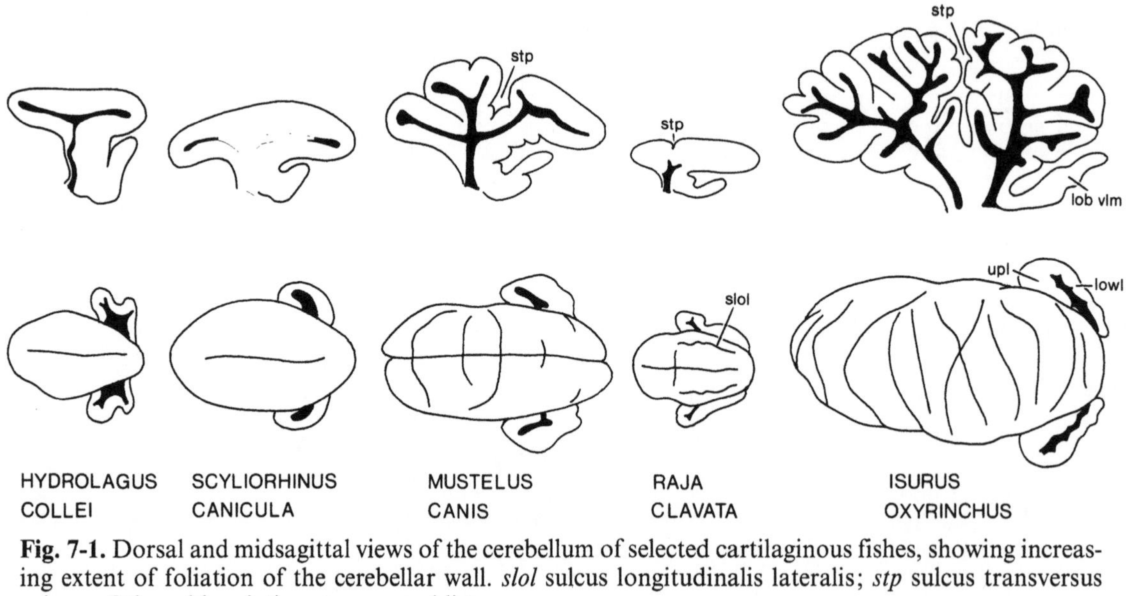

Fig. 7-1. Dorsal and midsagittal views of the cerebellum of selected cartilaginous fishes, showing increasing extent of foliation of the cerebellar wall. *slol* sulcus longitudinalis lateralis; *stp* sulcus transversus primus. Other abbreviations see general list

oidea of the fourth ventricle and can be subdivided on each side into a rostromedial upper leaf and a caudolateral lower leaf (Fig. 7-1). The lower leaf is contiguous caudally with the areae octavolaterales while the upper leaves of both auricles unite medially as the "lower lip", which bridges the fourth ventricle. The upper leaves and the lower lip are separated from the corpus cerebelli by a deep groove. Voorhoeve (1917) designated the lateral parts of this groove as the sulci para-auriculares and the intermediate part (i.e. the part marking the boundary between the lower lip and corpus cerebelli) as the sulcus postremus.

The unpaired corpus cerebelli encloses a large ventricular cavity (small in batoids) and extends rostrally over the roof of the midbrain and caudally over the lower lip. The basal part of the corpus forms a distinct stalk, the peduncle, which joins rostrally with the tectum, via a very short velum medullare anterius, and caudally with the lower lip and upper leaves of the auricles. The basal parts of the peduncle protrude medially into the ventricle as the eminentiae ventrales cerebelli, which contain on each side an aggregation of cells, first described by Edinger (1901) under the name of nucleus lateralis cerebelli. This nucleus, now generally called the nucleus cerebelli, is a large cluster of diffusely arranged medium-sized cells. Edinger and several later students considered this nucleus to be a subcerebellar structure but studies on its ontogenetic development by Rüdeberg (1961) have shown that it originates from the cerebellar anlage and lies within the cerebellum.

The extensive comparative studies of Voorhoeve (1917) emphasized that the chondrichthyan corpus cerebelli varies considerably in external form and size between species (Fig. 7-1). In primitive sharks such as *Hexanchus* and *Heptranchias,* and in some other sharks (e.g. *Scyliorhinus*) and in holocephalians also, the cerebellar surface is smooth or has only a slender median groove. However, in most sharks and in all batoids the corpus cerebelli displays transverse grooves. The sulcus transversus primus divides the corpus into anterior and posterior lobes of about equal size. In small-bodied species (e.g. *Squalus*) this groove is shallow and single, but in larger animals it is deep and associated with two other sulci. These sulci are situated before and behind the sulcus transversus primus and divide the anterior and posterior lobes into two further lobuli. In the larger sharks and rays accessory sulci divide these four cerebellar lobuli into numerous sublobuli and in some (e.g. *Dasyatis*) the corpus becomes markedly asymmetrical. The cerebellum of batoids differs from that of the sharks in having a well developed sulcus longitudinalis lateralis. It is important to emphasize that the convolution of the chondrichthyan cerebellum differs from that of amniotes in that the entire wall is folded, whereas only the external surface is convoluted in amniotes.

Voorhoeve introduced the idea that the variations in size of the cerebellum were in some way related to the size of the body of the fish. In fact such a simple correlation between cerebellar size and body size does not stand examination for, as noted by Northcutt (1978), specimens of different species of the same size (e.g. *Squalus acanthias, Mustelus canis, Sphyrna tiburo*) can exhibit the total range of cerebellar complexity seen among chondrichthyans. Further investigation reveals that the size of the cerebellum relates to the overall size of the brain. Hence the question of the relative development of the cerebellar corpus in different species becomes the much wider problem of accounting for differences in overall brain development. The size of the brain presumably reflects different levels of sensory analysis and motor performance and relates to the relative complexity of the behaviour pattern of a particular species. But correlations of this type are very difficult to make even for species for which there are complete descriptions of behaviour, and are not yet possible for elasmobranchs.

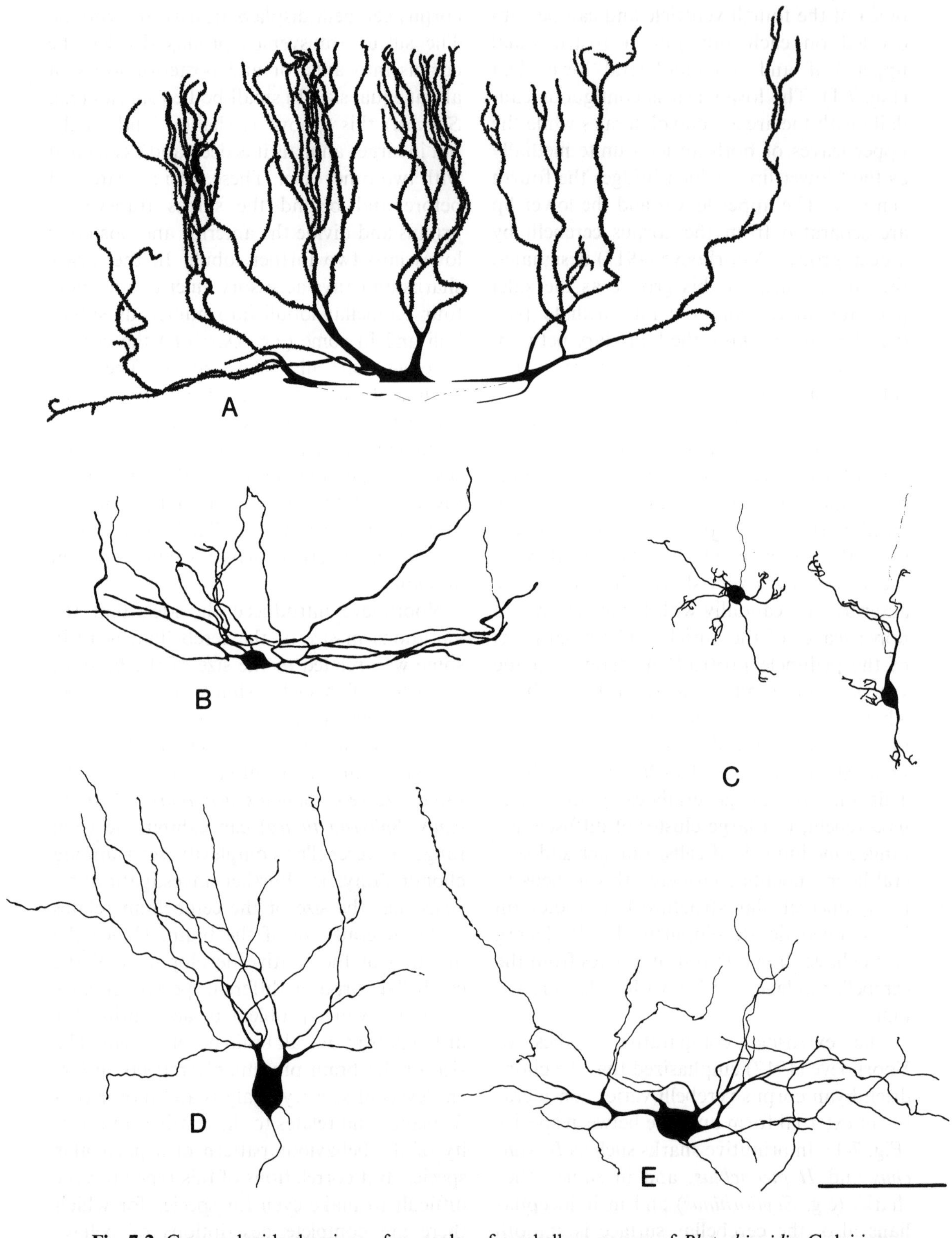

Fig. 7-2. Camera lucida drawings of examples of cerebellar neurons of *Platyrhinoidis*. Golgi impregnation. **A** Purkinje cell; **B** stellate cell; **C** granule cells; **D** and **E** Golgi cells. *Bar scale* = 100 μm. (Combined figures from Nicholson et al. 1969).

In its internal organization the chondrichthyan cerebellar corpus is very similar to that of other gnathostomes. Its wall is differentiated into four layers: the innermost granular layer, the fibre zone, the layer of Purkinje cells, and the outer molecular layer. The granule cells are concentrated in two longitudinal ridges, the prominentiae granulares of Nieuwenhuys (1967a), which are situated on either side of the median plane. They begin immediately behind the trochlear decussation, curve along the inside of the wall of the corpus cerebelli, and terminate as two slight ventral protrusions of the lower lip. The granule cells are small and densely packed, as in the cerebellum of other vertebrates. The Golgi studies of Schaper (1898), Houser (1901) and Nicholson et al. (1969) have shown that these elements have from three to six short dendrites that emerge randomly from the somata, sometimes bifurcate once or twice, and terminate as large, convoluted claws. The axon of the granule cell arises either from the soma or from the proximal region of a dendrite (Fig. 7-2) and then passes to the molecular layer where it divides as a T to give parallel fibres, lying in the transverse plane of the cerebellum. Nicholson et al. (1969) confirmed from Golgi material and electrophysiological studies that half of the parallel fibres cross the midline so that a given Purkinje cell receives parallel fibre inputs from groups of granule cells on each side of the median plane.

In addition to granule cells, the prominentiae granulares also contain larger neurons, assumed to be Golgi cells (Nicholson et al. 1969; Eccles et al. 1970) on the basis of comparison with similar cells found in higher vertebrates. Golgi cells probably receive input from mossy fibres and the more superficial elements of this type extend into the molecular layer, where they are in contact with parallel fibres.

In the lateral parts of the corpus cerebelli, where granule cells are absent, a zone of fibres can be distinguished beneath the Purkinje cells. This zone includes incoming afferent fibres and outgoing Purkinje cell axons that are passing ventrally to the cerebellar nucleus.

The Purkinje cells form a regular layer, one or two cells thick, which extends throughout the entire wall of the corpus cerebelli, with the exception of the paramedian regions. In *Hydrolagus,* however, the arrangement of Purkinje cells is less clear. In this species the Purkinje cells do not form a layer but are rather scattered throughout the molecular layer (Fig. 7-3). The studies of Schaper (1898), Houser (1901) and Nicholson et al. (1969) have revealed that the Purkinje cell has an elongated body from which several thick spineless processes emerge. Each process gives rise to a number of ascending spiny dendrites, which form together a dendritic tree that is quite markedly flattened in the parasagittal plane.

The molecular layer contains several elements: (1) axons ascending from granule cells, and their branches (i.e. parallel fibres); (2) Purkinje cell dendrites; (3) possibly climbing fibres, and (4) stellate-type interneurons.

A thorough discussion of whether climbing fibres are present in the cerebellum of fishes has been provided recently by Paul (1982). The inferior olivary nucleus, the chief source of climbing fibres in the mammal, has long been recognized in many elasmobranchs (see p. 74). Recent studies in *Scyliorhinus* have shown this nucleus to label retrogradely with HRP injected into the cerebellum (Paul and Roberts unpublished), suggesting that axons of olivary neurons pass to the cerebellum. It remains uncertain, however, how they terminate there. Although Ariëns Kappers et al. (1936) thought that all cerebellar afferents form climbing fibres, and Nicholson et al. (1969) have reported climbing-fibre-like synapses on Purkinje cell dendrites, other authors have failed to recognize climbing fibres in Golgi preparations. For teleosts, Pouwels (1978) and Finger (1978b) have reported that climbing fibre terminals are confined to the Purkinje somata.

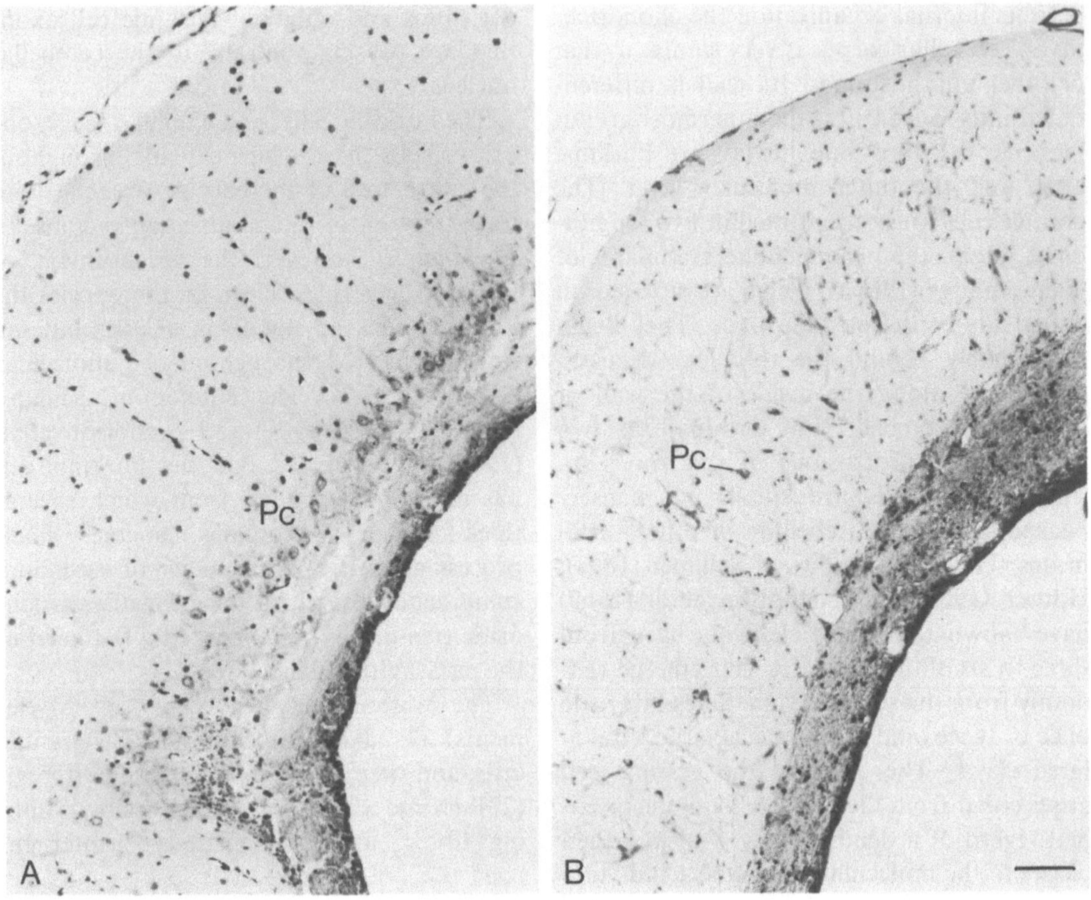

Fig. 7-3. Photomicrographs of transverse sections showing the arrangement of Purkinje cells in the corpus cerebelli of *Raja* **A** and *Hydrolagus* **B**. *Pc* Purkinje cell. *Bar scale* = 100 µm

Stellate cells have been described in detail in the dogfish, *Mustelus*, by Schaper (1898) and Houser (1901), and in the thornback ray, *Platyrhinoidis*, by Nicholson et al. (1969). The cells found by these authors display some morphological variations but can be divided into two categories according to their location in the molecular layer. The dendrites of the more superficially located cells, although passing in all directions, predominantly lie parallel to the sagittal plane. They are rather sparsely covered with small spines. The axon of the superficial stellate cell emerges from the soma or from one of the dendrites to ramify within the territory of the dendrite. It is characterized by its delicate structure and bears small swellings. The more deeply located stellate cells tend to have a fusiform body and their dendritic trees are flattened horizontally. Unlike mammalian basket cells they do not form pericellular baskets.

Electrical stimulation of the surface of the corpus has been carried out for several elasmobranch species and evokes a series of potentials that can be recorded with microelectrodes inserted into the molecular layer. These compound potentials are thought to result from a volley of parallel fibre activity, conducting at around 0.2 ms^{-1} (Paul 1969), and the subsequent excitation of the Purkinje cell and stellate cell dendrites. The precise interpretation of these potentials has caused some controversy (see Paul 1982). For example, complex responses of the type attributed in mammals to climbing fibre activity have been seen in elasmobranchs in response to surface stimulation. Nicholson et al. (1969) have shown, however, that these responses could be blocked by intracellular hyperpolarizing current, and so were not synaptic potentials, whilst similar responses could be directly evoked by the injection of depolarizing current into the cell, suggesting that they are a property of the Purkinje cell membrane. Some of the potentials evoked by surface stimulation have been interpreted as resulting from stellate cell inhibition of the Purkinje cells (Nicholson and Llinás 1969).

A recent study by Montgomery (1978) has extended our knowledge of the organization of the auricles. The three component regions, the upper and lower leaves and the lower lip (pars medialis of Larsell 1967), differ in their internal organization and connections. The lower lip shows a typical cerebellar construction with a molecular layer, conspicuous Purkinje cells and granular layer. The granule cells not only form a layer beneath the Purkinje cells but also cluster as median accumulations, the granular cords or prominentiae granulares, which join with the prominences of the cerebellar corpus. The lateral extent of the upper leaf is composed exclusively of granule cells, and perhaps, Golgi cells and is not covered by the molecular layer. Only a few Purkinje cells are found in the lower leaf but in the lower lip they form a well defined layer (except for *Hydrolagus*). Their dendritic arborizations are oriented transversely except for those cells in the most rostral part of the lower lip which lie in the sagittal plane (as they do in the corpus).

7.3 Cerebellar Afferents and Their Centres of Origin

Data on afferent fibre connections to the chondrichthyan cerebellum have been presented by a great number of authors using normal material (e.g. Voorhoeve 1917; Ariëns Kappers 1921; Ariëns Kappers et al. 1936; Larsell 1967; Nieuwenhuys 1967a). In more recent years our knowledge of cerebellar afferents has been extended by the experimental studies of Hayle (1973b), Boord and Campbell (1977), Montgomery (1978), Boord and Roberts (1980), Paul and Roberts (1981) and Ebbesson and Hodde (1981); these findings are summarized in Fig. 7-4. The following afferents to the chondrichthyan cerebellar regions have been described: (a) primary fibres of the lateral line nerves; (b) primary fibres of the octavus nerve to the lower lip; (c) fibres originating in the area octavolateralis; (d) trigeminocerebellar fibres; (e) a tractus spinocerebellaris; (f) an olivocerebellar system; (g) a tractus mesencephalocerebellaris, and (h) a tractus lobocerebellaris. The first three projections, associated with the octavolateralis system, have already been described.

Trigeminocerebellar fibres have been described by Haller (1898) and Edinger (1901) although Ariëns Kappers et al. (1936) and Larsell (1967) were uncertain whether direct root fibres of the trigeminal nerve reached the cerebellar corpus or the auricles. A connection between the mesencephalic trigeminal nucleus and the anterior lobe of the corpus cerebelli has been reported by Witkovsky and Roberts (1975, *Scyliorhinus*) who traced axons of the mesencephalic trigeminal neurons from the caudal part of the tectum to the anterior lobe of the corpus cerebelli, where they terminated in close proximity to Purkinje cells. Some electrophysiological evidence for the existence of such a pathway was also obtained (Roberts and Witkovsky 1975).

A *spinocerebellar system* (tractus spinocerebellaris) comprising a smaller rostral and a larger caudal portion was described in normal material by Ariëns Kappers (1921), Ariëns Kappers et al. (1936) and Larsell (1967). According to these authors the rostral portion reaches the cerebellum by way of the velum medullare anterius and is believed to be homologous with the ventral spinocerebel-

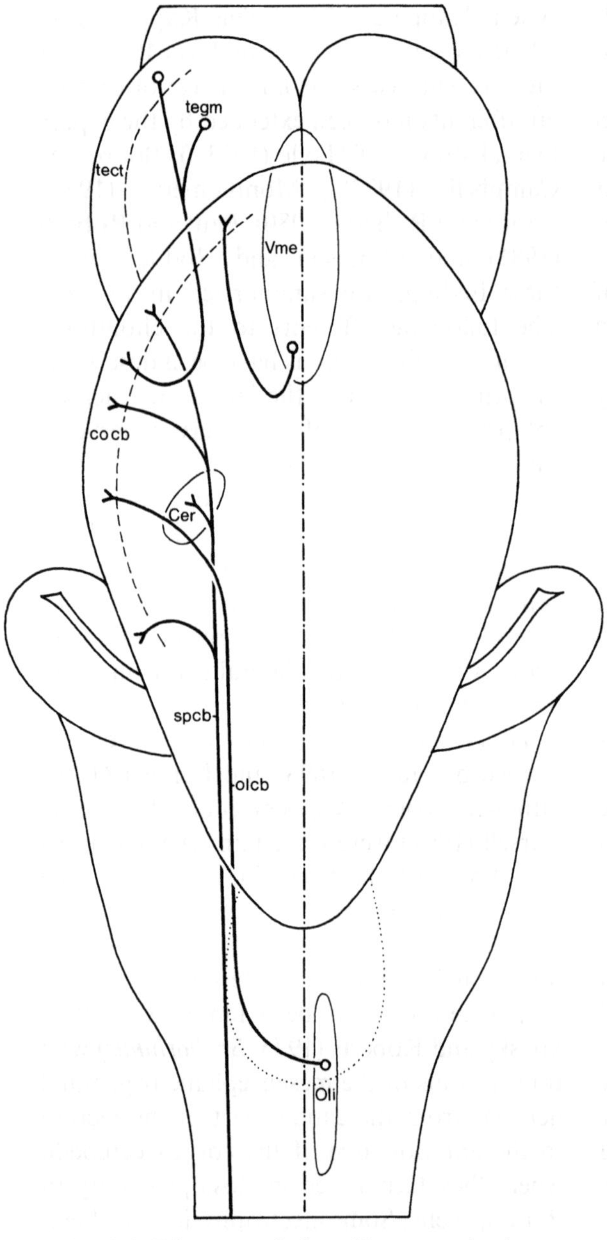

Fig. 7-4. Diagram of the afferent projections to the cerebellum, as viewed from the dorsal surface of the brain

lar tract of mammals. The caudal portion, which would represent the dorsal spinocerebellar tract of mammals, ascends toward the corpus cerebelli behind the eminentia ventralis cerebelli through the basal part of auricles. The greater part of the spinocerebellar system terminates ipsilaterally, but some of its fibres decussate through the commissural plate of the corpus cerebelli to the other side (Ariëns Kappers et al. 1936). A degeneration study by Hayle (1973b) has shown that in the dogfish *Scyliorhinus* the spinocerebellar projections cannot be divided into dorsal and ventral spinocerebellar tracts and that degenerating spinocerebellar fibres cannot be clearly distinguished from spinobulbar fibres, although they tend to lie more dorsally in the spinal lemniscus. The spinocerebellar fibres terminate as mossy fibres bilaterally in the prominentiae granulares. According to Hayle (1973b), the density of the degeneration involving terminal arborizations was greatest throughout the intermediate portion of the corpus cerebelli and rather less within its caudal and rostral poles. There was also some evidence of a preferential distribution to the dorsal rather than to the ventral granular eminence. In *Ginglymostoma* Ebbesson and Hodde (1981) confirmed that the spinocerebellar pathway terminates in the granular layer. They also observed some termination in the cerebellar nucleus.

An *olivocerebellar system* i.e. tractus olivocerebellaris, accompanies the spinocerebellar tract and contains fibres that originate from the contralateral inferior olive and terminate in all parts of the body of the cerebellum (Ariëns Kappers 1921). The oliva inferior is an elongated nucleus, which occupies a paramedian position in the caudal part of the rhombencephalon and comprises small, round and ellipsoid cells. This nucleus has been observed in a variety of cartilaginous fish and equated to the mammalian inferior olive by numerous authors, among them Rohon (1878), Kooy (1917) and Ariëns Kappers (1920). Gerlach (1947) questioned the homology with the inferior olive of amniotes and preferred the non-committal name, nucleus paraseptalis. The mammalian inferior olive is the source of climbing fibres to the cerebellum. The debate about whether cartilaginous fish have a climbing fibre projection has already been considered (Sect. 7.2).

A *tractus mesencephalocerebellaris* has been described by Voorhoeve (1917) and Ariëns Kappers (1921) as running caudally through the midbrain tegmentum and the isthmus region. These fibres arch dorsally in the rostral part of the rhombencephalon and pass rostral to the eminentia ventralis. Ariëns Kappers (1921) thought that at least a part of this system arose from the rostral part of the tectum mesencephali, but so far the exact origin of this bundle has not been determined.

A *tractus lobocerebellaris,* originating from the lobus inferior hypothalami, was reported by Voorhoeve (1917) and Ariëns Kappers (1921) as passing from the caudal part of the inferior lobe of the hypothalamus to the ipsilateral anterior part of the corpus cerebelli. Sterzi (1909) described both direct and crossed lobocerebellar tracts, but Ariëns Kappers et al. (1936) were not convinced of the presence of such a system. In sagittal sections of the four species studied numerous fibres can be recognized which connect the inferior lobes with the caudal brain parts. The existence of a lobocerebellar tract, therefore, cannot be excluded.

7.4 The Efferent Projections from the Cerebellar Corpus and Auriculae

The only output from the cerebellar cortex of mammals is provided by Purkinje cell axons which project onto deep cerebellar nuclear neurons and inhibit their activity. It is a question of considerable comparative interest as to whether axons of Purkinje cells of elasmobranchs terminate in a similar fashion, or pass directly to other regions of the brain. A recent attempt by Ebbesson and Campbell (1973) to answer this question for the shark *Ginglymostoma* indicated that the Purkinje cell axons do not project to the brain stem but terminate in the cerebellar nuclei located in the cerebellar peduncle. A similar result has also come from an experimental study in *Scyliorhinus* (Paul and Roberts unpublished) where it has been demonstrated that the Purkinje cells inhibit the activity of cerebellar nuclear neurons (Paul and Roberts 1976).

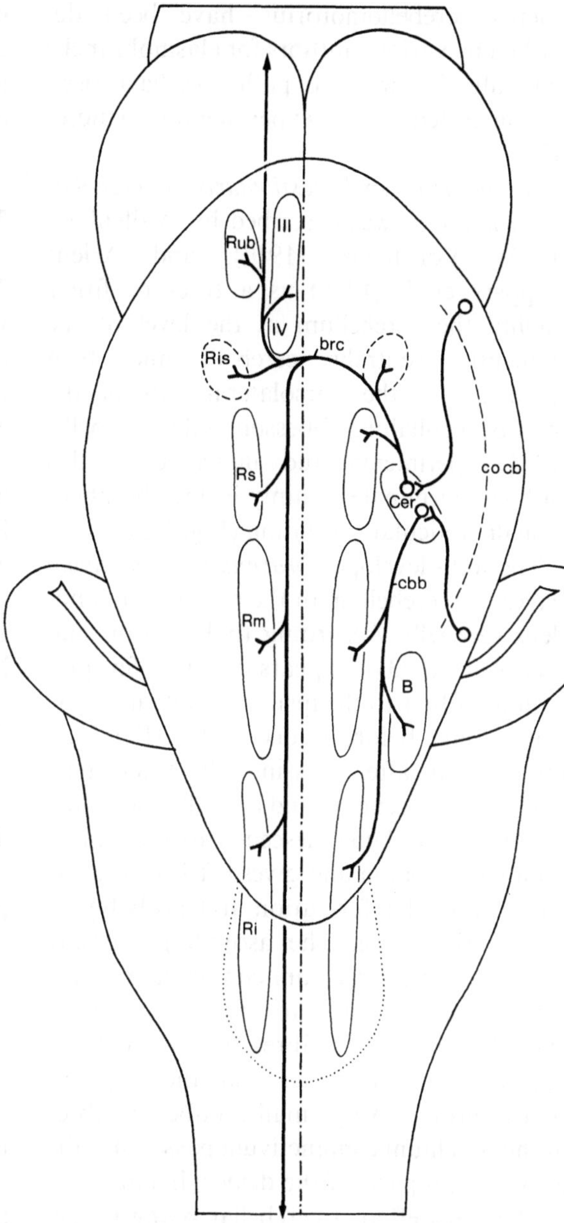

Fig. 7-5. Diagram of the efferent projections of the cerebellum, as viewed from the dorsal surface of the brain

Three cerebellofugal systems—a tractus cerebellovestibularis et cerebellobulbaris rectus, a brachium conjunctivum and a tractus cerebellomotorius—have been described by various authors for elasmobranchs but only the first two pathways have been demonstrated by experimental studies (Fig. 7-5).

A *tractus cerebellovestibularis et cerebellobulbaris rectus* was described by Wallenberg (1907), Voorhoeve (1917), and Ariëns Kappers et al. (1936) as a tract of fibres leaving the cerebellum at the level of the eminentia ventralis cerebelli and then running into the dorsolateral part of the rhombencephalon. Ebbesson and Campbell's (1973) experimental study indicated that this pathway terminates mainly in the rhombencephalic reticular formation (Fig. 7-5).

The well-developed *brachium conjunctivum* leaves the cerebellum in the region of the nucleus cerebelli. According to Ebbesson and Campbell (1973) it appears to issue a few preterminal fibres to the ipsilateral reticular formation just before it decussates (Fig. 7-5). Immediately after crossing, the tract splits into a large ascending and a small descending pathway. The latter descends to the caudal rhombencephalon and gives off fibres to the medial reticular formation and probably to the inferior olive. The ascending crossing pathway of the brachium conjunctivum provides a massive input to the nucleus ruber, and also distributes fibres to the trochlear and oculomotor nuclei and the adjacent central grey. A very small number of fibres of the brachium conjunctivum passes into the dorsomedial parts of the diencephalon.

The third efferent cerebellar pathway, i.e. the *tractus cerebellomotorius,* has already been dealt with in connection with the components, which constitute together the medial longitudinal fascicle (Sect. 6.3).

The output of the auricles is less well known. At least some of the granular cells of the upper and lower leaves send their axons as parallel fibres into the crest that covers the dorsal and intermediate nuclei (Montgomery 1981; Roberts 1981b). The projections from the Purkinje cells of the lower lip have not been determined experimentally but there is some electrophysiological evidence suggesting that they terminate in, and inhibit the activity of, the octaval nuclei (Montgomery 1978).

7.5 Functional Correlations

There have been few neurophysiological investigations of the elasmobranch cerebellum and the majority of these have been designed to elucidate interneuronal connections. Nevertheless the physiological studies that have been performed, when taken in conjunction with anatomical work, show that the elasmobranch cerebellum contains the same kind of elements as the mammalian cerebellum, and has similar interactions (Nicholson et al. 1969). The basic organization of the cerebellum was evidently established early in vertebrate history.

Although it is widely held that the cerebellum plays an important role in motor control it still remains unclear how this is achieved. The profound locomotor disturbances seen in mammals following cerebellectomy are primarily the result of loss of postural control, and of course this problem is considerably simplified for aquatic animals because of the support supplied by the dense medium. Why, therefore, is the cerebellum so well developed in elasmobranch fishes which perform relatively stereotyped movements, uncomplicated by postural problems?

Some attempts have been made to determine cerebellar function by observing the behavioural changes that follow cerebellar lesions. The results of such surgery are hard to interpret for it is always uncertain whether the reported modification is a direct or an indirect effect of the ablation. Lesions of the auricles of various elasmobranchs were found by Ten Cate (1930) to lead to tonic muscular contractions, bending towards the lesioned side and spiral swimming. He also thought that removal of the corpus alone had no effect but other workers have noted that ablation of the corpus results in a decline in the strength and performance of body and fin movements. Lesions that include damage to just one of the cerebellar nuclei are followed by asymmetry in body and fin muscle tone (Polimanti 1911; Tilney 1923; Dijkgraaf 1949; Paul and Roberts 1979).

A recent study on the consequence of cerebellar ablation for a reflex pectoral fin movement in *Scyliorhinus* gives some insight into the significance of the cerebellum (Paul and Roberts 1979). This work showed that the reflex was considerably attenuated in cerebellectomized fish and suggested that the cerebellum plays a modulating rather than an initiating role in locomotor control. The motor loops that generate movements of the fins and body are located in the spinal cord and the "gain" of these circuits is regulated by the rhombencephalic reticular formation which, in its turn, is controlled by cerebellar action. This might mean that different regions of the body will be regulated by selected areas of the cerebellum which will effectively hold a map of the body's motor activities. Consequently, complex movements might require elaborated cerebella. This may help us to understand why elasmobranch cerebella are relatively so much larger than that of most teleosts, for the latter, in possessing neutral buoyancy, do not have the problem of matching thrust against lift which confronts the swimming elasmobranch.

8 The Tectum Mesencephali

8.1 Introductory Notes

In the majority of cartilaginous fishes the tectum mesencephali is strongly developed and differentiated into two bilateral lobes, which surround expansions of the ventricular cavity. In the species studied this cavity is most spacious in *Squalus* and *Hydrolagus*, but much reduced in *Raja*.

The tectum mesencephali is the chief centre of termination for retinofugal fibres and, as in other fishes, generally appears to be well developed in those species with large eyes. Thus, the tecta of *Squalus* and *Hydrolagus*, (essentially deep-water forms with large eyes) are relatively larger than those of *Scyliorhinus* and *Raja*. As well as having a clear relationship to vision the tectum mesencephali also receives afferent inputs from several non-visual systems so the widely used term tectum opticum is inappropriate and should be avoided. The efferent pathways from the tectum pass to all main regions of the brain, except for the telencephalon.

8.2 Structure

The chondrichthyan tectum differs between species not only in the size of the enclosed ventricular cavity but also in the degree of internal lamination. Several tectal laminae can be distinguished although there is presently disagreement about the exact number in the different species (Table 8-1). For the present analysis we have subdivided the tectum into six tectal zones (Fig. 8-1). Passing from meningeal to ventricular surfaces these are: (1) stratum medullare externum; (2) zona externa of the stratum cellulare externum; (3) zona interna of the stratum cellulare externum; (4) stratum medullare internum; (5) stratum cellulare internum, and (6) stratum fibrosum periventriculare. *Squalus* and *Scyliorhinus* exhibit a particularly clear lamination. A similar, although less distinct, lamination is seen in *Raja*, whereas the tectum of *Hydrolagus* differs considerably from that of the elasmobranchs. In the latter, most of the tectal neurons occupy a periventricular position (Fig. 8-1 D) and hence, a deep fibre layer (tectal layer 6) is absent. This periventricular cell layer, seven to eight cells thick, is capped by three to four additional cell layers, each one to two cells thick, separated by thin fibre layers.

We will survey briefly the characteristics of the neurons and fibres of the several tectal layers as revealed by the Golgi studies of Houser (1901; *Mustelus canis*), Leghissa (1962; *Squalus acanthias, Raja clavata, Torpedo ocellata, Scyliorhinus stellaris, Mustelus laevis*), Schroeder and Ebbesson (1975; *Ginglymostoma cirratum*), and Witkovsky et al. (1980; *Raja clavata* and *Raja eglanteria*).

Layer 1, the stratum medullare externum, consists of a few neurons, scattered between a network of unmyelinated fibres. The neurons in this marginal layer are bipolar with horizontally radiating dendrites and an axon that terminates in layers 2 and 3. Some

Fig. 8-1. Photomicrographs of transverse sections showing the tectal lamination in *Squalus* (**A**), *Scyliorhinus* (**B**), *Raja* (**C**) and *Hydrolagus* (**D**). *Left half* Klüver and Barrera staining; *right half* Nissl staining. *Bar scale* = 100 μm

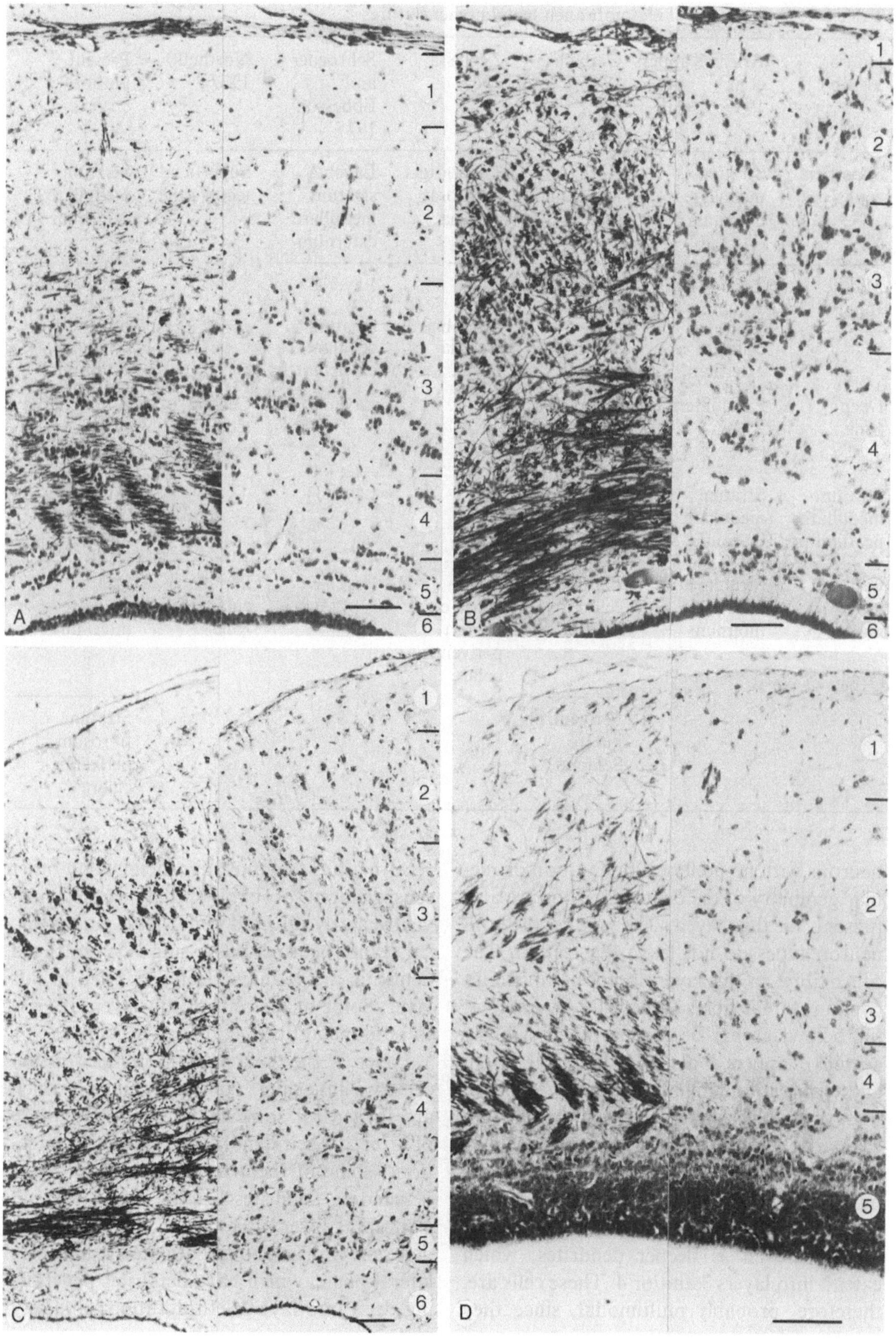
A
1
2
3
4
5
6
B
1
2
3
4
5
6
C
1
2
3
4
5
6
D
1
2
3
4
5

Table 8-1. Comparison of elasmobranch tectal nomenclatures

Houser 1901	Ariëns Kappers et al. 1936	Gerlach 1947	Leghissa 1962	Schroeder and Ebbesson 1975	Northcutt 1977	Present authors
Superficial zone	Stratum medullare externum	Kernarme Molekularschicht	Zona grigia e fibrosa externa	Layer A stratum medullare externum	Superficial tectal zone	Stratum medullare externum
Middle zone	Zona externa	Opticusfasern		Layer B		Zona externa
	Stratum cellulare externum		Zona biana e gigria media	Stratum cellulare externum		Stratum cellulare externum
Deep zone	Zona interna	Hauptzellenschicht		Layer C	Central tectal zone	Zona interna
Stratum medullare profundum	Stratum medullare internum	Lamina medullaris interna		Layer D		Stratum medullare internum
Central gray matter	Stratum cellulare internum	Lamina cellularis interna	Zona gigria e fibrosa periventricolare	Layer E	Periventricular tectal zone	Stratum cellulare internum
		Periventrikuläre Schicht				Stratum fibrosum periventriculare

neurons with no preferred orientation or regular geometry of the dendritic arbor are also present in this layer. The axons of both neuron types do not leave the tectum. The other fibres in this layer do not comprise incoming retinal fibres (except in *Hydrolagus*, see below) as has been reported previously (Ariëns Kappers et al. 1936) but their origin and termination remain unknown.

Layer 2, the zona externa of the stratum cellulare externum, consists of numerous pyramidal and bipolar cells. Both types of cell have superficial dendrites which branch repeatedly into layer 1, and receive retinal input, as well as deeper dendrites, which extend into layers 3 and/or 4. These cells are, therefore, probably multimodal, since they receive a retinal input superficially and other sensory input (probably spinal) via their periventricularly directed dendrites. In addition, small interneurons possessing very restricted dendritic arbors have been identified in this layer. Some of the cells of this layer give rise to retinopetal fibres (Luiten 1981 a).

Layer 3, the zona interna of the stratum cellulare internum, contains neurons which are in general somewhat larger than those of layer 2. Their dendrites extend in all directions, some of them reaching layer 2. Their descending dendrites usually do not extend as deep as layer 4. However, there are some neurons, with cell bodies oriented in a mediolateral plane, which have dendrites penetrating deep into layer 4. In addition, a variety of cells with long, radially oriented processes have been described in tectal layer 3. It is generally assumed that the axons of the neurons of tectal layer 3 contribute to the main tectal efferent pathways. There is experimental evidence that this layer receives a small retinal input, an input from the contralateral tectal half (Smeets 1982) and a telencephalic input (Ebbesson and Schroeder 1971).

Layer 4, the stratum medullare internum contains only a few cells. The main efferent tectal fibres course through this layer (Smeets 1981 b).

Layer 5, the stratum cellulare internum, contains various cell types. The most notable cells of this layer are the very large neurons of the mesencephalic trigeminal nucleus (see p. 35). Tectal neurons in this layer have piriform perikarya and processes that often arborize horizontally. An HRP study by Smeets and Timerick (1981) has revealed that at least some of these neurons project directly to the spinal cord.

Layer 6, the stratum fibrosum periventriculare, has not previously been distinguished as a separate entity in the chondrichthyan tectum by most authors (Table 8-1). The origin and termination of the unmyelinated fibres in this layer are unknown.

8.3 Connections

8.3.1 Afferent Projections to the Tectum

The bulk of the input to the tectum comes from the retina which, as is well known, is an embryological extension of the brain and consists of visual receptor cells and a network of neurons.

In terms of its main cell types and layers the retina is remarkably constant in all vertebrates (see Stell 1972a for review). The elasmobranch retina was reinvestigated by Stell (1972b, *Squalus acanthias*), Stell and Witkovsky (1973a, b, *Mustelus canis*) and Witkovsky and Stell (1973, *Mustelus canis*). It was shown that the elasmobranch retina can be subdivided into three cellular and two synaptic or plexiform layers. The receptor cell bodies constitute the outer nuclear layer. Both rods and cones are present although rods predominate (e.g. 100:1 in *Mustelus;* Witkovsky and Stell 1973). The retina of *Raja,* however, is thought to be without rods (Dowling and Ripps 1970). The outer plexiform layer is formed by the dendrites of horizontal and bipolar cells as well as the telodendria of photoreceptors. The inner nuclear layer contains the perikarya of horizontal, bipolar and amacrine cells, the three main categories of intrinsic retinal neurons. The inner plexiform layer consists of the terminals of bipolar and amacrine cells and the dendrites of ganglion cells. The innermost cellular layer is formed by the giant ganglion cells.

The axons of the ganglion cells constitute the optic nerve and carry information which represents the activity of photoreceptor cells, transformed by convergence and synaptic interactions of intermediary neurons, to the brain. In addition to these retinofugal fibres the optic nerve also contains retinopetal fibres. Witkovsky (1971, *Mustelus*) observed myelinated fibres coming from the optic nerve and making synapses with both amacrine and bipolar cells. Recently Luiten (1981 a), after intraocular HRP-injections in the nurse shark *Ginglymostoma,* found labelled tectal neurons in the zona externa of the stratum cellulare externum, contralateral to the injection side. As Luiten suggests, these cells may, therefore, be considered as cells of origin of retinopetal fibres.

The retinotectal projection, which constitutes the majority of the retinofugal fibres, has been studied in recent years with experimental hodological techniques, viz. selective silver impregnation and autoradiographic tracing (Ebbesson and Ramsey 1968, *Ginglymostoma cirratum* and *Galeocerdo cuvieri;* Graeber and Ebbesson 1972, *Negaprion brevirostris;* Northcutt 1979a, *Squalus acanthias;* Ebbesson and Meyer 1980, *Rhinobatos productus;* Northcutt and Wathey 1980, *Platyrhinoidis triseriata;* Witkovsky et al. 1980, *Raja clavata* and *Raja eglanteria;* Smeets

1981 a, *Scyliorhinus canicula* and *Raja clavata*). It appears from these experiments that the optic fibres destined for the tectum decussate totally in the optic chiasm and proceed in a dorsocaudal direction along the periphery of the diencephalic wall and, together with fibres which terminate in the thalamus and in the pretectal area, constitute the tractus opticus marginalis. Upon attaining the rostral pole of the tectum this bundle separates into a tractus opticus medialis and a tractus opticus lateralis. The fibres of the former course dorsomedially and spread over the rostral and dorsomedial portions of the tectum, while those of the latter terminate in the lateral and caudal parts of the tectum. All studies using selective silver impregnation or autoradiography have revealed only contralateral retinotectal projections, except for the autoradiographic study of Northcutt and Wathey (1980) in *Platyrhinoidis* which revealed the presence of ipsilateral as well as contralateral retinotectal pathways. The sparse ipsilateral projection appeared to be restricted to the rostral half of the tectum.

Considerable differences in the tectal termination of optic fibres are seen between species. In both *Ginglymostoma* (Ebbesson and Ramsey 1968) and *Scyliorhinus* (Smeets 1981 a) most optic fibres terminate in the stratum medullare externum, whereas in *Negaprion* (Graeber and Ebbesson 1972), *Platyrhinoidis* (Northcutt and Wathey 1980) and *Raja* (Smeets 1981 a) the zona externa of the stratum cellulare externum receives the bulk of retinal input. In *Galeocerdo* (Ebbesson and Ramsey 1968), *Squalus* (Northcutt 1979 a), and *Rhinobatos* (Ebbesson and Meyer 1980) the stratum medullare externum and zona externa of the stratum cellulare externum receive about the same number of terminations.

In addition to the visual input, the tectum mesencephali has been reported to receive fibres from all main parts of the brain and spinal cord (Fig. 8-2). The presence of spinotectal fibres was suggested by Ariëns Kappers et al. (1936) on the basis of normal material.

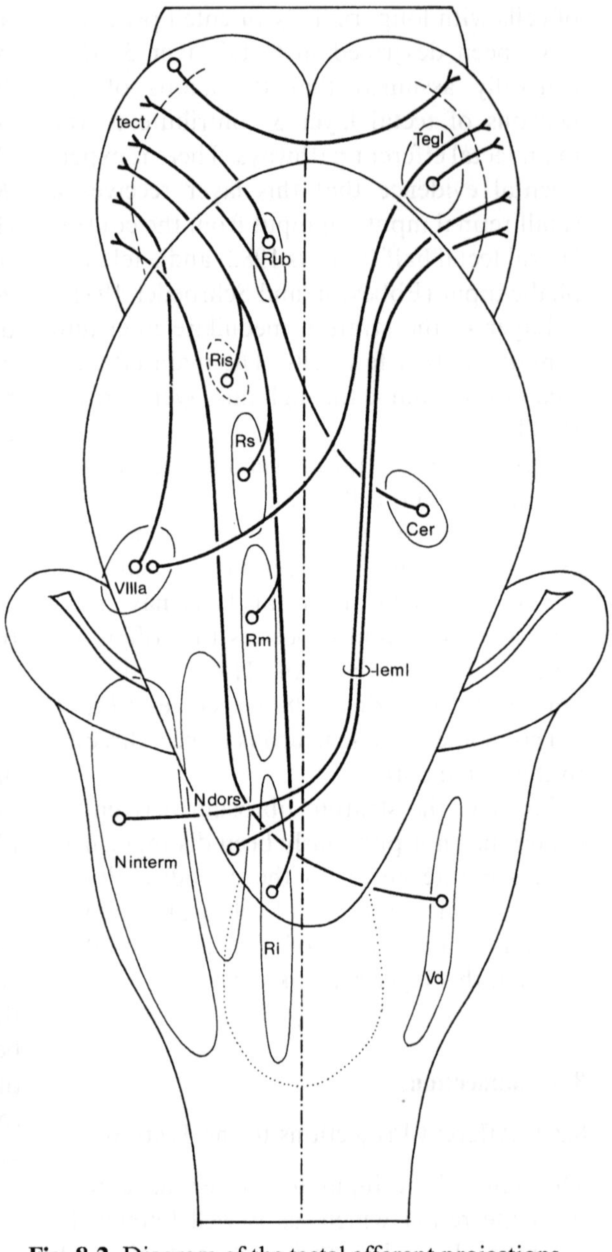

Fig. 8-2. Diagram of the tectal afferent projections originating from the brain stem, as viewed from the dorsal surface of the brain

Hayle (1973 a) traced degenerating axons to the ipsilateral zona externa of the stratum cellulare externum, following spinal cord lesions in *Scyliorhinus*. This degeneration appeared to be distributed to the intermediate portion only and was absent in both the rostral and caudal poles of the tectum. A degeneration study by Ebbesson and Hodde

(1981) on *Ginglymostoma* has also demonstrated the presence of a spinal projection to the deep layers of the tectum. In a recent study with unilateral HRP injections in the tectum of *Scyliorhinus* and *Raja* (Smeets 1982) retrogradely labelled cells were found in the spinal cord. These labelled cord neurons were sited both ipsilateral and contralateral to the injection side and were confined to the dorsal half of the "cervical cord".

The rhombencephalic cell masses that are known to project to the tectum mesencephali, are the reticular formation, the nucleus tractus descendens nervi trigemini, the nucleus dorsalis and intermedius areae octavolateralis and the nucleus octavus ascendens (Fig. 8-2). Following HRP injection in the tectum of *Scyliorhinus* labelled neurons could also be found in the nucleus cerebelli and the nucleus funiculi lateralis (Smeets 1982).

Reticulotectal connections have also been reported in teleosts (Luiten 1981c; Grover and Sharma 1981) and in the frog (Wilczynski and Northcutt 1977). Whereas HRP injections into the spinal cord of *Raja* and *Scyliorhinus* label large and very large reticular neurons, injections into the tectum mark only medium-sized reticular neurons.

A crossed tectal afferent pathway, originating from the nucleus tractus descendens nervi trigemini, has been described by Wallenberg (1907) in *Scyliorhinus stellaris* on the basis of Marchi experiments. The HRP-study by Smeets (1982) has confirmed the existence of such a pathway in *Scyliorhinus* and *Raja;* the projection was found to be bilateral, however.

Afferent connections of the tectum which originate from the area octavolateralis have been described by Boord and Northcutt (1979, 1982) for the ray *Raja eglanteria.* These authors have traced degenerating fibres, resulting from lesions in the nucleus dorsalis and intermedius areae octavolateralis, which ascend within the lateral lemniscus to mesencephalic levels particularly on the contralateral side. Most of these fibres terminate in the dorsolateral part of the tegmentum, but a substantial number continues beyond, to terminate within the tectum. The HRP-study of Smeets (1982) has confirmed the existence of a direct connection between the octavolateral area and the tectum and also the predominantly contralateral character of this projection.

The afferent tectal connections reported by Smeets (1982) as originating from the nucleus funiculi lateralis, the nucleus octavus ascendens and the nucleus cerebelli have not been previously described in cartilaginous fishes, nor in fact in other vertebrates. The labelling of the nucleus cerebelli is especially strong after a tectal injection. However, the paucity of experimental data about tectal afferents in other vertebrates would render it premature to draw conclusions from these findings.

The contralateral tectal half, the nucleus tegmentalis lateralis, the ventrolateral tegmental area, and the nucleus ruber have all been shown to project to the tectum (Fig. 8-2). It has been found (Smeets 1982) in *Scyliorhinus* and *Raja* that the input from the contralateral tectum terminates in the zona interna of the stratum cellulare externum. The fact that labelled cells can be recognized in a nucleus, which we have previously designated as nucleus ruber (Smeets and Nieuwenhuys 1976), indicates that this cell mass is more complex than in amniotes.

In the diencephalon the pretectal area, the thalamus dorsalis pars medialis, the thalamus ventralis pars lateralis and the nucleus medius hypothalami were seen to be the sources of pathways projecting to the tectum mesencephali in *Raja,* as well as in *Scyliorhinus* (Smeets 1982). Labelled neurons could also be seen in the ipsilateral lateral geniculate body after an HRP-injection in the tectum of *Scyliorhinus.* The projections originating from the pretectal area, the thalamus dorsalis pars medialis and the thalamus ventralis pars lateralis, which are apparently confined to the rostral part of the tectum, are predominantly ipsilateral. A substantial input to the tectum also appears to originate from the nucleus medius infundibuli, and this is also a predominantly ipsilateral connection.

A projection from the telencephalon to the tectum has been described in *Ginglymostoma* by Ebbesson and Schroeder (1971). Large, unilateral telencephalic lesions resulted in the degeneration of several descending tracts, one of which crossed in the decussatio tractus pallii to the opposite side and proceeded in a dorsolateral and caudal direction to terminate in the zona interna of the stratum cellulare externum. In a recent study Smeets (1982) has shown that in *Scyliorhinus* and *Raja* the telencephalic neurons which project to the tectum are located in the pars lateralis of the central dorsal pallium. In contrast to *Ginglymostoma* this projection is bilateral in *Scyliorhinus*. In *Raja*, however, the projection is small and appears to be only contralateral.

8.3.2 Tectal Efferents

The efferent connections of the tectum mesencephali have been experimentally investigated by Ebbesson (1972, *Ginglymostoma*), Northcutt and Boord (1981, *Raja*) and, in a more detail, in *Scyliorhinus* and *Raja* by Smeets (1981b). The ensuing survey is based on the latter study.

After a unilateral lesion had been made through all six tectal layers four distinct degenerating pathways could be observed: (1) an ascending projection both ipsi- and contralateral to the pretectal area, the dorsomedial region of the thalamus and the lateral geniculate body; (2) a commissural projection to the contralateral tectal half, some fibres of which could be traced to the contralateral tegmentum mesencephali in *Scyliorhinus;* (3) a descending projection which could be subdivided into two main pathways, one crossed, the other remaining ipsilateral; and (4) an additional rather diffuse ipsilateral tecto-tegmental system.

The fibres of the ascending group invade the pretectal area as well as the dorsomedial regions of the thalamus (Fig. 9-6). A few fibres run farther lateralward to the corpus geniculatum laterale, torus lateralis (*Raja*) and lobus inferior. Finally, some fibres run parallel and medial to the optic tract. These ascending projections pass both ipsi- and contralateral and reach the opposite side by

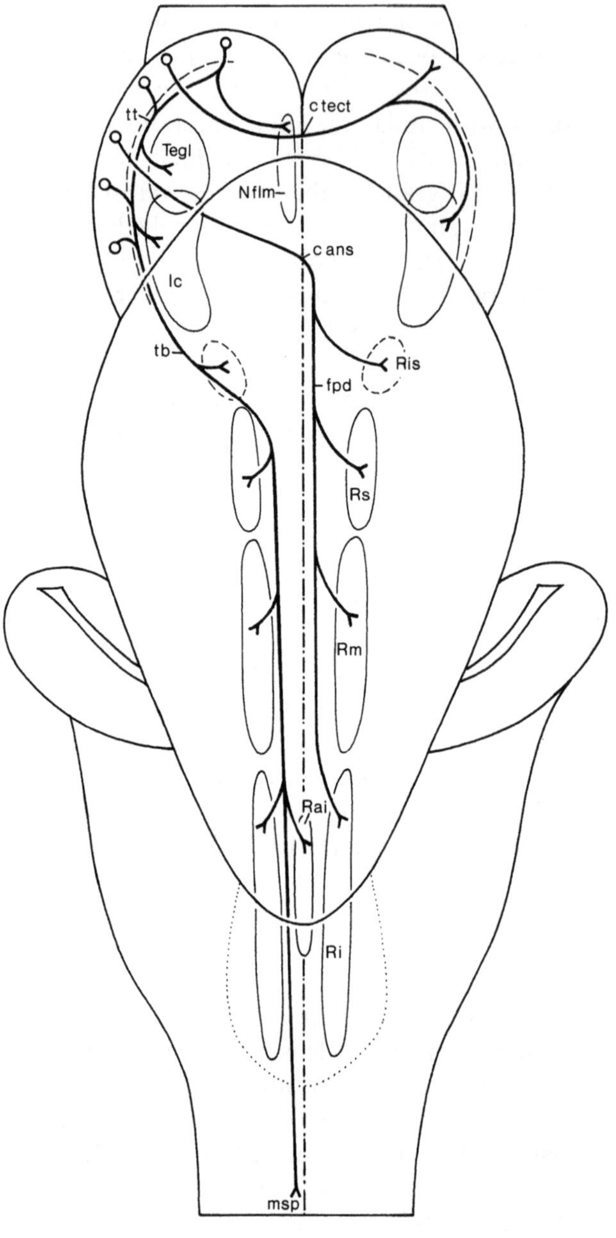

Fig. 8-3. Diagram of the tectal efferent projections to the brain stem and spinal cord, as viewed from the dorsal surface of the brain

way of the tectal commissure, just as in anurans (Rubinson 1968).

The commissural fibres also reach the contralateral tectal half by way of the tectal commissure and are found to project to an area roughly homotopic to the lesion site (Fig. 8-3). In *Scyliorhinus* some degenerating fibres course caudally, together with the contralateral ventral tectobulbar tract, distributing fibres to the ventrolateral isthmus region and the rhombencephalic reticular formation.

The descending pathways of *Scyliorhinus* and *Raja* can be divided into an ipsilateral and a contralateral component (Smeets 1981b). The ipsilateral descending pathway can be subdivided into fasciculi tectotegmentales ventrales and intermedii. A number of these fibres terminate in the tegmentum. The majority of the fasciculi tectotegmentales, however, continue farther caudalward as tractus tectobulbaris ventralis and intermedius. The tractus tectobulbaris ventralis distributes fibres to the reticular formation of the isthmus region, the nucleus reticularis superior, medius and inferior, and to the nucleus raphes inferior. In actinopterygians (Grover and Sharma 1979; Ebbesson and Vanegas 1976; Sligar and Voneida 1976), amphibians (Rubinson 1968) reptiles (ten Donkelaar 1976b) and mammals (Martin 1969; Rafols and Matzke 1970) the ipsilateral tractus tectobulbaris ventralis projects to the lateral part of the reticular formation, while the contralateral component distributes mainly to the medial part of the reticular formation. In cartilaginous fish, however, no distinct separation between tectal projections to the lateral and medial reticular formation can be made. In the rostral parts of the rhombencephalon, following tectal lesions, degenerating fibres have been traced to the more lateral regions of the reticular formation, but the medial reticular formation, which contains many very large neurons, also receives a tectal input.

The tractus tectobulbaris intermedius, which is very prominent in *Raja* (Smeets 1981b), also distributes fibres to the ipsilateral medial reticular formation and to the nucleus raphes inferior. The tectal projection to the nucleus last mentioned is contralateral in reptiles (ten Donkelaar 1976b).

The contralateral descending pathway is formed by the fasciculus tectotegmentalis dorsalis. A small number of these fibres terminate ipsilaterally in the nucleus of the fasciculus longitudinalis medialis (Northcutt and Boord 1981; Smeets 1981b), but most decussate in the commissura ansulata and continue caudalward as the fasciculus predorsalis. This fascicle is situated ventrolateral to the fasciculus longitudinalis medialis and distributes fibres to the medial reticular formation.

A small, mainly ipsilateral, tectospinal projection, originating from the stratum cellulare internum, was demonstrated in *Scyliorhinus* and *Raja* after an HRP injection in the "cervical" cord (Smeets and Timerick 1981).

Several additional tectal efferent systems have been described in the older literature: a tractus tectocerebellaris (Johnston 1905), a tractus tectolobaris (Ariëns Kappers et al. 1936) and a tractus tectohabenularis (Addens 1945). However, none of these efferent connections has been confirmed experimentally, although after a tectal lesion, degenerating fibres coursing to the inferior lobes of the hypothalamus may represent a tractus tectolobaris (Smeets 1981b).

8.4 Comparative Considerations

Although tectal lamination in cartilaginous fishes is not as distinctive as reported for teleosts (e.g. Meek and Schellart 1978; Pinganaud and Clairambault 1979; Kishida 1979; Schroeder et al. 1980), and anurans (for review see Székely and Lazar 1976), it is generally possible to distinguish six tectal layers in these fishes. The most primitive condition occurs in the holocephalian *Hydrolagus* where most of the tectal neurons are confined within a wide periventricular layer (Fig. 8-1D). Among elasmobranchs the squalomorph shark *Squalus* probably exhibits the most primitive tectal pattern, though extensive cellular migration away from the ventricular surface has taken place (Fig. 8-1A). However, in the galeomorph shark *Scyliorhinus* and, especially, in the batoid *Raja*, tectal cells have migrated still further away from the ventricular surface (Fig. 8-1B, C). This result supports Northcutt's suggestion (1979a) that the evolution of tectal visual circuitry in cartilaginous fish may have occurred by the migration and increase of neuronal number, rather than by development of new visual pathways.

Several experimental studies have shown that superficial layers 1, 2 and, to a lesser extent, 3 receive input from the ganglion cells of the retina. Layer 2 receives, additionally, a small direct spinal projection. Tectal layer 3 also contains the cells of origin of tectotectal connections. Layer 4 can be considered as the zone along which the main efferent tectal pathways leave the tectum, while tectal layer 5 contains neurons which project directly to the "cervical" spinal cord. The origin and termination of the fibres in tectal layer 6 are unknown.

At the moment we do not know how the tectal lamination of *Hydrolagus* conforms to the general tectal pattern just outlined. Two striking differences can be easily observed in our normal material of this species: (1) most tectal neurons are situated in a periventricular plate, and (2) some optic nerve fibres do not enter the tectum via the deep tectal zone (layer 3), but via the superficial layer 1. However, experimental studies will be needed to determine to what extent the tectal organization in Holocephali is similar to that in elasmobranch fish.

Though few studies with modern experimental neuroanatomical techniques have been devoted to afferent and efferent tectal connections in cartilaginous fish (Ebbesson 1972; Smeets 1981b, 1982) it is evident that the basic organization seen in other vertebrates is present. A telencephalotectal pathway has been identified which originates from the pallium dorsale pars centralis lateralis. Telencephalotectal connections have also been demonstrated in teleosts (Ito and Kishida 1977; Luiten 1981c; Grover and Sharma 1981) but in amphibians such a direct connection could not be identified (Wilczynski and Northcutt 1977). Diencephalic cells of origin of pathways projecting to the tectum mesencephali have been found in several thalamic areas, in the pretectum and in the nucleus medius hypothalami. It is noteworthy that the latter nucleus exerts an inhibitory action on the release of the melanocyte stimulating hormone (Wilson and Dodd 1973c).

Tectotectal and tegmentotectal connections comparable to those found in other vertebrates have been seen in chondrichthyans, but a rubrotectal connection has, so far, been reported only in teleosts (Luiten 1981c). In the rhombencephalon, labelled cells have been found in the reticular formation, the nucleus tractus descendens nervi trigemini and the nucleus dorsalis and intermedius areae octavolateralis. In other vertebrates these cell masses also contain some labelled cells after tectal HRP injections. However, direct projections originating from the nucleus octavus ascendens, as demonstrated in *Raja* and *Scyliorhinus*, and from the nucleus cerebelli and the nucleus funiculi lateralis, as shown in *Scyliorhinus* (Smeets 1982), have not been identified in any other vertebrate. Conversely, reciprocal connections between tectum and nucleus isthmi, as reported in actinopterygian fish (Ebbesson and Vanegas 1976; Sligar and Voneida 1976; Grover and Sharma 1979, 1981; Northcutt and Butler 1980; Luiten 1981c; Ito et al. 1981, 1982) and amphibians (Wilczynski and Northcutt 1977; Gruberg and Udin 1978; Glasser and Ingle 1978), have not been observed in the chondrichthyan brain.

The efferent pathways from the tectum show the basic vertebrate pattern, consisting of ascending, commissural and descending pathways. The ascending fibres project to the pretectal area, the dorsomedial region of the thalamus and the lateral geniculate body, both ipsi and contra-laterally. According to Smeets (1981 b) the ascending tectal efferents reach to the contralateral side by way of the tectal commissure. A distinct postoptic commissure could not be identified. Similar ascending projections have been experimentally confirmed in actinopterygians (Ebbesson and Vanegas 1976; Sligar and Voneida 1976; Grover and Sharma 1979; Northcutt and Butler 1980), amphibians (Rubinson 1968), reptiles (Butler and Northcutt 1971; Braford 1972; Foster and Hall 1975; ten Donkelaar 1976b; Schroeder 1981) and mammals (e.g. the opossum: Martin 1969; Rafols and Matzke 1970). The commissural and descending pathways are also, in general, similar to those found in other vertebrates. However, experimental studies in chondrichthyan fish (Smeets 1981 b, 1982) do not give any evidence of a reciprocal tectal connection with the nucleus isthmi, nor of a distinct segregation with respect to laterality of tectoreticular connections as found in other vertebrates.

The identification of tectal cells which project to the retina (Luiten 1981 a) indicates that in chondrichthyan fish such cells apparently play an essential role in the retinotectal system as they seem to do in other vertebrate classes (Rubinson 1968: anurans; Vanegas et al. 1973: teleosts). However, the location of neurons projecting to the retina differs considerably not only between classes, but even within one class. In the lamprey Vesselkin et al. (1980) recognized neurons projecting to the retina predominantly from the contralateral mesencephalic tegmentum, whereas in urodeles (Fritzsch and Himstedt 1981) and mammals (Itaya 1980; Jaeger and Benevento 1980) these neurons are situated within the pretectal area. In some reptiles the tegmentum mesencephali directly rostral to the isthmus contains cells that project to the retina (Ferguson et al. 1978). However, in other reptiles a diencephalic origin (Halpern et al. 1976; Repérant et al. 1980) or telencephalic origin (Hoogland and Welker 1981) of the efferent fibres to the retina has been identified. In teleosts, Meyer et al. (1981, *Tetradon fluviatilis*) found that in addition to the tectal source of retinopetal fibres, thalamic and pretectal areas also contain cells of origin of these fibres. Münz and Claas (1981), Münz et al. (1982), and Ebbesson and Meyer (1982), however, identified an additional telencephalic origin of retinopetal fibres in cichlid and poecilid fishes.

It is too early to make any general statement about retinopetal projections for cartilaginous fishes before these have been investigated in more species.

8.5 Functional Correlations

Its central position, its multimodality and its extensive connections with other centres, emphasize that the tectum plays an important co-ordinating role in the behaviour of fishes. Nevertheless there are few ideas about the exact function of this region or of its behavioural significance, although the studies by Sanders (1940) on the effects of tectal ablations in the goldfish indicate that it has an important role in learning. Ablation studies in elasmobranchs have given confusing results, some workers finding that tectal ablations produce motor deficits, while others report little change (see Healey 1957). More recently Graeber and Ebbesson (1972) reported that removal of parts of the tectum in *Ginglymostoma* had only an initial impact on locomotor performance and that even after extensive tectal ablation the fish could still learn visual discriminations. Indeed there was little evidence to suggest any substantial post-operative visual discrimination deficit (Graeber 1978).

The anatomical studies have shown that the chondrichthyan tectum receives a major visual projection and also has inputs from other brain regions and sensory modalities. Extensive studies on teleosts have shown that the retinotectal pathway is highly ordered so that the dorsal half of the visual field projects to the dorsal tectal surface and the ventral field to the ventral tectum. This orderly retinotectal topography has been found sofar in all vertebrate classes but there is no evidence yet for this arrangement in chondrichthyans.

There have been many electrophysiological studies on the tectum of teleost fishes, particularly in relation to the question of tectal mapping and reorganization after ablation, but the tectal regions of elasmobranchs have been much less well explored. Witkovsky et al. (1980) have reported on the field potentials that can be recorded contralaterally in the tectum of *Raja* in response to optic nerve stimulation. They observed at the surface a strong negative wave, with a peak around 16 ms after the stimulus, which reversed at around 500 μm depth in the tectum. This electrophysiological study supports the anatomical work in showing that optic afferent projections are restricted to the more superficial parts of the tectum and that the cells of the deep tectal layer do not receive a direct optic nerve projection.

Current-source-density profiles, determined from the field potentials, suggest that the optic nerve fibres are made up of three groups which conduct at different velocities. The two fastest groups of fibres seem to project to tectal neurons near to the surface of the tectum, while the slowest fibres pass more deeply (Witkovsky et al. 1980).

Some evidence of the nature of other inputs to the deeper layers of the elasmobranch tectum has come from the study by Platt et al. (1974) on *Torpedo*. These authors obtained responses in the tectum following stimulation of the electrosensory nerves and also demonstrated a short latency response in the tectum after spinal cord stimulation.

9 The Diencephalon

9.1 Introductory Notes

The structure and function of the chondrichthyan diencephalon is poorly understood. There have been few detailed studies on this region of the brain in these fishes except for the work of Bergquist (1932), who carefully analyzed the diencephalon of *Squalus* at various developmental stages.

Unlike other regions of the brain, the diencephalon in cartilaginous fishes has retained essentially the embryonic tube-like shape. Its thickened walls surround a narrow ventricle that expands ventrocaudally within the hypothalamus to form the lobi inferiores hypothalami.

The location of the precise boundary between the telencephalon and diencephalon is disputed. Kuhlenbeck (1929a, 1977) and Gerlach (1947) recognize the rostral border as passing from the point of attachment of the velum transversum to the commissura anterior. We follow Johnston (1911), Bäckström (1924), and Bergquist (1932) and draw the boundary as passing from the velum transversum to the caudal border of the optic chiasm.

Although the chondrichthyan diencephalon is dominated by the entrance and crossing of the optic nerve and by the broad course of this nerve to the tectum mesencephali, other prominent fibre streams such as the fasciculus basalis telencephali, tractus pallii, stria medullaris and thalamotegmental fibres are present in the diencephalon. Fibres ascending from the brain stem and spinal cord are thought to terminate within the brain stem and the diencephalic area, for no experimental study has revealed ascending fibres passing beyond the diencephalic level. Conversely, the majority of fibres descending from the telencephalon seem to terminate at the level of the diencephalon, though a number of fibres reach to more caudal levels (Ebbesson and Schroeder 1971; Smeets 1983). Clearly, the diencephalon has to be considered as an important relay-centre between the telencephalon and other regions.

9.2 Morphological Pattern

The diencephalon of the anamniotes has served as one of the major battle grounds on which the problem of homology has been fought by numerous authors including Herrick (1910), Kuhlenbeck (1929a), Haller (1929), Bergquist (1932), and more recently, Nieuwenhuys and Bodenheimer (1966).

The problem of establishing homologies is central to comparative neuroanatomy. The giving of a name to a neuronal cell mass implies that it has certain similarities with nuclei of the same name in the brains of other species. Thus a pre-existing anatomical name should only be used for a nucleus when we are certain of its homology. On the other hand the creation of a novel name for a region denies homologies while continual proliferation of novel names inevitably leads to confusion in the literature.

The following procedures have been used by various authors to establish homologies in the nervous system, particularly with emphasis on the diencephalon (for details see Nieuwenhuys and Bodenheimer 1966).

1. Herrick (1910) divided the diencephalon into four longitudinal zones: epithalamus,

pars dorsalis thalami, pars ventralis thalami and hypothalamus. These zones are separated from each other by the sulcus dorsalis thalami, the sulcus medius thalami and the sulcus ventralis thalami respectively. Herrick held that the sulci defined regions which are *functionally* as well as *morphologically* distinct.

2. Kuhlenbeck (1929a) also used ventricular sulci to divide the diencephalon of all vertebrates into "Grundbestandteile", which form together the structural plan (Bauplan). In Kuhlenbeck's opinion regions in different brains that occupy a corresponding position in the Bauplan were to be considered homologous, regardless of their structure or fibre connections. Haller (1929) also considered the ventricular sulci as a basis for the homologization of parts of the diencephalon. However, this author believed that the sulci are related to the embryonic transverse neuromeres, and that they imply a vertical rather than a horizontal subdivision of the diencephalon. The nearly horizontal appearance of these sulci was accounted for by the rotation of the diencephalic portion of the neural tube about the axis of the tuberculum posterius.

3. A totally different way of determining homologies was established by the Swedish school of neuroanatomists (Bergquist 1932; Källén 1951) who attempted to homologize areas of the brain on the basis of a study of the early embryological development of cytoarchitectonic units. Thus Bergquist subdivided the diencephalon into areas in the ventricular wall, "Grundgebiete", where the mitotic activity is higher than in intervening areas during early stages in development. He found that a sulcus usually forms in the middle of each "Grundgebiete" just at the point of highest mitotic density and therefore he called these sulci "Proliferationsfurchen".

Thus we have three different procedures for the establishing of homologies in the diencephalon. Herrick felt that sulci, nuclear structure, and fibre connections are all closely related so that a subdivision based on any one of these criteria will conform with subdivisions derived from the others. Kuhlenbeck believed that, at least for anamniotes, the sulci are the important determinants of brain organization, rather than the cytoarchitecture or disposition of fibre tracts. Finally, Bergquist and Källén held that nuclear structure as seen in the embryonic brain is primarily important, with sulci and fibre connections being of no assistance in the establishment of homologies.

More recently, Nieuwenhuys and Bodenheimer (1966) enunciated the following working definition of neural homology: "entities in different species which, within an obviously similar structural plan common to all species compared, occupy a corresponding topological position, should be considered homologous" (Nieuwenhuys and Bodenheimer 1966, p. 426). They tested the practical validity of the three criteria for establishing homologies—the ventricular sulci, nuclear boundaries and fibre connections—in the diencephalon of bony fish and concluded that there is no pan-diencephalic parallelism among the results of all three, or between any two, of the criteria used to establish homologies. They concluded that similarity in the topological position of cytoarchitectonic entities is the most useful criterion.

We shall test the practical validity of these criteria for establishing homologies in the diencephalon of the four chondrichthyan fish studied (Sect. 9.6), after first providing a description of sulcal pattern, cell masses and fibre connections.

Apart from a subdivision of the diencephalon into epithalamus, dorsal thalamus, ventral thalamus and hypothalamus, the cell masses of this brain region can also be subdivided according to their degree of migration into: (1) a zone situated immediately lateral to the ventricular ependyma; (2) a zone lateral to the inner ventricular zone but still a part of the ventricular grey, called the outer ventricular zone, and (3) migrated nuclei which are either completely separate from, or which retain only a slight connection with, the ventricular zones.

9.3 Ventricular Sulci

The ventricular sulci in the diencephalon of cartilaginous fishes are generally very distinct when compared with those of other vertebrates. However, their nomenclature is confusing because most previous authors did not investigate the diencephalic sulcal pattern systematically. A selective list of the various names in the literature is given in Table 9-1.

The following sulci could be recognized in the four species examined in the present study.

The *sulcus intrahabenularis* is a short sulcus within the habenular ganglion. It is always present on the side of the large habenula, but it is not a constant feature of the small habenula.

A very distinct *sulcus subhabenularis*, present on both sides, generally extends into the posthabenular recess under the subcommissural organ. The sulcus subhabenularis separates the epithalamus from the thalamus in the rostral part of the diencephalon.

A *sulcus organi subcommissuralis* could be identified in all species examined and forms the ventral border of the organon subcommissurale. Together wit the *sulcus accessorius organi subcommissuralis*, it also forms the dorsal boundary of the fasciculus retroflexus.

The *sulcus diencephalicus dorsalis* is represented not by a single groove but by several sulci, which pass from the level of the habenular commissure rostrally to the di-mes-encephalic boundary caudally. In *Squalus* and *Raja* there are two sulci, of which the most dorsal one (sdd1) is very short in *Raja*. In *Scyliorhinus* only the ventral component (sdd2) is present, whereas in *Hydrolagus* only the dorsal component can be found. The dorsal component of the sulcus diencephalicus dorsalis forms the ventral boundary of the fasciculus retroflexus or bundle of Meynert, which causes a slight eminence of the ventricular surface during its course to the interpeduncular nucleus. The ventral component, if present, always lies in the middle of a cell mass, the thalamus dorsalis pars medialis, which sometimes can be divided into a pars dorsomedialis and a pars ventromedialis. In *Squalus* the boundary between these two parts coincides with this groove.

A distinct *sulcus diencephalicus medius* varies in length in the different species studied. In *Squalus* and *Hydrolagus* it continues less far rostralward than in *Scyliorhinus* and *Raja*. However, in the latter two species sulcus f (see below) is absent and the rostral part of the sulcus diencephalicus medius of *Scyliorhinus* and *Raja* may be equivalent to sulcus f of *Squalus* and *Hydrolagus*.

The sulcus diencephalicus medius is an important structure since it has been used by many authors (Kuhlenbeck 1929a, 1973; Addens 1945; Gerlach 1947) to subdivide the thalamus into a pars dorsalis and a pars ven-

Table 9-1. Comparison of the nomenclature of diencephalic ventricular grooves in cartilaginous fishes

Johnston (1911)	Bergquist (1932)	Ariëns Kappers et al. (1936)	Addens (1945)	Kuhlenbeck and Niimi (1969)	Present study
–	sih	–	sih	–	sih
ssh	ssh	ssh	s dorsalis	s dorsalis	ssh
c	sos	–	–	s synencephalicus	sos
–	–	–	–	s synencephalicus	sosac
b	sd	sm	sm	s synencephalicus	sdd1
a	sm	sm	sm	s synencephalicus	sdd2
s Monroï	sv	sm	sv	sv	sdm
s Monroï	–	s th hyp	–	–	sdv
–	shyp	–	–	ssi	shyp

tralis. According to our observations this sulcus only coincides with the boundary between the dorsal and ventral thalamus in *Raja* and *Hydrolagus*. In *Squalus* and *Scyliorhinus* the sulcus diencephalicus medius appears to course within the rostral part of the dorsal thalamus.

The *sulcus diencephalicus ventralis* is a short and shallow groove, appearing only in the diencephalon of *Squalus*, and according to some authors (Kuhlenbeck 1929a, 1973; Gerlach 1947) separating the ventral thalamus from the hypothalamus. The sulcus diencephalicus ventralis of *Squalus* lies in the dorsal part of the hypothalamic area, but forms no clear boundary between the ventral thalamus and hypothalamus.

In the hypothalamic area a number of sulci are found which have not been separately labelled. These *sulci hypothalami* are distinct and are, as far we can judge from our material, rather constant features. One of these grooves is surrounded by a thickened and often vascularized ependyma, which probably represents the ependymal organ of Kappers-Charlton (Charlton 1928; Kuhlenbeck 1970, 1977).

A *sulcus preopticus*, found only in *Squalus*, is very distinct. Johnston (1909) and Ariëns Kappers et al. (1936) considered this sulcus to be the rostral end of the sulcus limitans of His, but according to our observations this sulcus is in fact not continuous with any sulcus caudal to the optic chiasm. The sulcus preopticus bears no clear relation to cytoarchitectonic fields.

Sulcus f occurs only in *Squalus* and *Hydrolagus*. It passes from the level of the anterior commissure to the frontal pole of the habenular region. In the same position in the diencephalon of *Scyliorhinus* and *Raja* only the rostral continuation of the sulcus diencephalicus medius is seen. Sulcus f can be compared in position and extension to the rostral part of the sulcus diencephalicus medius in *Scyliorhinus* and *Raja*. Sulcus f and its presumed equivalent pass over a cell mass, the eminentia thalami.

9.4 Cell Masses

In chondrichthyans the cell masses of the diencephalon have a more diffuse arrangement and contain neurons of more uniform size than those in the brain stem. About 20 cell masses could be delineated in this area in the four species studied.

9.4.1 Epithalamus

The epithalamus comprises the epiphysis and the ganglia habenulae. The *epiphysis* or *pineal organ* of *Scyliorhinus*, investigated by Rüdeberg (1969), is a long thin, tubular structure, the closed distal end of which rests against the roof of the brain case (Fig. 9-1). The proximal end opens into the third ventricle, between the habenular and posterior commissures. Using light- and electron-microscope techniques, Rüdeberg concluded that the pineal parenchyma consists of receptor cells, supporting cells and ganglion cells. The latter give rise to the diffuse tractus pinealis, which passes to the posterior commissure and, to a lesser extent, to the habenular commissure. However, its exact termination site has not been identified. A pineal tract could be recognized in all four species studied, the course of which is in full agreement with the description of Rüdeberg (1969). This author also found that the receptor cells have well developed outer segments similar to that of retinal photoreceptors. Hamasaki and Streck (1971) demonstrated electrophysiologically that the epiphysis of *Scyliorhinus* is a very sensitive photoreceptor organ and a preliminary study by Wilson and Dodd (1973a) has provided some evidence that in *Scyliorhinus* the pineal gland evokes a paling response in darkness either directly through the secretion of e.g. melatonin, or indirectly through nervous connections (perhaps via the MSH-system; see Sect. 9.7).

The *ganglia habenulae*, which are well developed in chondrichthyans, constitute the rostral part of the epithalamus. In all species studied the left habenular ganglion was

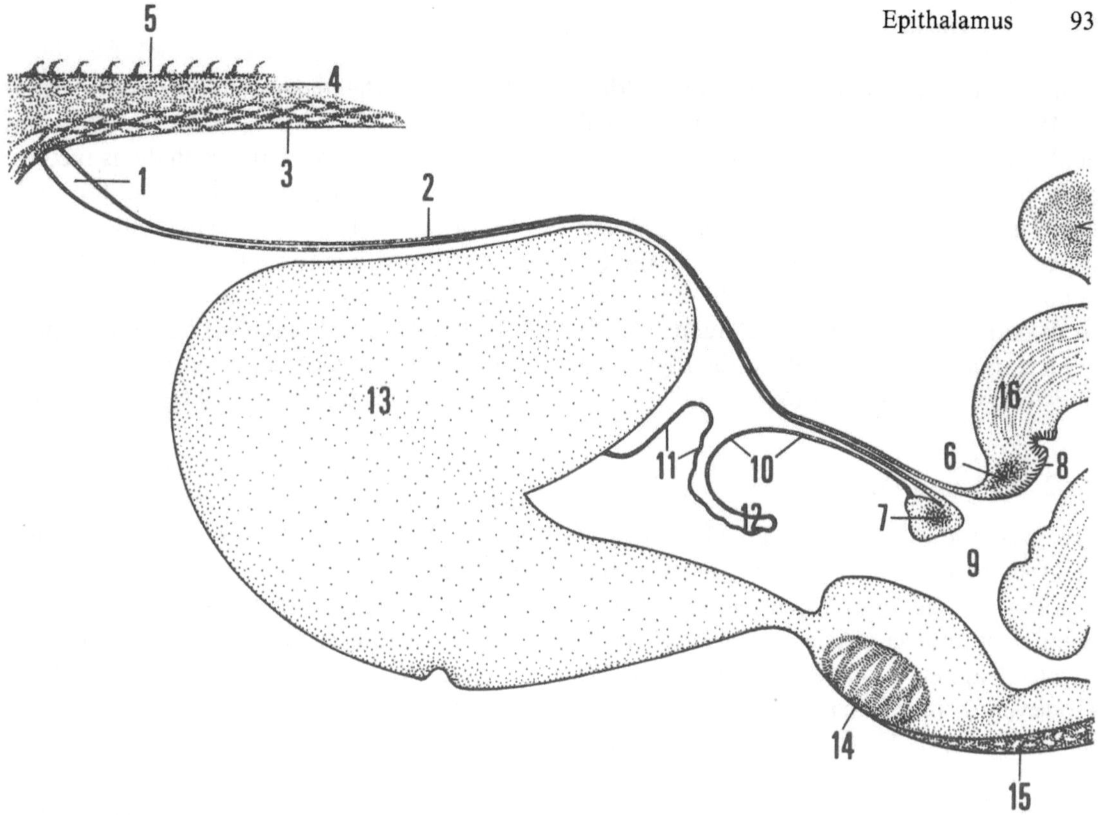

Fig. 9-1. Diagram of sagittal section through the anterior part of the brain of *Scyliorhinus canicula* *1* pineal end-vesicle; *2* pineal stalk; *3* connective tissue braincase; *4* dermis; *5* epidermis; *6* commissura posterior; *7* commissura habenulae; *8* organon subcommissurale; *9* third ventricle; *10* saccus dorsalis; *11* paraphysis; *12* velum transversum; *13* telencephalon; *14* chiasma opticum; *15* hypophysis; *16* tectum mesencephali. (After Rüdeberg 1969)

found to be larger than the right, in agreement with Ariëns Kappers and Carpenter (1911), Bergquist (1932), Ariëns Kappers et al. (1936), Gerlach (1947), Kuhlenbeck (1977), Farner (1978), Kemali et al. (1980) and Miralto and Kemali (1980). The difference in size, mainly produced by the somewhat more frontal extension of the left ganglion, is less in *Squalus* and *Hydrolagus*, than in *Scyliorhinus* and *Raja*. Differences in histological structure between the two ganglia, however, are more considerable. In Klüver-Barrera sections it is particularly striking that the dorsolateral region of the left ganglion is provided with a dense network of myelinated fibres. In the left ganglion, cells lying within the myelinated network are somewhat larger and more loosely arranged than are the other cells of the ganglion. The latter cells, just like all the cells of the right ganglion, are of the characteristic small habenular type and lie tightly packed together. These findings are in full agreement with those of Kemali et al. (1980) in *Scyliorhinus stellaris*.

In many groups of vertebrates the ganglia habenulae can be subdivided into medial and lateral nuclei. In chondrichthyans, however, such a subdivision of the ganglia habenulae is only vaguely suggested at some habenular levels.

Caudally the two ganglia habenulae fuse across the midline in the area below the habenular commissure. This commissure comprises a cell mass which consists of small cells of the same size and type as the ordinary habenular cells, although they are more loosely arranged. This nucleus has been termed the nucleus of Bellonci by Addens (1945) and was also recognized by Kemali et al. (1980) in *Scyliorhinus stellaris;* however, the latter considered it to be a midline habenular nucleus. In *Scyliorhinus* and *Hydrolagus* we have found a cell mass in the habenular commissure which may be equivalent to the nucleus of Bellonci of Addens (1945). Such a nucleus could not be distinguished, however, in *Squalus* and *Raja.*

9.4.2 Thalamus Dorsalis

Previous studies of the diencephalon of cartilaginous fishes merely describe major subdivisions of this brain area without giving details of nuclei (Ariëns Kappers 1906; Ariëns Kappers et al. 1936; Gerlach 1947). However, those cell masses that receive a retinal input have been recently described in more detail by Northcutt (1979a) and Smeets (1981a).

The dorsal thalamus can be subdivided into two parts, a pars medialis that consists of small densely packed periventricular cells and a pars lateralis which contains slightly larger and more diffusely arranged cells. The thalamus dorsalis pars medialis extends from the level of the optic chiasm to the di-mesencephalic boundary. This cell mass can be further subdivided in *Squalus* and *Hydrolagus* into a pars dorsalis and a pars ventralis because of the greater density and slightly more intensive staining of the cells of the pars ventralis.

The position of the thalamus dorsalis pars lateralis varies somewhat in the four species studied. In *Scyliorhinus, Raja* and *Hydrolagus* the lateral dorsal thalamus lies at the level of a zone connecting the habenular commissure with the optic chiasm, whereas in *Squalus* this cell mass has a more caudal position.

9.4.3 Thalamus Ventralis

According to Ariëns Kappers et al. (1936), Gerlach (1947), and Kuhlenbeck (1929a, 1973, 1977) the *thalamus ventralis* of chondrichthyans is separated from the thalamus dorsalis by the sulcus diencephalicus medius. However, it can be concluded from the study of Northcutt (1979a: *Squalus*) and our own observations that this sulcus cannot be used to identify the exact boundary between the dorsal and ventral thalamus because its position varies rostrocaudally with reference to the cell groups that run the length of the thalamus.

The ventral thalamus, like the dorsal thalamus, is composed of two parts, a periventricular pars medialis and a migrated pars lateralis. The former consists of some loosely arranged laminae of small cells, the latter of diffusely arranged small cells. The position of these two cell masses differs little in the species studied. The thalamus ventralis pars lateralis caps the fasciculus basalis telencephali.

Rostral to the thalamus ventralis pars medialis a distinct nucleus could be delineated which, following Bergquist (1932) and Kuhlenbeck (1977), has here been termed eminentia thalami. This cell mass could be distinguished on account of the larger size and the more intensive staining of its constituent neuronal elements. The existence of such a nucleus could be confirmed in all species studied except *Hydrolagus* and is particularly strongly developed in *Raja.* In *Hydrolagus,* in the middle region of the thalamus dorsalis pars medialis, a number of cells are seen that are of the same character as the neurons of the eminentia thalami in the other three species (Fig. 9-2).

Before turning to the description of the migrated thalamic cell masses it should be noted that the external part of the ventricular layer contains, in addition to the cell masses described, numerous diffusely scattered cells.

The *corpus geniculatum laterale* is the only nucleus in the chondrichthyan thalamus which can be included in the category of migrated cell masses. This nucleus is embedded in the stream of optic nerve fibres and can be easily recognized because it is the only

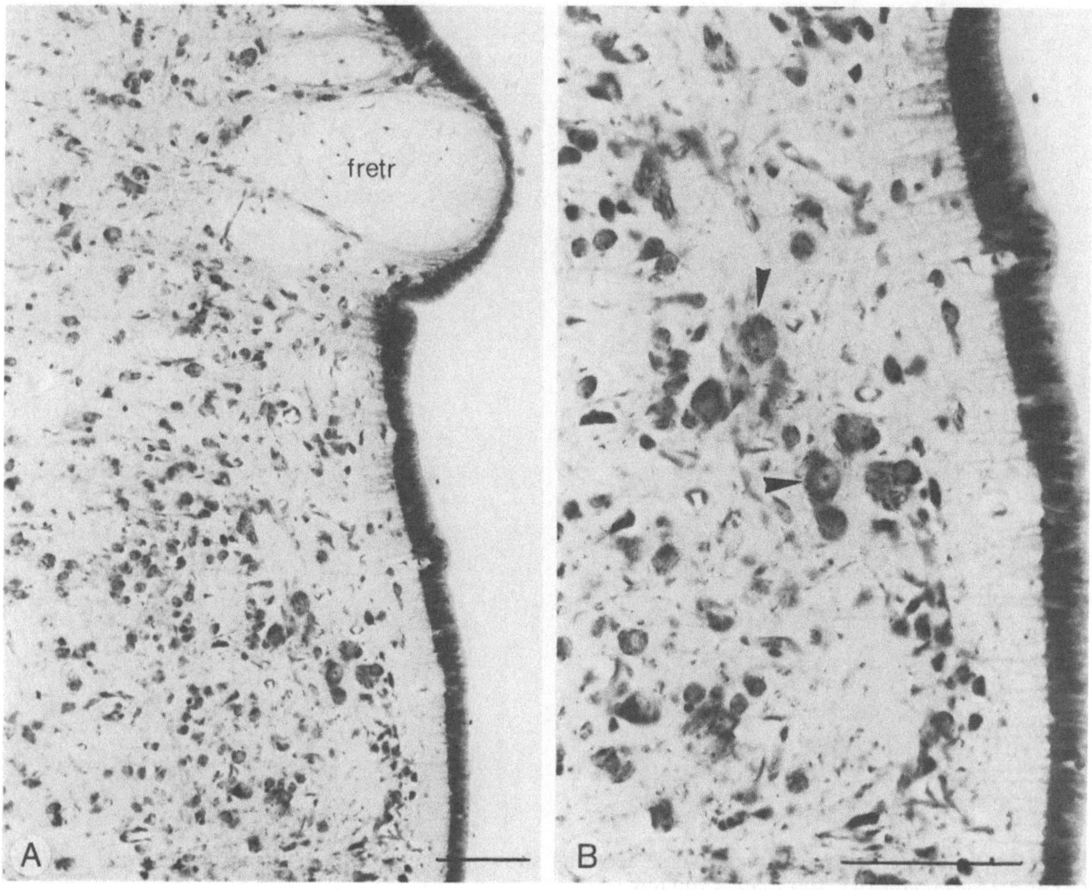

Fig. 9-2 A. Photomicrograph of a transverse section showing the thalamus dorsalis pars medialis of *Hydrolagus collei*. Nissl staining. **B** Higher power view of part of the same section shown in **A**. *Spear-heads* point to neurons which may be equivalent to the eminentia thalami of the elasmobranchs studied. In **A** and **B** *bar scale* = 100 μm

cell mass within the diencephalon that comprises medium-sized cells (Fig. 9-3).

Comparisons of the lateral geniculate body between species showed that it contains more neurons and has a more compact arrangement in *Scyliorhinus, Raja* and *Hydrolagus* than in *Squalus*. Another striking difference is that the lateral geniculate body of the first three species has migrated further lateralward in harmony with the suggestion of Northcutt (1979a) that "evolution of primary visual pathways in cartilaginous fishes occurs by migration and an increase in neuronal number, rather than by occurrence of new visual pathways" (Northcutt 1979a, p. 219).

Although this nucleus has been termed the corpus geniculatum laterale in the present analysis, following the terminology of Ariëns Kappers et al. (1936), this does not imply that this cell mass is equivalent to the lateral geniculate body of amniotes. Recent experimental studies of Northcutt (1979a: *Squalus*), Luiten (1981a, b: *Ginglymostoma*) and Smeets (1981a, b: *Scyliorhinus, Raja*) have revealed that there is considerable disagreement concerning the question of which nucleus should be designated as the lateral geniculate body. Northcutt (1979a) avoided this term and labelled the lateral geniculate body of the present authors as the nucleus pretectalis superficialis. Luiten (1981a, b)

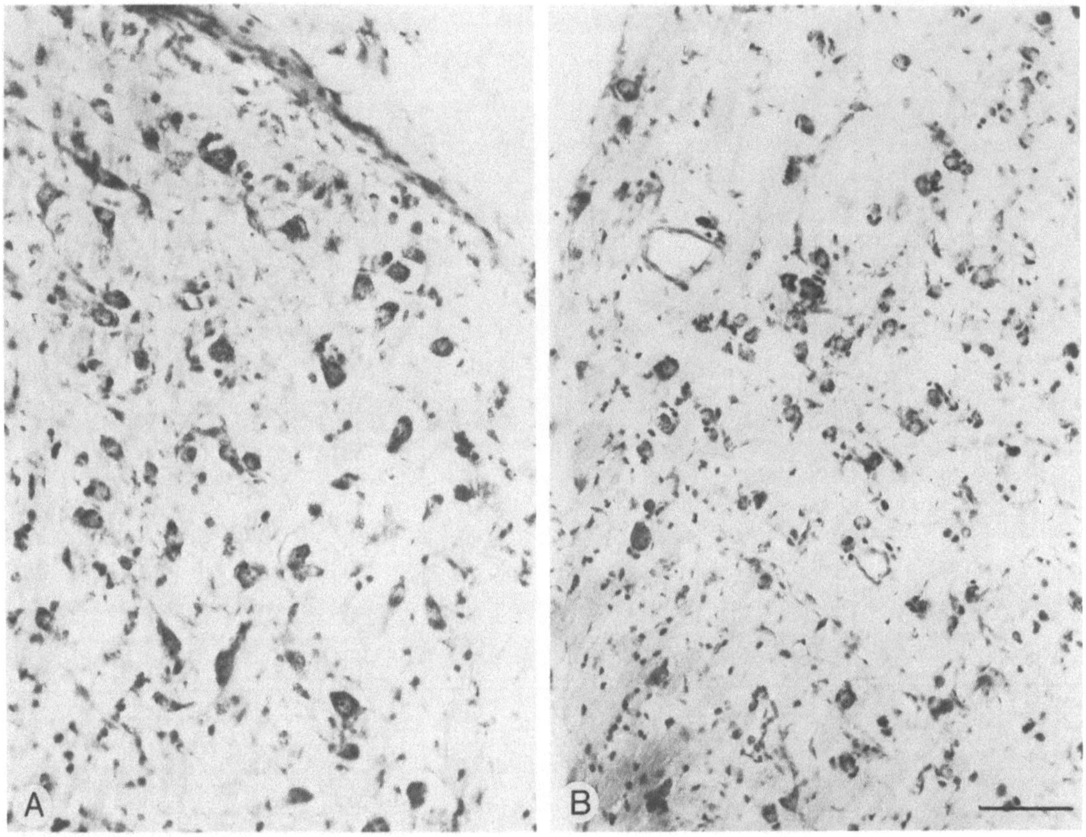

Fig. 9-3. Photomicrographs of transverse sections showing the corpus geniculatum laterale of *Raja* (**A**) and *Hydrolagus* (**B**). *Bar scale* = 100 μm

divided a cell mass, which has about the same position in the diencephalon of the nurse shark *Ginglymostoma,* into two, calling only the dorsal part the lateral geniculate body. This portion does not receive direct retinal projections but does have tectal input, whereas the ventral part, i.e. the ventrolateral optic nucleus receives a strong retinal input. Both (sub)nuclei, project to the caudal part of the telencephalic pallium. The degeneration studies of Smeets (1981a, b) in *Scyliorhinus* and *Raja* revealed both retinal and tectal inputs to the nucleus, labelled here as corpus geniculatum laterale. These species differences may be explained by considering the brain of *Ginglymostoma* to be more highly developed (Northcutt 1977, 1978) than that of *Scyliorhinus* and *Raja.* According to the parcellation theory of Ebbesson (1980a, p. 183) "neural systems evolve not by the mixing of systems, but by differentiation and parcellation, which involves competition of inputs, the redistribution of inputs, and the loss of connections". Experimental studies, particularly HRP-injections in the caudal part of the pallium of *Scyliorhinus* and *Raja,* will be needed to test the validity of this theory with regard to the lateral geniculate body in elasmobranchs.

9.4.4 The Hypothalamus

The anatomy of the hypothalamus is complicated, since many of the boundaries between cellular condensations are rather ill-defined, and because relatively large hypothalamic regions consist of diffusely arranged neurons.

In the species studied we have delineated the following cell masses: the nucleus preopticus; the nucleus suprachiasmaticus; the nucleus medius hypothalami; the nucleus periventricularis hypothalami; the nucleus lateralis tuberis, and the nucleus lobi lateralis hypothalami.

Although it belongs morphologically to the telencephalon, we shall include the *nucleus preopticus* here because of its functional relationship to the hypothalamohypophyseal system, to be described in Sect. 9.7. Its boundaries are not very distinct and, except in *Hydrolagus,* it merges rostrally into the interstitial nucleus of the commissura anterior. The caudal boundary is marked ventrally by the optic chiasm and dorsally by differences in the cytoarchitecture of the preoptic area and thalamus. Charlton (1932) found that, just as in other vertebrates, the cells of the chondrichthyan preoptic area lining the preoptic recess differ in size and could be subdivided on that account into a pars magnocellularis and a pars parvocellularis. In the present study such a difference in size of neurons within the preoptic area could also be recognized (Table 4-2) but because large and small neurons intermingle a subdivision into magnocellular and parvocellular parts could not be made, although the larger neurons lie mainly at the level of the preoptic recess. In *Raja,* a huge accumulation of these neurons is observed at the level where the optic nerve enters the brain (Fig. 89). The

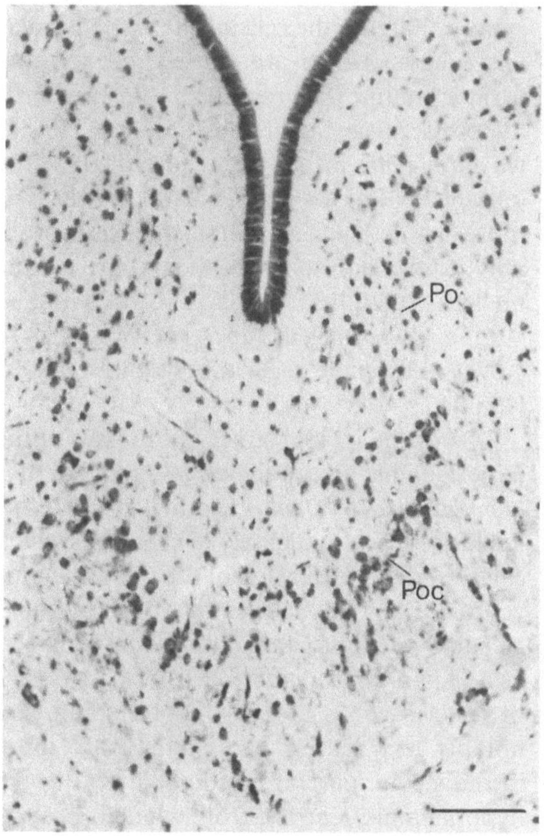

Fig. 9-4. Photomicrograph of a transverse section showing the different components of the preoptic nucleus in *Hydrolagus collei.* Nissl staining. *Bar scale* = 100 μm

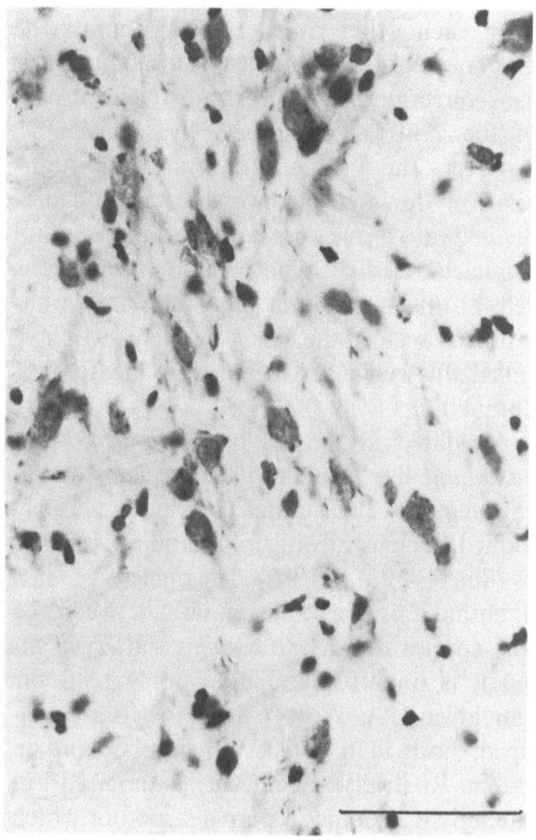

Fig. 9-5. High power photomicrograph of a transverse section showing the nucleus suprachiasmaticus of *Squalus acanthias.* Nissl staining. *Bar scale* = 100 μm

caudal part of the preoptic nucleus in *Hydrolagus* contains a zone of cells which stain darker and are somewhat larger than the surrounding preoptic neurons (Fig. 9-4). This subdivision of the nucleus preopticus we have termed the pars compacta.

A *nucleus suprachiasmaticus* has been recognized in the elasmobranchs. The existence of such a nucleus in cartilaginous fish has never been mentioned, although the nucleus labelled nucleus chiasmaticus by Ebbesson and Ramsey (1968: *Galeocerdo, Ginglymostoma*) may be homologous. The rostral part of the suprachiasmatic nucleus consists of cells which are scattered among the fibres of the decussating optic tract and the commissura postoptica (Fig. 9-5). Caudally this nucleus is replaced by the nucleus medius hypothalami of Wilson and Dodd (1973c). These two nuclei can be easily distinguished from each other by the orientation of the bipolar cells of the nucleus suprachiasmaticus, converging toward the most ventral part of the ventricular surface. In *Hydrolagus,* however, the postchiasmatic area contains relatively few neurons, which cannot be subdivided into a nucleus suprachiasmaticus and a nucleus medius hypothalami. Because the cells of this area do not show the same orientation as seen in the elasmobranchs, we have called this cell mass the nucleus medius hypothalami.

As stated above, the *nucleus medius hypothalami* lies caudal to the nucleus suprachiasmaticus. This nucleus was previously termed nucleus medius infundibuli by one of us (Smeets 1981b, 1982). The nucleus medius hypothalami varies substantially between the four species studied. In *Squalus* and *Hydrolagus* it is unpaired and does not extend far caudalward. In *Scyliorhinus* it is better developed, both in terms of cell number and in extent. Rostrally this nucleus is unpaired but caudally it becomes a paired structure which extends over some distance into the median eminence. In *Raja* the nucleus medius hypothalami shows the greatest degree of development, in cell number, extent, and cell type. As the different cell types are not collected into well-defined nuclei we have considered them together as the nucleus medius hypothalami.

Apart from the nucleus suprachiasmaticus and the nucleus medius hypothalami, the postchiasmatic region also contains the *nucleus periventricularis hypothalami* and the *nucleus lateralis tuberis.* These latter nuclei surround the infundibular recess and lie within the periventricular zone. The nucleus periventricularis hypothalami occupies the dorsolateral periventricular zone, whereas the nucleus lateralis tuberis takes rostrally a ventrolateral position but shifts to a ventral position at more caudal levels. These nuclei differ not only in topological position but also in size, intensity of staining, and, especially, in cell density. The cells of the nucleus periventricularis hypothalami are more loosely arranged than are the cells of the nucleus lateralis tuberis, which are densely packed and form a lamina. The nucleus periventricularis hypothalami is rather indistinct and, as far we know, has never been described before (except for Smeets and Timerick 1981). The more distinct nucleus lateralis tuberis is probably equivalent to the caudal part of the nucleus lateralis tuberis of Sathyanesan (1965), to the nucleus tuberis of Wilson et al. (1974) and to the regio tuberis of Bergquist (1932).

One of the most distinct cell masses of the hypothalamic area is the *nucleus lobi lateralis hypothalami.* In *Squalus* and *Scyliorhinus* this nucleus consists of densely packed cells which cap the lateral expansion of the hypothalamic ventricle. In both species this nucleus could be subdivided into pars dorsalis, pars medialis and pars ventralis. The cells of pars dorsalis nuclei lobi lateralis hypothalami are notably less densely packed than are those in the medial part. It lies dorsolateral to the hypothalamic ventricle from which it has migrated a short distance. The medial part of the nucleus is the largest portion of the nucleus and lies lateral and ventral to the adjacent hypothalamic ventricle. At more caudal levels a pars ventralis of the nucleus lobi lateralis can be distinguished, which contains loosely arranged cells and has migrated a short distance from the ventricle, like the dorsal part. The pars ventralis occupies a ventromedial position in the infundibular wall. The nucleus lobi lateralis hypothalami cannot in *Raja* be partioned into three because the bilateral recesses of the hypothalamic ventricle are almost completely obliterated and the cells are diffusely scattered in the inferior hypothalamic lobe. There are nevertheless differences in size and shape of the neurons of the nucleus lobi lateralis of *Raja* but it is impossible to group those into different parts. Contrary to the sharks, in *Raja* the cells have migrated far away from the ventricular surface.

The nucleus lobi lateralis of *Hydrolagus* stands with regard to its size and degree of differentiation between that of the sharks and rays. The recesses of the hypothalamic ventricle are less well developed, compared with those of sharks, but are much larger than those of rays. The nucleus lobi lateralis, with different cell types, caps the lateral recess in the same way as in sharks. However, the cells have migrated farther away from the ventricular surface and do not form distinct subdivisions. On the basis of differences in cell density and staining a dorsal and a ventral part can be recognized. The cells of the dorsal part stain darker than those in the ventral part and constitute a lamina, whereas the cells of the ventral part are more diffusely arranged. At more caudal levels a group of cells lies lateral to the recess in a submeningeal position. This cell mass has been indicated as the nucleus lobi lateralis pars lateralis by the present authors. We are unable to homologize the subdivisions of the nucleus lobi lateralis of *Hydrolagus* with those of sharks.

Finally it is noteworthy that Braak (1963) has described a highly vascularized ependyma at rostral optic chiasmatic levels in *Chimaera* which he called the organon vasculosum praeopticum. A similar structure can be seen in our material of *Hydrolagus* (Fig. 124, ovp).

9.4.5 Synencephalic Region

During ontogenesis the synencephalic region forms the transition region between the archencephalon and deuterencephalon (for review see Kuhlenbeck 1973). Though initially belonging to the deuterencephalon rostral parts of it become incorporated in the diencephalon during further development of the brain. We have subdivided this transitional zone into a dorsal pretectal region and a ventral di-mes-encephalic transition zone.

The Pretectal Region. The pretectal region lies, as its name implies, rostral to the tectum mesencephali. The region has distinct boundaries in the median plane, but laterally it merges with other areas of the brain. The habenular commissure forms the rostral boundary, the sulcus organi subcommissuralis marks the ventral boundary, and the dorsal border is provided by the surface of the brain. On its lateral border the pretectal region merges ventrally with the thalamus and dorsally with the tectum. The caudal extent is limited ventrally by the rostral wall of the wide ventricle of the tectum, and dorsally is continuous with the tectal grey.

The pretectal region can be subdivided into three longitudinal areas that lie one on another: (1) the area pretectalis sensu strictiori; (2) the commissura posterior, and (3) the organon subcommissurale.

The *area pretectalis sensu strictiori,* which constitutes the most dorsal zone of the pretectal region, extends throughout the length of the latter. Northcutt (1979a) divided the pretectal area in *Squalus* into three cellular groups—the periventricular pretectal nucleus, the central pretectal nucleus and the superficial pretectal nucleus—all of which receive retinofugal fibres. He believed that these three nuclei correspond topologically to the rostral continuation of the periventricular, central and superficial tectal layers.

Using normal material of the four species studied we were unable to delineate the cell masses described by Northcutt.

A distinct retinal target area in the pretectal region of *Scyliorhinus* and *Raja*, labelled nucleus pretectalis, has been found with experimental neuroanatomical techniques (Smeets 1981 a). This nucleus probably corresponds to the periventricular pretectal nucleus of *Squalus* (Northcutt 1979a) and to the pretectal area of *Ginglymostoma* and *Galeocerdo* (Ebbesson and Ramsey 1968; Luiten 1981a) and *Negaprion* (Graeber and Ebbesson 1972). A central pretectal nucleus as observed in *Squalus* (Northcutt 1979a) could not be recognized in *Scyliorhinus*, but is possibly represented in *Raja* (Pretc, Fig. 84). A distinct superficial pretectal area as described in *Squalus* (Northcutt 1979a) could not be recognized in *Scyliorhinus* and *Raja* (Smeets 1981a). Graeber and Ebbesson (1972, *Negaprion*) and Luiten (1981a, *Ginglymostoma*) identified a posterior optic nucleus which probably corresponds to the superficial pretectal nucleus of Northcutt (1979a). As the pretectum is a complex and poorly understood region of the vertebrate brain, it would be premature to attempt to compare cell masses between elasmobranchs.

The middle layer of the pretectal region, the bed of the commissura posterior, consists chiefly of decussating fibres, but also contains the *nucleus commissurae posterioris*. This nucleus is formed of diffusely spread cells that lie among the ventrolaterally directed fibres of the commissure. Because of the diffuse arrangement of the neurons, and the consequent difficulty in delineating this nucleus, we have only indicated its position in the transverse sections of the diencephalon.

The third and most ventral layer of the pretectal region is made up of the *organon subcommissurale*, which forms the roof of the ventricle from the caudal end of the habenular commissure to the caudal end of the posterior commissure. The epiphyseal stalk opens into the ventricle through the rostral part of the subcommissural organ. This organ is formed of tall, thin columnar cells, from which Reissner's fibre originates. This fibre, non-nervous in character, consists of a muco-polysaccharide-protein complex that passes caudalward through the ventricles of the brain and the central canal of the cord and attaches to a mass of mesenchymatous tissue at the end of this canal (for reviews see Gabriel 1970; Ziegels 1976). The function of this specialized epithelium is still unknown, although its most important feature seems to be its secretory activity (Ziegels 1979).

The Ventral Di-Mes-Encephalic Region. The cell masses which will be discussed in this section are: the basal optic nucleus; the torus lateralis; the nucleus tuberculi posterioris; the nucleus sacci vasculosi, and the nuclei labelled provisionally K and L.

We have recognized a presumed nucleus opticus basalis consisting of relatively large cells in our normal material of *Squalus, Raja* and *Hydrolagus*, but could not identify such a nucleus in *Scyliorhinus*. Its position has been indicated in the topographical reconstructions of the diencephalon. Experimental studies on the retinal efferents have provided evidence for the existence of a basal optic nucleus in elasmobranchs (Ebbesson and Ramsey 1968: *Galeocerdo;* Northcutt 1979a: *Squalus;* Ebbesson and Meyer 1980: *Rhinobatos;* Witkovsky et al. 1980: *Raja;* Luiten 1981a: *Ginglymostoma;* Smeets 1981a: *Raja*). It appears, however, that in addition to the basal optic nucleus delineated by the present authors, another target area for the basal optic tract can be identified in the midbrain tegmentum (Northcutt 1979a; Witkovsky et al. 1980; Luiten 1981a). This area does not form a distinct cell mass, however, in normal material.

In *Raja* a prominent cell mass has been found in front of the nucleus tegmentalis lateralis. This cell mass, which we have termed *nucleus tori lateralis*, has a spherical shape and consists of small bipolar and triangular cells. This nucleus, which has also been recognized by Northcutt (1978) in *Raja eglanteria* and labelled by that author as the mesencephalic lateralis nucleus, strongly resembles in its shape and position the nucleus tori lateralis of some actinopterygian fish. The nucleus tori lateralis in *Raja* is caudally continuous with the nucleus tegmentalis lateralis from which it can be distinguished, however, because of differences in density and arrangement of the neurons.

Bergquist (1932) described a regio tori lateralis in *Squalus acanthias*. It is noteworthy that the caudal part of this region is comparable to our nucleus tegmentalis lateralis, whereas its rostral part could not be delineated in our adult *Squalus* material.

The *nucleus tuberculi posterioris* is a rather ill-defined cell mass, situated in the tuberculum posterius and in the area immediately rostral to this conspicuous encephalic landmark. Rostrally it is continuous with the thalamus ventralis pars medialis and caudally it merges into the tegmental grey. Its position is nevertheless rather constant in the four species studied. The nucleus tuberculi posterioris contains ellipsoidal cells which are in general somewhat larger than those of the thalamus ventralis pars medialis.

The *nucleus sacci vasculosi* is a distinct cell mass in cartilaginous fish that consists of bipolar or triangular, relatively large neurons and lies in the ventral part of the tuberculum posterius, just ventral to the commissura postinfundibularis pars inferior. There are some minor differences with regard to the nucleus sacci vasculosi between the four species studied. For example, in *Squalus* the nuclei of both sides are fused to form an unpaired nucleus sacci vasculosi, a feature also reported by Bergquist (1932), whereas the nucleus is paired in the other three species.

At the level of the nucleus sacci vasculosi in *Scyliorhinus* and *Hydrolagus*, an accumulation of cells can be identified which has been tentatively labelled K. In *Scyliorhinus* this cell mass takes a somewhat more medial position than in *Hydrolagus*. A comparable cell mass has been recognized in *Raja*, though the position of this nucleus K is somewhat more caudal than that in *Scyliorhinus* and *Hydrolagus*. In all species this cell mass consists of small, ellipsoidal neurons which are in general smaller than those of the neighbouring lobus lateralis hypothalami. Nucleus K lies between fibres which probably form the so-called tractus lobocerebellaris and so it is possible that this nucleus functions as a relay centre. The nucleus, labelled L, identified only in *Hydrolagus* and consisting of small, ellipsoidal cells, lies caudal to the presumed basal optic nucleus which contains larger neurons.

9.5 Fibre Tracts

The fibre systems in the diencephalon of cartilaginous fishes have been studied in normal material by several authors among whom Edinger (1892), Ariëns Kappers (1906), Ariëns Kappers and Carpenter (1911), Ariëns Kappers et al. (1936) and Addens (1945) should be mentioned. Recent experimental studies have confirmed the presence of most of the previously described fibre systems. These experimental studies have revealed, in addition, the presence of a number of previously unknown connections.

In this section we shall first describe the fibre tracts of the diencephalon that we could distinguish in normal material and then discuss the results of experimental studies. The fibre tracts of the diencephalon will be subdivided into the following categories: (1) connections between diencephalon and brain stem, (2) connections between diencephalon and telencephalon, (3) intrinsic diencephalic connections, (4) diencephalic commissures and (5) the tractus opticus.

9.5.1 Connections Between Diencephalon and Brain Stem

Under this heading the following fibre tracts will be discussed: the fasciculus retroflexus;

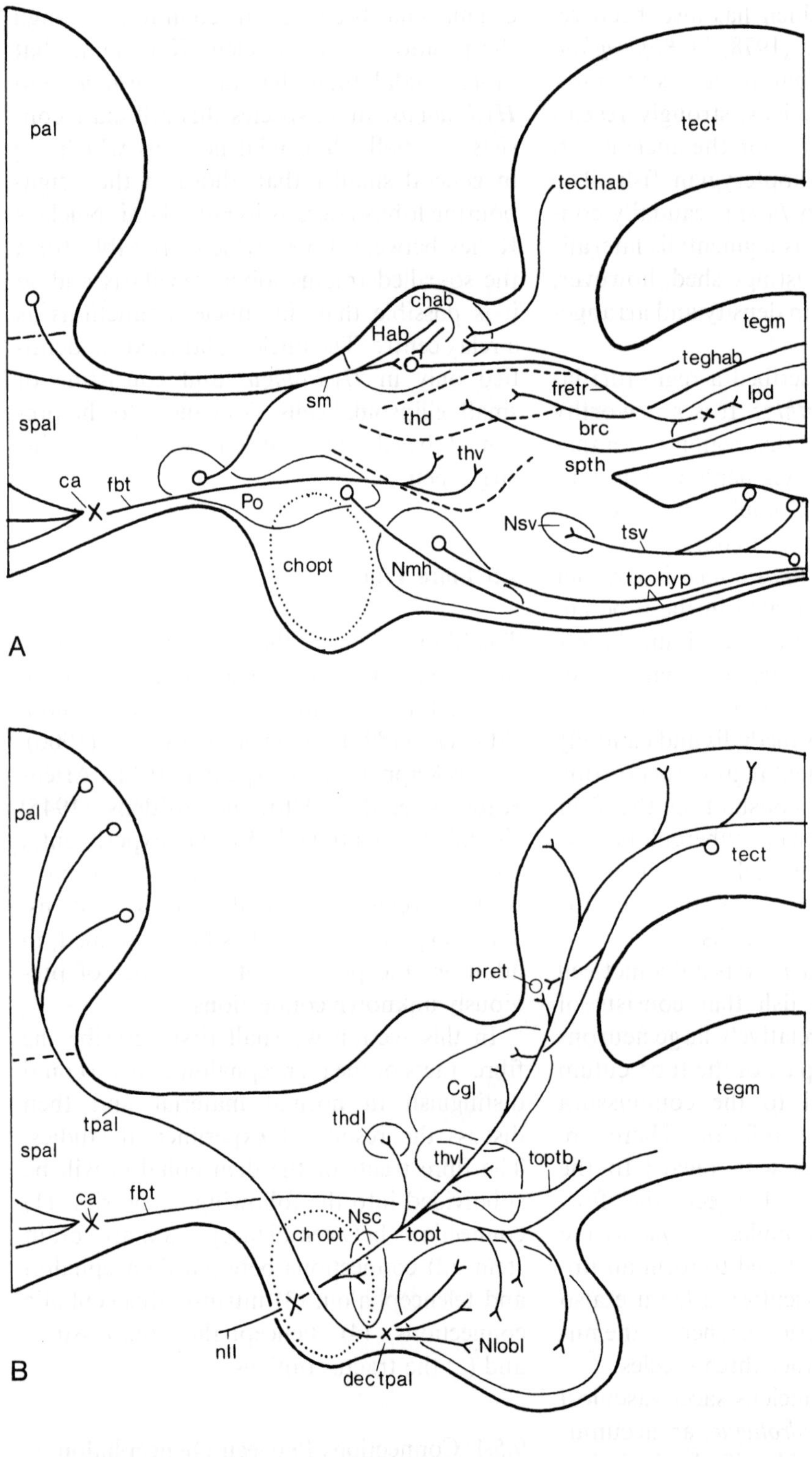

Fig. 9-6. A, B Diagrammatic projection on a sagittal plane of diencephalic afferent and intrinsic connections, occupying a medial (**A**) or lateral (**B**) position. **C, D** Diagrammatic projection on a sagittal plane of diencephalic efferent projections occupying a medial (**C**) or lateral (**D**) position

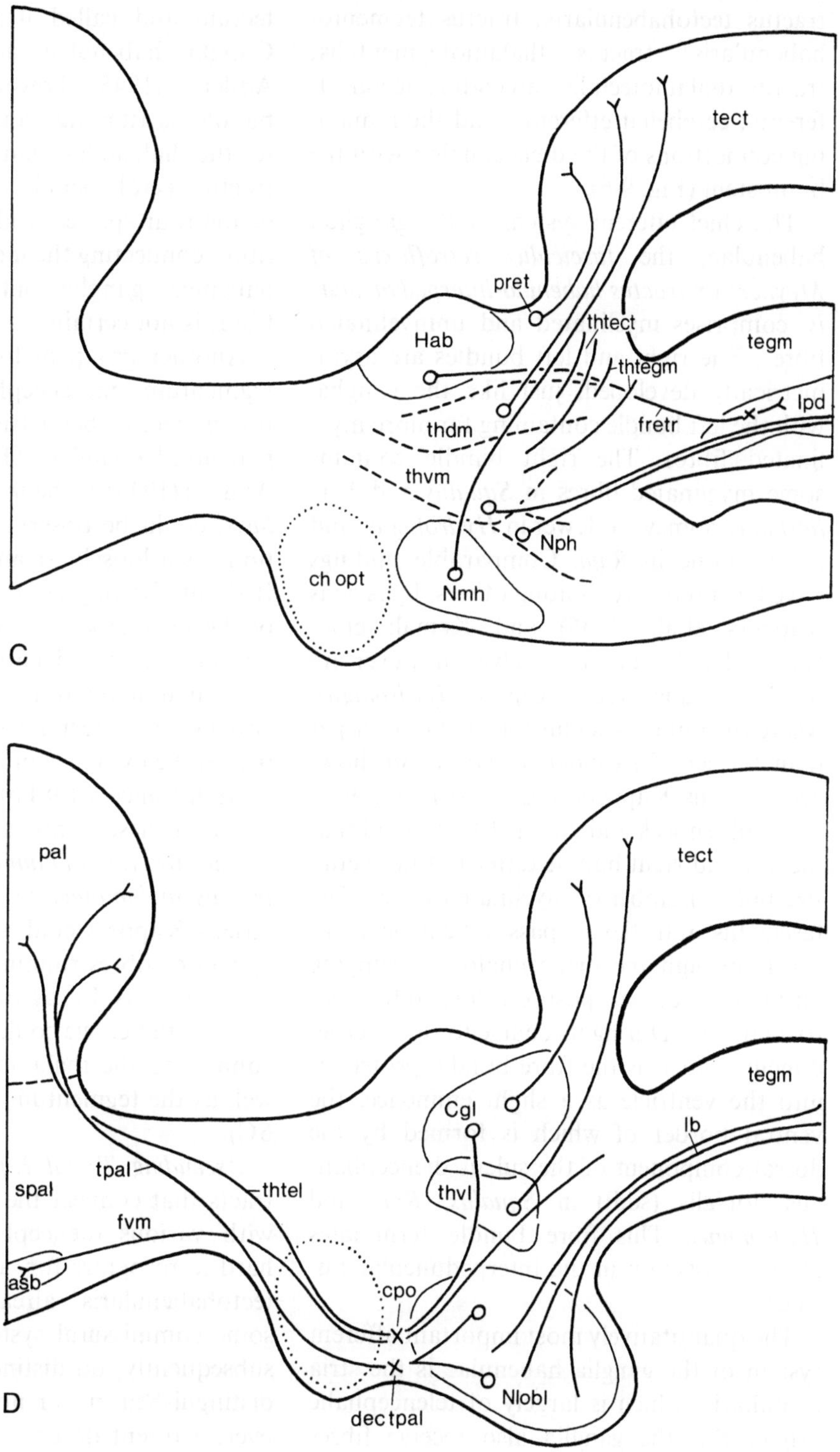
tect
pret
thtect
Hab
tegm
thtegm
lpd
thdm
fretr
thvm
Nph
ch opt
Nmh
C
pal
tect
tegm
Cgl
lb
tpal
thtel
spal
thvl
fvm
asb
cpo
Nlobl
dec tpal
D

tractus tectohabenularis; tractus tegmentohabenularis; tractus thalamotegmentalis; tractus thalamotectalis; ascending tectal efferents; cerebellar efferents, and the remaining connections of the diencephalon with the brain stem (Fig. 9-6).

The chief efferent system of the ganglion habenulae, the *fasciculus retroflexus of Meynert* or *tractus habenulo-interpeduncularis,* comprises myelinated and unmyelinated fibres. The right and left bundles are asymmetrically developed, just like the ganglia, with the left bundle containing far more myelinated fibres. The right bundle contains some myelinated fibres in *Squalus* and *Scyliorhinus,* somewhat fewer in *Hydrolagus,* and almost none in *Raja.* Comparable findings have been reported, among others, by Ariëns Kappers et al. (1936) and Kemali et al. (1980). The bundles themselves, however are of about equal size, except in *Hydrolagus,* where the right fasciculus retroflexus is larger than the left. In *Chimaera monstrosa,* however, Ariëns Kappers and Carpenter (1911) and Kuhlenbeck and Niimi (1969) found that the left and right bundle differ not in overall size but in number of myelinated fibres. The fasciculus retroflexus passes ventrocaudalward through the diencephalon, occupying during this course a position close to the ventricular wall. During its course to the interpeduncular nucleus the fibre bundle protrudes into the ventricle as a slight eminence, the ventral border of which is formed by the dorsal component of the sulcus diencephalicus dorsalis (sddl) in *Squalus, Raja* and *Hydrolagus.* This fibre bundle terminates after decussation in the interpeduncular nucleus.

The quantitatively most important afferent system of the ganglia habenulae is the stria medullaris, which is largely of telencephalic origin. But the ganglia also receive fibres from the mesencephalon, the majority of which arise from the tectum mesencephali and constitute the *tractus tectohabenularis.* This bundle was first recognized by Edinger (1892), who regarded it as running to the tectum and called it accordingly "tractus Ganglii habenulae ad Mesencephalum". Addens (1945) however, recognized this bundle as running from the mesencephalon to the habenular ganglion and termed it tractus tectohabenularis. In sagittal sections of the four species studied we can recognize fibres connecting the mesencephalon with the habenular ganglia, but the direction of the fibres is not certain.

Another group of fibres originates in the tegmentum mesencephali and joins the tractus tectohabenularis before reaching the ganglion habenulae. This tract, described by Addens (1945) as the *tractus tegmentohabenularis,* could be observed in our sagittal sections as a loosely arranged bundle of fibres. It should be emphasized that the description of the tectohabenular and tegmentohabenular tracts is based entirely on observations in normal material. Experimental studies, involving large tectal lesions, have not confirmed the existence of a tectohabenular connection (Smeets 1981b), so that a habenular origin of these fibres cannot be excluded.

The *tractus thalamotegmentalis* and the *tractus thalamotectalis* mentioned briefly by Ariëns Kappers et al. (1936) form a diffuse system of fibres running between brain stem and thalamus. In sagittal sections of the four species studied we could also recognize fibres connecting the thalamus with the tectum as well as the tegmentum mesencephali (Fig. 9-6C).

Ascending Tectal Efferents. Distinct fibre tracts that connect the tectum mesencephali with various diencephalic cell masses are hard to recognize for, apart from the tractus tectohabenularis already mentioned and some commissural systems to be dealt with subsequently, no distinct fibre tracts can be distinguished in our normal material. However, a recent degeneration study by one of us (Smeets 1981b) on the tectal efferent pathways in *Scyliorhinus* and *Raja,* has revealed that the diencephalon receives a substantial tectal input. After lesions that included all six tectal layers, ascending fibres could be traced to the pretectal area and to the dorsomedial region of the thalamus. A few fibres appeared to run further lateralward to the corpus geniculatum laterale and the torus lateralis (in *Raja*). These ascending projections are both ipsi- and contra-lateral. The majority of the ascending efferent tectal fibres probably reach the contralateral pretectal and thalamic regions by way of the tectal commissure. However, some fibres run parallel and medial to the optic tract and probably reach to the opposite side, via the commissura postoptica.

Cerebellar Efferents. An important connection between cerebellum and mesencephalic and diencephalic regions, is provided by the *brachium conjunctivum* or *tractus cerebello-mesencephalicus et diencephalicus* of Ariëns Kappers et al. (1936). This tract arises from the cerebellar nucleus and swings ventromedialward and forward to cross in the most caudal part of the tegmentum mesencephali, ventral to the fasciculus longitudinalis medialis. The majority of its fibres terminate in the nucleus ruber; however, some run to the diencephalic region and terminate reportedly in the thalamus and hypothalamus (Wallenberg 1907; Ariëns Kappers et al. 1936; Gerlach 1947). An experimental study of the cerebellar efferent pathways in the nurse shark, *Ginglymostoma,* (Ebbesson and Campbell 1973) has confirmed the existence of this pathway with the exception of a cerebellohypothalamic connection (Fig. 9-6A).

Remaining Connections of the Diencephalon with the Caudal Regions. Under this heading we will mention some fibre tracts which have been described in the older literature and which can also be recognized at some levels in our material. However, the polarity of these fibre systems is unknown and the present authors are convinced that only experimental studies can unravel this network of diffuse fibre systems. According to Ariëns Kappers et al. (1936) and Gerlach (1947), fibre tracts which arise chiefly in the lobus inferior hypothalami form a tract, *tractus lobobulbaris,* which can be traced to the most caudal regions of the rhombencephalon, where it probably becomes associated with the visceral efferent nuclei of the region, particularly those of trigeminal and facial nerve nuclei. Another tract, the *tractus mamillo-interpeduncularis* (Gerlach 1947), courses together with the tractus lobobulbaris from the region of the recessus posterior or recessus mamillaris to the nucleus interpeduncularis in the midbrain tegmentum.

Our material, especially in sagittal sections, does not exclude the existence of these two tracts. Numerous fibres, both myelinated and unmyelinated, run from the caudal part of the lobus inferior hypothalami and also from the recessus mamillaris to the midbrain tegmentum (Fig. 9-6D). Probably these fibres include the *tractus lobocerebellaris,* described by Ariëns Kappers et al. (1936).

Ebbesson and Hodde (1981) desdribed a direct spinothalamic connection in *Ginglymostoma.* After a hemisection in the third spinal segment a small number of degenerated fibres could be traced to the dorsal thalamus. Although in our normal material, such a pathway could not be recognized, a direct projection from the thalamus ventralis and nucleus periventricularis hypothalami to the spinal cord has also been experimentally confirmed by Smeets and Timerick (1981) in *Scyliorhinus* and *Raja.*

9.5.2 Connections Between Diencephalon and Telencephalon

The following tracts will be discussed in this section: (1) the stria medullaris; (2) the tractus pallii; (3) the fasciculus basalis telencephali, and (4) the tractus thalamotelencephalicus.

The *stria medullaris,* the most important connection of the ganglion habenulae, passes through the most dorsal part of the telencephalon impar and swings upward into the habenular ganglion to form the habenular commissure. The stria medullaris consists of (1) fibres which cross in the habenular commissure to synapse with habenular cells and (2) fibres which constitute a true telencephalic commissure, the commissura superior telencephali of Ariëns Kappers et al. (1936). According to these authors this commissure contains unmyelinated axons which are surrounded by the myelinated fibres of the decussating olfactohabenular tract. The stria medullaris can be easily recognized in the four species studied and is of approximately equal size on the two sides, collecting fibres from both the pallial and subpallial regions of the telencephalon. It consists of a thin sheet of fibres that surrounds a bundle of unmyelinated fibres. The various tracts that make up the stria medullaris are known principally from the work of Johnston (1911) who recognized as many as seven components. A more detailed description of the several components of the stria medullaris complex will be given later in Chap. 10.

An important fibre tract, the *tractus pallii,* originates in the roof of the telencephalon and was first described by Edinger (1892) as "Mantelbündel". This tract, which has been noted by all later students of the chondrichthyan brain (e.g. Johnston 1911; Bäckström 1924; Ariëns Kappers et al. 1936; Gerlach 1947), connects the telencephalic pallial regions with the hypothalamic areas. According to Johnston (1911) the tractus pallii constitutes a system relating visceral and gustatory centres of the hypothalamus to the olfactory centres of the telencephalon. As he stated (Johnston 1911, p. 18): "this is equivalent to saying that the massive roof in selachians is an olfactogustatory correlation centre".

In the species studied by us, the tractus pallii is a well defined tract of predominantly myelinated fibres. It passes immediately ventrolateral to the stria medullaris, through the telencephalon impar, keeping a position lateral to the basal forebrain bundle and then runs superficial to, or intermingles with, lateral fibres of the optic tract. Finally the fibres of the tractus pallii after decussating in part, in the decussatio tractus pallii, caudoventral to the chiasma opticum, mostly terminate in the rostral half of the nucleus lobi lateralis inferior. The caudal part of the tractus pallii is quite similar in the four species studied. The features of the more rostral part will be discussed later in Chap. 10.

There are several possible sites of origin and termination of the fibres of the tractus pallii. Catois (1901) using Golgi material concluded that the tract contains descending fibres, whereas Ariëns Kappers and Theunissen (1907) with normal material, and Wallenberg (1907) with the Marchi technique, held that this pathway consists of ascending fibres. More recent preliminary studies using axon degeneration techniques have revealed that in fact the tractus pallii contains both descending and ascending fibres (Ebbesson 1972; Smeets unpublished), and that most of the fibres cross in the decussatio tractus pallii.

The largest fibre system that connects the telencephalon with more caudal, and particularly diencephalic centres is the *fasciculus basalis telencephali* or *basal forebrain bundle.* This tract was first recognized by Edinger (1892) who called it "das basale Vorderhirnbündel". It has subsequently been called the tractus striothalamicus (Ariëns Kappers 1906), tractus striohypothalamicus (Ariëns Kappers and Carpenter 1911), medial forebrain bundle (Johnston 1911) and tractus striothalamicus et hypothalamicus (Ariëns Kappers et al. 1936).

In most groups of lower vertebrates two basal bundles can be distinguished, namely the fasciculus medialis telencephali and the fasciculus lateralis telencephali. It has to be emphasized that the medial and lateral forebrain bundles are named according to their topographic position in the telencephalic peduncle and they do not necessarily share similar origins and terminations in different groups (Nieuwenhuys 1969). In teleosts, two parts of the basal forebrain bundle can be recognized, whereas in primitive actinopterygians and in cartilaginous fishes this separation is not distinct.

As a detailed desciption of the components and origin of the basal forebrain bundle will be given in the next chapter we will confine ourselves here to its course and site of termination within diencephalic and brain stem regions.

After partly crossing in the commissura anterior the basal forebrain bundle enters the diencephalon to distribute its fibres mainly to the more superior and caudal parts of the lobi inferiores hypothalami. However, some fibres can be traced to the ventral thalamus and to the midbrain tegmentum. Experimental studies by Ebbesson (1972) and Smeets (1983) confirm this route for the basal forebrain bundle.

The *tractus thalamotelencephalicus*. Ebbesson and Schroeder (1971) and Schroeder and Ebbesson (1974) reported that in the nurse shark, *Ginglymostoma,* following thalamic lesions a strong, largely crossed bundle of degenerating fibres can be traced to the pallium. The fibres of this thalamotelencephalic tract collect in the rostral thalamic region and pass ventrally to decussate mostly in the postoptic commissure, directly dorsal to the decussation of the tractus pallii. Further rostrally the tract moves laterally and finally swings dorsalward to terminate in an area known as the pallium dorsale pars centralis (Fig. 9-6D).

A recent HRP-study by Luiten (1981b) in *Ginglymostoma* has revealed that the cells of origin of this thalamotelencephalic pathway are mainly located in three nuclei: the lateral geniculate body; the ventrolateral optic nucleus, and a centre, which he designated as the central thalamic nucleus. The two nuclei first mentioned probably constitute together the corpus geniculatum laterale of our description, but the central nucleus of *Ginglymostoma* is hard to compare with any of the thalamic cell masses of the four species studied by us.

A thalamotelencephalic tract could not be distinguished as a separate entity in our normal material, although it is probable that the lateral part of the basal forebrain bundle contains the fibres of this tract. Johnston (1911) recognized a thalamocortical tract, which connects the telencephalon with the lateral geniculate body, nucleus pretectalis and possibly the tectum mesencephali. However, we consider it very unlikely that the latter two connections are actually present, at least as ascending systems.

9.5.3 Intrinsic Diencephalic Connections

Many diffusely arranged fibres pass dorsoventralward and rostrocaudalward within the diencephalon. Amongst these fibres earlier students of the chondrichthyan brain (Dammerman 1910; Johnston 1911; Ariëns Kappers et al. 1936) recognized a tractus habenulothalamicus, a tractus thalamolobaris, a tractus saccothalamicus, a tractus tuberoposterior, a tractus sacci vasculosi and a tractus preopticohypophyseus. However, only the last two of these can be clearly distinguished in our normal material (Fig. 9-6A).

A *tractus habenulothalamicus,* mentioned by Johnston (1911), was stated by Ariëns Kappers et al. (1936) to arise from the subhabenular, rather than from the habenular region and to surround partly the fasciculus retroflexus. These authors were unable to determine its area of termination.

According to Ariëns Kappers et al. (1936) a *tractus thalamolobaris* is formed by diffusely arranged fibres which in more caudal diencephalic regions connect the thalamus dorsalis with the lobi inferiores.

A *tractus saccothalamicus* has been reported by Dammerman (1910) as consisting of those fibres of the tractus sacci vasculosi, that pass beyond the nucleus sacci vasculosi, together with processes from the cells of the latter nucleus. This tract extends frontalward and dorsalward toward the central grey of the dorsal thalamus.

According to Dammerman (1910) another fibre system, the *tractus tubero-posterior,* arises from the nucleus sacci vasculosi and passes caudalward from the nucleus through the tuberculum posterius. The termination of this tract is at present unknown.

A fibre system which could be identified in all species examined is the *tractus sacci vasculosi*. This is a well-developed unmyelinated tract formed by the processes of the cells of the saccus vasculosus. The wall of the saccus vasculosus consists of cubic or cylindric cells (coronet cells) which have sensory hairs with knob-shaped terminations turned toward the lumen of the sac. Between these cells lie smaller, supporting elements. The axons of the coronet cells run in the wall of the sac, forming an evident fibre layer. Frontalward they form a fascicle of fibres. In the region where the wall of the sac passes over into that of the recessus mamillaris this fibre tract enters the grey substance of the tuberculum posterius as a distinct little bundle which ascends to the level of the commissura post-infundibularis pars inferior. After decussating, the majority of the fibres terminate in the nucleus sacci vasculosi, which has already been mentioned.

There is little or no agreement about the function of the saccus vasculosus. According to Dammerman (1910) the fact that the sensory hairs of the crown or coronet cells protrude into the ventricular cavity of the saccus suggests that it is concerned with perceptions related to the ventricular fluid. A secretory function of the epithelium of the saccus vasculosus, based on the presence of inclusions in the coronet cells, has been proposed by Bargmann (1954), van de Kamer and Verhagen (1954) and Altner (1964b, 1965), whereas Mellinger (1964) and Altner (1964a) have stated that these inclusions could not be interpreted as a sign of a secretion process, but are due to differences in endoplasmic structure. Watanabe (1966) and von Harrach (1970) favoured a receptor function and suggested that the saccus vasculosus could be an organ for chemoreception or for the perception of liquor pressure. A third possibility is that in chondrichthyan fish the coronet cells are functionally involved in a transcellular ion transport process between the cerebrospinal fluid and the blood, in the same manner as proposed by Jansen (1969) for actinopterygians.

A more recent electronmicroscopic study by van de Kamer et al. (1973) has revealed that large quantities of glyocogen are present in the coronet cells and that these cells are filled with an extensive system of vertically oriented, smooth endoplasmic reticulum tubules which might facilitate transcellular ion transport.

The second intrinsic diencephalic tract, recognized in all species examined, is the *tractus preopticohypophyseus*. Earlier students of the diencephalon have considered this tract to be the afferent connection of the saccus vasculosus and therefore termed it the tractus thalamosaccularis (Dammerman 1910; Ariëns Kappers et al. 1936; Gerlach 1947). However, more recent studies (e.g., Meurling and Björklund 1970; Wilson et al. 1974) have revealed that this tract passes to the pituitary and, hence, deserves the name tractus preopticohypophyseus (see Sect. 9.7).

9.5.4 Diencephalic Commissures

The *commissura habenulae* consists of myelinated and unmyelinated fibres of the stria medullaris. The unmyelinated fibres are probably true commissural fibres that form a commissura superior telencephali (Ariëns Kappers et al. 1936), whereas the myelinated fibres arise from the stria medullaris, which decussates in the commissura habenulae and terminates in the contralateral ganglion habenulae. The commissura habenulae also receives fibres from the pineal tract.

The *commissura posterior* consists of a large mass of fibres forming a dorsally convex arch just ventral to the pretectal region. It is compact in the midplane but sprays out laterally, with most of its fibres going in a ventrocaudal direction. The most ventrally directed fibres run near the periventricular grey toward the tuberculum posterius. The most caudally running fibres are lost in the tegmental fibre stream. No definite connections of the posterior commissure could be distinguished but the fibres possibly relate to all of the nearby cellular regions. Connections with the tectum, as stated by Ariëns Kappers et al. (1936), have not been confirmed by the experimental study of Smeets (1981 b). The posterior commissure also receives fibres from the pineal tract.

The *commissura postoptica.* In addition to the crossing optic fibres, decussations of other pathways are seen ventrocaudal and dorsocaudal to the optic chiasm. The ventrocaudal part of the commissura postoptica is formed by crossing fibres of the tractus thalamotelencephalicus, as described above. It can be easily distinguished from the more ventral decussation of the tractus pallii. The dorsal part of the postoptic commissure probably represents a connection between the two sides of the rostral parts of the hypothalamus and may also contain crossing fibres of the commissura transversa.

The *commissura transversa,* sometimes infelicitously termed commissura postoptica (Ariëns Kappers 1936), crosses immediately behind the optic chiasm in the dorsal part of the commissura postoptica. Little agreement exists about the origin and termination of these fibres. Edinger (1892), who considered it more a decussation than a commissure, saw its fibres end under the tectum mesencephali, while Catois (1901) described it as a commissure between the posterior and lateral regions of the mesencephalon. Haller (1898), who called it "commissura postoptica superior", claimed that it contains only decussating fibres of the tectum mesencephali. Ariëns Kappers et al. (1936) stated that the fibres of this commissure pass along the lateral surface of the diencephalon to tectal and tegmental centres with which they are functionally related.

Although in our material the commissura transversa could easily be traced over some distance, its origin and termination could not be determined with certainty. It seems probable that these fibres are subtectal rather than tectal decussating fibres, though lesions of the tectum show some degenerating fibres which course to the postoptic commissure.

Two other commissures, the *commissura preinfundibularis* and the *commissura postinfundibularis,* can be recognized in the four species studied. The commissura preinfundibularis lies caudal to the decussation of the tractus pallii in the ventral part of the hypothalamus and consists partly of myelinated dorsally-running fibres, the origin and termination of which are still unknown. The commissura postinfundibularis can be subdivided into a pars superior and a pars inferior. The pars superior of the commissura postinfundibularis lies somewhat more rostral in the tuberculum posterius than the inferior part and probably forms a connection between the caudal cell masses of the hypothalamus. In *Squalus* and *Scyliorhinus* this commissure extends rather far in the rostrocaudal direction. The pars inferior of the commissura postinfundibularis takes a more ventral position in the tuberculum posterius and has a much smaller rostrocaudal extent. This part is formed by the crossing fibres of the tractus sacci vasculosi. In *Scyliorhinus* the preinfundibular commissure is poorly developed, whereas in *Raja* the postinfundibular commissure is less distinct.

9.5.5 The Tractus Opticus

The tractus opticus of cartilaginous fish has been the subject of several recent experimental studies (Ebbesson and Ramsey 1968; Graeber and Ebbesson 1972; Northcutt 1979a; Ebbesson and Meyer 1980; Northcutt and Wathey 1980; Luiten 1981a; Smeets 1981a). These studies show that the retinal efferents are not restricted to the corpus geniculatum laterale and the tectum mesencephali, as was previously believed (Houser 1901; Ariëns Kappers et al. 1936), but project to several additional diencephalic and mesencephalic regions (Fig. 9-6B).

In *Scyliorhinus* and *Raja* (Smeets 1981a) the optic nerve fibres decussate completely in the chiasma opticum with the exception of a very small ipsilateral projection to the caudal part of the preoptic nucleus or to an area immediately ventral to this nucleus, which corresponds to our nucleus suprachiasmaticus. The decussating optic nerve fibres also terminate for a very small part in the contralateral preoptic area, in the caudal region of which are scattered cells among the decussating optic nerve fibres. The crossed optic nerve fibres form a lateral or marginal optic tract which courses both dorsally and caudally along the lateral wall of the diencephalon. Some of the optic nerve fibres leave the main stream and curve medialward to terminate in the thalamus dorsalis pars lateralis. At somewhat more caudal levels, the optic nerve fibres or their collaterals terminate within the area of the lateral geniculate body and this is the most prominent diencephalic projection. At the same level the thalamus ventralis pars lateralis also receives a retinal projection from fibres which run in the ventral part of the main stream of optic fibres.

A medial optic tract splits off from the marginal tract and courses dorsomedially, where it divides into fascicles which terminate for a small part in a pretectal nucleus and for the greater part distribute their fibres over the rostral and medial parts of the tectal lobe. In *Raja* also, a central pretectal terminal field was found (Fig. 84). The remainder of the marginal optic tract, the tractus opticus lateralis, courses dorsolaterally and supplies the lateral and caudal parts of the tectal lobe.

Caudal to the division into medial and lateral optic tracts, fibres were observed in *Raja* leaving the lateral optic tract and these fibres could be traced to the midbrain tegmentum, where they terminate in the basal optic nucleus, thus forming a tractus opticus basalis.

Comparisons between the retinal projections found by other authors are difficult because there is little agreement about the diencephalic nomenclature. However, it can be concluded that in all elasmobranch species studied so far, the preoptic area, the ventral and dorsal thalamus, the pretectal areas, as well as the tectum mesencephali, receive a contralateral retinal input. Ipsilateral projections to the preoptic area or nucleus suprachiasmaticus have been reported by Graeber and Ebbesson (1972: *Negaprion*), Northcutt (1979a: *Squalus*), Ebbesson and Meyer (1980: *Rhinobatos*), Northcutt and Wathey (1980: *Platyrhinoidis*) and Smeets (1981a: *Scyliorhinus, Raja*). As far as we know, only Northcutt and Wathey (1980) have observed ipsilateral projections to the other diencephalic and the mesencephalic retinal targets.

A tractus opticus basalis has not been recognized in all chondrichthyan fish investigated. In *Galeocerdo* (Ebbesson and Ramsey 1968), *Negaprion* (Graeber and Ebbesson 1972) and *Scyliorhinus* (Smeets 1981a), following enucleation of an eye, degenerating fibres are observed leaving the lateral optic tract ventrally and coursing caudalward to the midbrain tegmentum. Their site of termination could not be ascertained, except in *Galeocerdo*. In *Squalus* (Northcutt 1979a), *Platyrhinoidis* (Northcutt and Wathey 1980) and *Ginglymostoma* (Luiten 1981a) a distinct basal optic tract can be recognized, which partly terminates in a small cell group which is comparable to the basal optic nucleus of the present authors, and partly continues caudally and medially to terminate in an ill-defined tegmental area, which was termed the ventral accessory optic nucleus by Northcutt and Wathey (1980). According to Luiten (1981a), the close proximity of this terminal area to the oculomotor nuclei strongly suggests that this nucleus is the homologue of the accessory or basal optic nucleus of amniotes. It would be very interesting to know whether the presumed basal optic nucleus in cartilaginous fish projects to the oculomotor nucleus and the cerebellum, as has been demonstrated for some teleosts (Finger and Karten 1978), birds (Brauth and Karten 1977; Brecha et al. 1980) and reptiles (Reiner and Karten 1978).

9.6 The Diencephalon and the Problem of Homology

In Sect. 9.2 of the present chapter, several ways to solve the problem of establishing homologies were presented. Here we will test the practical validity of the three chosen criteria—ventricular sulci, cell masses and fibre tracts—for establishing homologies in the diencephalon of the four cartilaginous fish studied.

9.6.1 Ventricular Sulci

Eleven sulci have been recognized in the diencephalon and preoptic region of the dogfish *Squalus acanthias*. Two of these, the sulcus preopticus and the sulcus diencephalicus ventralis, could not be identified in any of the other three species and, hence, will not be discussed. The remaining nine sulci however, are rather constant features in every species examined.

The sulcus intrahabenularis is sometimes considered to separate dorsomedial and ventrolateral habenular nuclei (Addens 1945). In our material there are no distinct cytoarchitectonic differences between the parts of the habenula dorsal and ventral to the sulcus, and therefore we have qualified this groove as a sulcus within a nuclear mass rather than as a boundary sulcus.

The sulcus subhabenularis, also called the sulcus dorsalis by some authors (Table 9-1), is a very distinct groove in the diencephalon in all species studied. This sulcus forms the boundary between epithalamus and dorsal thalamus in the rostral part of the diencephalon. For this reason, and also because of its distinctness, there are no difficulties in defining and homologizing the habenular ganglia of different animals.

The sulcus diencephalicus dorsalis consists of two components, i.e. sdd1 and sdd2, which are very distinct and run caudally from the level of the habenular commissure rostrally to the di-mes-encephalic boundary. They are not continuous with the sulcus subhabenularis. Both sulci are present in *Squalus* and *Raja,* whereas in *Scyliorhinus* and *Hydrolagus* only one of the two grooves can be recognized.

The dorsal component of the sulcus diencephalicus dorsalis (sdd1) always forms the ventral boundary of the fasciculus retroflexus. The ventral component sdd2 always lies in the middle of a cell mass, the thalamus dorsalis pars medialis. Sometimes this cell mass can be divided into a pars dorsomedialis and a pars ventromedialis. In *Squalus* the boundary between these subnuclei corresponds with the sdd2.

Evidently there are constant relationships between the two components of the sulcus diencephalicus dorsalis and the boundaries of diencephalic cell masses or fibre tracts and, consequently, these sulci have great value when defining and homologizing diencephalic structures at least in cartilaginous fish.

The sulcus diencephalicus medius is an extremely important structure since it has been considered by many authors as marking the boundary between the dorsal thalamus and the ventral thalamus. However, in *Squalus* and *Scyliorhinus* this sulcus does not coincide with the boundary between these two entities, whereas in *Raja* and *Hydrolagus* this sulcus clearly demarcates these two regions. In *Scyliorhinus* and *Raja* this sulcus extends farther rostrally than the same sulcus in the other two species. However, in *Squalus* and *Hydrolagus* a sulcus f could be recognized which may be considered to be homologous with the rostral part of the sulcus diencephalicus medius of *Scyliorhinus* and *Raja,* since this sulcus lies within the region of the eminentia thalami, and also terminates at the level of the commissura anterior.

The sulcus organi subcommissuralis is a distinct groove, present in all species studied, which forms the ventral border of the pretectal area in the midline. In all, except *Raja,* this sulcus is accompanied by another groove, the sulcus organi subcommissuralis accessorius and together they form the dorsal boundary of the fasciculus retroflexus or bundle of Meynert.

The sulcus hypothalamicus does not exhibit any distinct relationship with the boundary of cytoarchitectonic structures.

In conclusion, therefore, we can state that of the various sulci, the sulcus subhabenularis, the sulcus organi subcommissuralis, the sulcus organi subcommissuralis accessorius, the sulcus diencephalicus dorsalis 1, and the caudal part of the sulcus diencephalicus medius (in *Raja* and *Hydrolagus*), mark the boundaries of cytoarchitectonic units. The sulcus intrahabenularis, the sulcus diencephalicus dorsalis 2 and the sulcus f (equivalent to the rostral part of the sulcus diencephalicus medius) lie in the middle of a cytoarchitectonic field, whereas the sulcus hypothalamicus shows no distinct relations with any cytoarchitectonic structure in the four species examined.

It should be added that in the species studied most of the diencephalic sulci show an overall horizontal course and that it is only the boundaries of periventrically situated cell masses which correspond with these sulci.

9.6.2 Cell Masses

The habenular ganglia are similar in structure in all the species examined, the left ganglion being of the same size or larger than the right. Each ganglion consists of small, closely packed cells that line the ventricular surface and somewhat larger, more loosely arranged cells that lie in the dorsolateral region. At the level where the two habenular ganglia fuse a separate cell mass consisting of small granular cells could be recognized in *Scyliorhinus* and *Hydrolagus,* but not in *Squalus* and *Raja.*

We have seen that the *dorsal thalamus* can be divided into a pars medialis and a pars lateralis. The pars medialis thalami dorsalis shows great similarity in all species examined and consists of small, densely packed periventricular cells and has a constant topographical position. In *Squalus* and *Hydrolagus* it can be further subdivided into a pars dorsomedialis and ventromedialis on account of the greater density and a slightly darker staining of the cells of the ventral part. In *Squalus* this subdivision coincides with the sulcus diencephalicus dorsalis 2. As shown in Fig. 103 the thalamus dorsomedialis of *Hydrolagus* is situated between the sulcus diencephalicus dorsalis 1 and the sulcus diencephalicus medius, but this relation is less distinct in the other three species. The pars lateralis of the dorsal thalamus varies somewhat in position among the species studied, since in *Scyliorhinus, Raja* and *Hydrolagus* this cell mass lies at the level of the axis habenular commissure-optic chiasm, whereas in *Squalus* it takes a more caudal position (cf. Figs. 5, 38, 70, 104).

The *ventral thalamus,* which is also composed of a medial and a lateral part, is readily comparable between the four species studied, as only minor variations are seen, in relation to its rostrocaudal extent. An eminentia thalami could be distinguished rostral to the ventral thalamus in all species, except in *Hydrolagus.* This cell mass is extremely well developed in *Raja.* In all, the sulcus diencephalicus medius and/or the sulcus f lies in this cytoarchitectonic unit.

The lateral geniculate body, which is embedded in the optic tract fibres, contains medium-sized cells (Table 4-2). Comparing the lateral geniculate body of the four species it can be stated that: (1) in *Scyliorhinus, Raja* and *Hydrolagus* this cell mass contains more neurons and shows a more compact arrangement than the equivalent nucleus in *Squalus;* (2) the lateral geniculate body in the three species first mentioned has migrated farther lateralward; (3) in *Raja* the lateral geniculate body rostrally takes a ventral position, and caudally occupies a more dorsal position, and (4) in *Scyliorhinus, Raja* and *Hydrolagus* the lateral geniculate body extends to the pretectal region, whereas in *Squalus* this cell mass does not extend so far caudally.

A *preoptic nucleus* could be delineated in all species studied but a subdivision into a pars magnocellularis and parvocellularis, as described by Charlton (1932), could not be made. A nucleus suprachiasmaticus could be identified on the basis of the orientation of its constituent cells. The position of this cell mass is rather constant within the three species studied. A *nucleus medius hypothalami* could be recognized in the four species, although varying considerably in size. In *Squalus* and *Hydrolagus* it is small and does not extend far caudalward, whereas in *Scyliorhinus* and *Raja* it is better developed, both in number of cells and in extent.

A nucleus periventricularis hypothalami and a nucleus lateralis tuberis could be recognized in all species, except *Raja*. However, the caudal boundary of the nucleus periventricularis hypothalami is hard to distinguish.

The remaining hypothalamic cell mass, the *nucleus lobi lateralis hypothalami,* has been observed in all species studied. This nucleus could be subdivided in both sharks into a pars dorsalis, a pars medialis and a pars ventralis, but such a tripartitioning of the nucleus is not possible in *Raja*. However, a tripartitioning was possible in *Hydrolagus,* but in a different way. On the basis of differences in cell density and intensity of staining, a pars dorsalis and a pars ventralis could be distinguished. A third part, termed nucleus lobi lateralis pars lateralis, could be distinguished at more caudal levels.

The *pretectal region*. It is easy to indicate the general position occupied by this region in the anamniote brain, but it has been impossible to define precisely its boundaries. We have defined this region as that part of the diencephalon lying rostral to the tectum, but only close to the midplane. In contrast Kuhlenbeck (1956), who considered the pretectal region to be both di- and mes-encephalic in origin, believed that this region extends much farther lateral- and ventralward. He believed that there are four parts to the pretectal region in all lower vertebrates: (1) the posthabenular periventricular diencephalic region, dorsal to the sulcus synencephalicus (our sulcus diencephalicus dorsalis 1); (2) nucleus commissurae posterioris, probably corresponding to our nucleus of the same name; (3) a mesencephalic, migrated nucleus lentiformis which may be equivalent to the rostral part of our nucleus tegmentalis lateralis and (4) the undifferentiated part of the periventricular grey of the tegmentum. The name nucleus pretectalis has, however, been used to describe various cell groups in the pretectal area which receive a retinal input (Northcutt 1979a; Smeets 1981a). We could not delineate distinct cell masses in the pretectal area in our material.

In the ventral part of the di-mes-encephalic transition zone a nucleus tuberculi posterioris, a nucleus sacci vasculosi, a nucleus opticus basalis and a torus lateralis can be recognized. Of these, the first two nuclei have been identified in all four species, occupying the same position in the diencephalon. The torus lateralis could only be distinguished in *Raja*. It is clear from the topographical reconstruction that this nucleus is a rostral continuation of the nucleus tegmentalis lateralis and is not represented in the other three species. A basal optic nucleus could be identified in *Squalus, Raja* and *Hydrolagus,* but not in *Scyliorhinus*. This nucleus has a similar position in the three species first mentioned, though it lies somewhat more rostral in *Hydrolagus*.

9.6.3 Fibre Tracts

It is difficult to distinguish individual fibre tracts within the diencephalon in Bodian and Klüver-Barrera series. Consequently, it cannot be excluded that the habenula, thalamus dorsalis, thalamus ventralis and hypothalamus are all connected with the tectum and the tegmentum of the midbrain, with the cerebellum and with the rhombencephalon; in fact, in the literature many such connections have been described. We will confine ourselves, however, to two groups of fibres, the fasciculus retroflexus and a fibre system that connects the hypothalamus with the rhombencephalon and tegmentum mesencephali.

The *fasciculus retroflexus* or bundle of Meynert, like the stria medullaris, is one of the most stable fibre systems in cartilaginous fishes. It has been universally described as coursing from the habenular ganglion to the interpeduncular nucleus in the midbrain tegmentum. The bundle on the left side contains many more myelinated fibres than that on the right and a decussation prior to the site of termination can be recognized. The dorsal component of the sulcus diencephalicus dorsalis, if present, always forms the ventral border of the protrusion caused by the bundle, whereas the sulcus organi subcommissuralis and its accessory component mark the dorsal boundary.

The connections of the hypothalamus with more caudal regions of the brain constitute a vast diffuse system, which could not be clearly divided into separate components. We wish to emphasize that there are far more caudal associations of the hypothalamus than of the thalamus. Most of these caudal connections of the hypothalamus are uncrossed, but some may decussate in the tuberculum posterius.

At this point it is worth recalling Herrick's (1948) comparison of the hypothalamus and habenula in *Ambystoma;* he noted that whereas the habenula discharges through one compact bundle, the fasciculus retroflexus, the hypothalamus has more widespread and diffuse efferents. These observations by Herrick also apply to chondrichthyans.

The *stria medullaris* is one of the most constant fibre systems in the vertebrate nervous system. It connects the telencephalon with habenular ganglia in all chondrichthyan species we have studied. In the actinopterygian fish *Acipenser* Johnston (1901) found that the larger ganglion receives both ipsi- and contra-lateral fibres, whereas the smaller ganglion gets only contralateral fibres. This may also be true in the chondrichthyans where the striae are of equal size bilaterally, although the ganglia are not.

The *tractus pallii* is also a rather constant fibre system in Chondrichthyes with respect to its diencephalic part. We may recall that this tract connects the telencephalic pallial regions with hypothalamic areas, terminating mainly in the rostral half of the nucleus lobi lateralis inferior. A decussation of the pallial tract fibres caudoventral to the optic chiasm could be recognized in all species. The tractus pallii probably contains both ascending and descending fibres and occupies a constant position in the species studied.

The *fasciculus basalis telencephali* connects the telencephalon with more caudal and particularly diencephalic centres. Fibres running farther than the di-mes-encephalic boundary have been demonstrated in experimental studies by Ebbesson (1972) and Smeets (to be published). The fibres of the basal forebrain bundle decussate partly in the commissura anterior. The bundle shows, with regard to course and areas of termination, a great similarity in the four species studied.

Two distinct intrinsic connections within the diencephalon have been recognized in our material, i.e., the *tractus sacci vasculosi* and the *tractus preopticohypophyseus.* Of these, the former connects the neurosecretory cells of the saccus vasculosus with the nucleus sacci vasculosi in the tuberculum posterius, while the latter plays an imporant role in the inhibitory control of the melanocyte stimulating hormone production and release from the pars intermedia of the pituitary (see Sect. 9.7). In the four species studied these two intrinsic diencephalic connections do not show great variation in origin, course and area of termination.

The *tractus opticus* forms the final fibre tract to be considered in this section. From recent experimental studies it has become clear that the retinofugal pathways are quite similar in all elasmobranchs studied. Retinal input to the preoptic area, dorsal and ventral thalamus, corpus geniculatum laterale, pretectum and the superficial layers of the tectum has been confirmed in all species examined and only minor variations are seen in respect to the retinal input of the various tectal layers. A basal optic tract could not always be established, but this may be dependent on the technique used (Smeets 1981a).

9.6.4 Conclusions

We have concluded our comparative evaluation of the sulci, nuclei, and fibre tracts of the four species, and it is now time to relate this material to the problem of the establishment of homologies in the diencephalon. We must first determine whether the three criteria employed for the subdivision of the anamniotic diencephalon give the same result.

The *sulcal pattern* of the chondrichthyan diencephalon comprises at least nine sulci which are very distinct and can be readily compared in the four species studied. A sulcus which is a boundary sulcus in one species never lies in the middle of a cell mass in another. Only the relation of the sulcus diencephalicus medius to the nuclei is variable. In *Raja* and *Hydrolagus* it separates the dorsal thalamus from the ventral thalamus, but this separation is not so clear in the sharks.

An obvious limitation to the use of sulci is that they can only delimit periventricular regions for the relation to migrated nuclei is less distinct, as can be seen in Fig. 70, where, for example, in *Raja* the relation of the corpus geniculatum laterale to the sulci differs considerably from that in the other three species.

The *cell masses* always occupy the same position in the topographical reconstruction, except for the corpus geniculatum laterale, the thalamus dorsalis pars lateralis and the nucleus opticus basalis. However, in the four species these nuclei have been given the same names on the basis of cytoarchitectonic characteristics and their relationships with the optic nerve fibres. This last observation raises the possibility of using the *fibre tracts* as useful entities for establishing homologies. In the diencephalon of Chondrichthyes a great similarity is shown by the well developed fibre systems, with only minor variations. However, as has been mentioned before, it is unreasonable to make comparisons using only normal material, since the fibre systems intermingle to such an extent that it is impossible to define exactly their origins and sites of termination.

In conclusion we can state that sulci, cytoarchitectonic boundaries, as well as fibre tracts are useful for establishing homologies in the chondrichthyan diencephalon. The first two criteria appear to be the most practical ones, with the criterion of fibre tracts providing additional certainty.

9.7 The Hypophysis

9.7.1 Introductory Notes

The hypothalamus influences behaviour by way of direct neural action on other neural centres and because of its powerful endocrine function. It collates information from the external and internal environment and causes the release of hormones from the hypophysis that bring about appropriate responses to environmental change. In this way, processes such as skin colour and reproduction may be synchronized to environmental events.

The hypophysis, or pituitary gland, derived from ectodermal tissue, is always made up of two structurally distinct parts which develop independently. The *neurohypophysis* arises as a downgrowth towards the oral cavity from the region of the neural tube that is to form the hypothalamus; the *adenohypophysis* originates as an ectodermal upgrowth (Rathke's pouch) from the roof of the embryonic buccal cavity. The two parts meet and enclose a mesodermal rudiment which gives rise to the intrinsic blood vessels. The adenohypophysis is the site of synthesis, storage and release into the blood of several peptide hormones. It is divided into the pars distalis (where most of the hormones are secreted) and the pars intermedia. In elasmobranchs the large pars intermedia is closely entwined with the neurohypophysis and for this reason the two parts are designated as the "neurointermediate lobe". The neurohypophysis is suspended from the diencephalon by a stalk which carries the axons of neurosecretory cell bodies that lie in the hypothalamus. Material secreted by these neurons is transported along the stalk to the core of the neurohypophysis. Together, these components form the hypothalamo-neurohypophysial system. In mammals the hypothalamus regulates adenohypophysis function by producing releasing factors (small peptides) which are carried in the portal blood system from the median eminence to the adenohypophysis. Although such a portal system exists in elasmobranchs, and a median eminence can be distinguished, there is no evidence as yet for the existence of releasing substances in these fishes.

9.7.2 The Adenohypophysis

Anatomy. In adult chondrichthyans the adenohypophysis can be divided into a pars distalis and a pars intermedia. According to Knowles et al. (1975) the elongated pars distalis can be further subdivided into a dorsal lobe, which includes a narrow rostral part or rostral lobe, a wider caudal part or median lobe, and a ventral lobe (Fig. 9-7). The ventral lobe is attached to the dorsal lobe by an epithelial stalk. The arrangement is somewhat different in Holocephali. In this group a follicular glandular structure, the "Rachendachhypophyse" lying far ahead of the main gland outside the cranium in the roof of the mouth, replaces the ventral lobe. This structure was described first by Kolmer (1923) and Fahrenholz (1928), and, more recently, by Sathyanesan (1965) and Wingstrand (1966). The Rachendachhypophyse is usually regarded as homologous with the elasmobranch ventral lobe (Sathyanesan 1965), though its embryology is hardly known. However, a recent study (Dodd et al. 1980) in *Hydrolagus* has revealed that this lobe is the main source of the gonadotropic hormone just like the ventral lobe in elasmobranchs. The Rachendachhypophyse appears, therefore, to be at least functionally equivalent to the elasmobranch ventral lobe.

The pars distalis (excluding the ventral lobe) in *Scyliorhinus canicula* has been arbitrarily divided into eight zones by Knowles et al. (1975). The rostral lobe (Fig. 9-7) contains four anterior zones (1–4), and the median lobe four posterior zones (5–8). Though the proportions of the lobes and their exact spatial relations to the hypothalamus vary in different individuals, their limits in most cases are as indicated in Fig. 9-7.

The pars intermedia of the adenohypophysis is intimately connected with the pars nervosa of the neurohypophysis so that they constitute together the neuro-intermediate lobe. The degree of intermingling of the two components varies in different cartilaginous fish, but the pars intermedia is always penetrated to some extent by neural tissue.

For a detailed description of the cell types of the different lobes we refer the reader to the studies of Mellinger (1962a, 1964, 1976). Alluchon-Gérard (1971, 1978a) and Knowles et al. (1975) which have shown that the various classes of endocrine cells in the gland are characterized not only by their location and their staining and histochemical properties, but also by their granulometric characteristics. Four (Alluchon-Gérard 1978a) to six (Knowles et al. 1975) endocrine cell types have been identified in the adenohypophysis of *Scyliorhinus canicula,* on the basis of the diameter of the secretory granules.

Adenohypophysial Hormones and Their Place of Secretion. Experimental endocrinological studies on the chondrichthyan adenohypophysis have revealed the presence of the fol-

Table 9-2. Localization of the various hormones in the adenohypophysis

	Rostral lobe	Median lobe	Ventral lobe	Neuro-intermediate lobe
TSH	–	–	+++	+
ACTH	+++	+	+	(+)
GTH	–	+	+++	+
Prolactine	+++	–	–	–
MSH	–	–	–	+++

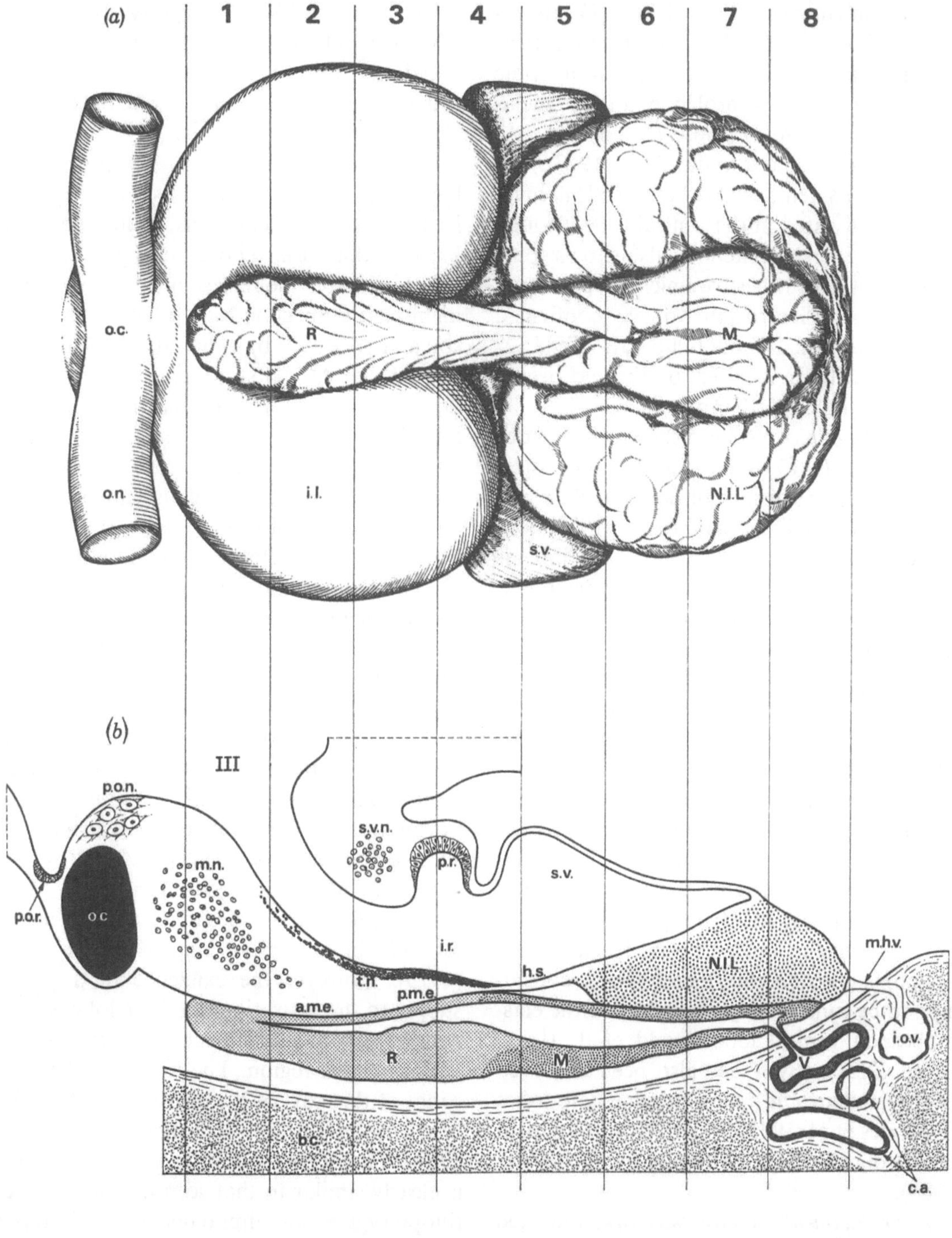

Fig. 9-7. The hypothalamus and hypophysis of *Scyliorhinus canicula*. **a** Ventral view. The regions mentioned in the text are indicated *1–8*. **b** Schematic sagittal section. *a.m.e.* eminentia mediana anterior; *b.c.* base of cranium; *c.a.* carotid arteries; *h.s.* hypophysial stem; *i.l.* lobus inferior; *i.o.v.* interorbital vein; *i.r.* recessus infundibularis; *M* median lobe; *m.h.v.* median hypophyseal vein; *m.n.* nucleus medius hypothalami; *NIL* neuro-intermediate lobe; *o.c.* chiasma opticum; *o.n.* nervus opticus; *p.m.e.* eminentia mediana posterior; *p.o.n.* nucleus preopticus; *p.o.r.* recessus preopticus; *p.r.* recessus posterior (mamillaris); *R* rostral lobe; *s.v.* saccus vasculosus; *s.v.n.* nucleus sacci vasculosi; *t.n.* nucleus lateralis tuberis; *III* third ventricle; *V* ventral lobe. (After Knowles et al. 1975)

lowing hormones (Table 9-2): (1) thyrotrophic hormone (TSH), (2) adrenocorticotrophic hormone (ACTH), (3) gonadotrophic hormone (GTH), (4) prolactin, and (5) melanocyte stimulating hormone (MSH).

A study by Jackson and Sage (1973a) on *Dasyatis* has demonstrated that *TSH activity* is mainly located in the ventral lobe, though the neuro-intermediate lobe also shows some activity; a recent electronmicroscopic study by Alluchon-Gérard (1978b) has confirmed these findings. Chondrichthyan TSH is highly specific in its effect on thyroid tissues (Jackson and Sage 1973b).

ACTH-activity has been shown by deRoos and deRoos (1967) to be located in the rostral lobe of the pars distalis of *Squalus, Raja* and *Hydrolagus*. Mellinger and Duois (1973) confirmed that in *Torpedo* the rostral lobe is indeed the main source of ACTH-activity. However, some activity was noted also in the neuro-intermediate lobe, but according to deRoos and deRoos (1967) this may be due to MSH, since there is a similarity in the chemical structure of the MSH and ACTH-principles (Lowry et al. 1974).

GTH, secreted mainly by the ventral lobe (Dodd et al. 1960; Mellinger 1964; Scanes et al. 1972; Dobson and Dodd 1977a, b) has recently been partially purified (Sumpter et al. 1978a, b). In *Hydrolagus* the highest GTH-activity is located in the Rachendachhypophyse which can, therefore, be considered to be functionally equivalent to the elasmobranch ventral lobe (Dodd et al. 1980). Some GTH-activity has also been observed in the median and neuro-intermediate lobes (Knowles et al. 1975; Mellinger 1972a; Firth and Vollrath 1973; Sumpter et al. 1978b, 1980).

A *prolactin-like* factor was first reported in the elasmobranch pituitary by Chadwick (1966). The presence of prolactin has been confirmed by Sage and Bern (1972) who showed the rostral lobe of the pars distalis to be the main source.

MSH is secreted by the neuro-intermediate lobe (Hogben 1936; Landgrebe and Waring 1941; Waring 1942). Recently Lowry and Chadwick (1970) have purified and elucidated the primary structure of the α-MSH found in the neuro-intermediate lobe of *Squalus acanthias*.

At present our knowledge of the physiological action of the various pituitary hormones in chondrichthyans is limited. It is known that the chondrichthyan thyroid gland increases the release of thyroxine when incubated with chondrichthyan TSH (Jackson and Sage 1973b). It has also been demonstrated that ACTH stimulates in vitro steroidogenesis in rajiform adrenal tissue (Macchi and Rizzo 1962; Klesch and Sage 1973, 1975), but not in shark adrenal (Bern et al. 1962). GTH plays a role in gametogenesis and gonad maturation (Dodd et al. 1960; Mellinger 1964; Dobson and Dodd 1977a, b), but the specific function of prolactin is still uncertain. MSH is concerned with the dark background response and the control of its release has attracted great interest (see Sect. 9.7.4).

Since there is no direct neurosecretory innervation of the pars distalis it seems that hypothalamic control is achieved through the portal system. Evidence for the existence of such a system has been provided by the studies of Meurling (1960, 1967a, b), Mellinger (1964), Sathyanesan (1965), Jasínsky and Gorbman (1966) and Knowles et al. (1975). However the existence of a portal supply to the neuro-intermediate lobe is puzzling in view of the rich neurosecretory innervation of this region. The functional significance of this part of the portal system, therefore, needs further investigation (Chevins 1968). The portal supply to the pars distalis is clearly similar to that seen in primitive actinopterygians and dipnoans and, as in those groups, is to be correlated with the absence of direct innervation of the pars distalis (Ball and Baker 1969).

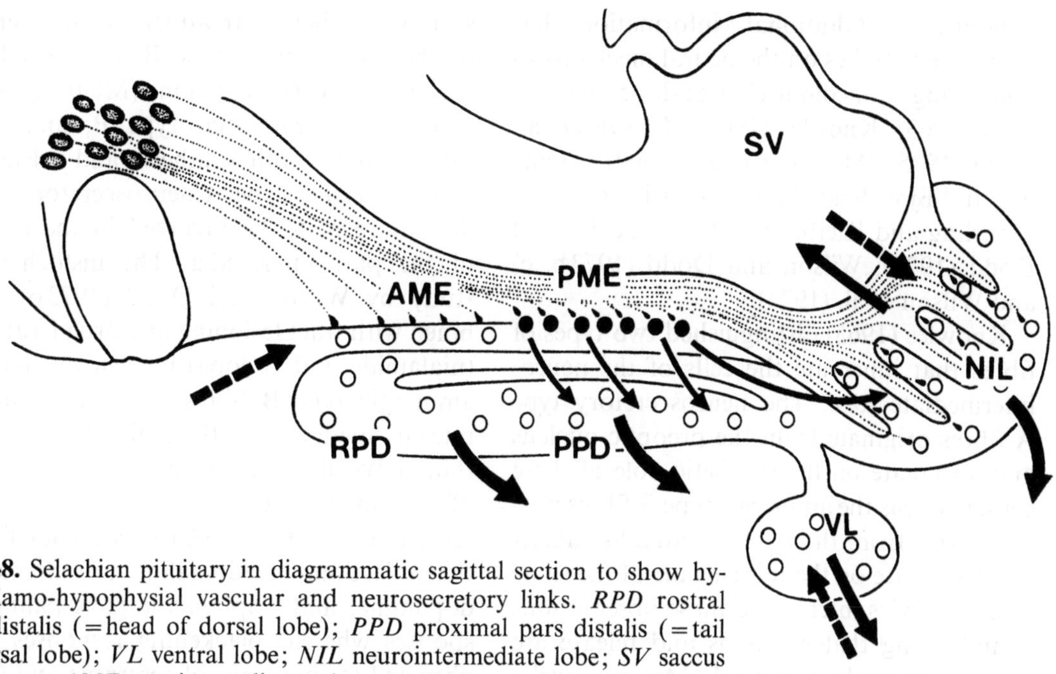

Fig. 9-8. Selachian pituitary in diagrammatic sagittal section to show hypothalamo-hypophysial vascular and neurosecretory links. *RPD* rostral pars distalis (= head of dorsal lobe); *PPD* proximal pars distalis (= tail of dorsal lobe); *VL* ventral lobe; *NIL* neurointermediate lobe; *SV* saccus vasculosus; *AME* anterior median eminence; *PME* posterior median eminence. *Dotted circles* neurosecretory cells of the PON; *dotted lines*, neurosecretory axons, with axon terminals marked; *filled circles* primary portal capillaries; *empty circles* secondary portal capillaries and other intrahypophysial capillaries; *interrupted arrows* arteries; *solid arrows* veins; *thin arrows*, portal veins. (After Jasinski 1969)

9.7.3 The Neurohypophysis

The neurohypophysis can be divided into the neural lobe, which is derived from the ventral wall of the saccus infundibuli and contacts closely the pars intermedia to form the "neuro-intermediate lobe"; the hypophysial stem carrying axons from the hypothalamus, and the median eminence.

According to Meurling (1967a) the median eminence of the chondrichthyan hypothalamus may be defined as that part of the ventral hypothalamic wall situated between the lobi inferiores that contains a specialized external zone formed by neurosecretory and glial processes. It contains a capillary plexus which drains towards the adenohypophysis. Differences in the arrangement of the capillaries led Knowles et al. (1975) to distinguish between anterior and posterior median eminences (Fig. 9-7). The posterior median eminence is continuous with the thin hypophyseal stem, which widens distally to form the neural component of the neuro-intermediate lobe. The blood supply to the rostral and median lobes originates in the anterior median eminence, the capillary supply of which forms the primary plexus of the pituitary portal system (Fig. 9-8).

The reader is referred to the review of Perks (1969) for a detailed description of the neurohypophyseal principles of chondrichthyans. At least two neurohypophyseal principles can be recognized—arginine vasotocin and an oxytocin-like principle (Sawyer 1965; Perks 1969)—and of these, the former has been demonstrated in most fish, and is presumably present in all.

9.7.4 The Hypothalamohypophyseal Control System

The morphology of the hypothalamohypophyseal system, first described by Scharrer (1952) and Bargmann (1953) in *Scyliorhinus*, consists of a well-developed neurosecretory preoptic nucleus and a long fibre tract that leads to the neuro-intermediate lobe of the hypophysis. Additional information has come from studies on the neural mechanisms underlying skin colour change done by Mellinger (1963), Knowles (1965), Knowles et al. (1970, 1975), Meurling et al. (1969), Meurling and Björklund (1970), Meurling (1972a), Meurling and Klefbohn (1974), Chevins and Dodd (1970), Wilson and Dodd (1973b, c) and Wilson et al. (1974).

Knowles (1965) distinguished two types of fibres that innervate the cells of the neuro-intermediate lobe. The neurosecretory type A fibres originate from the preoptic nucleus and terminate on the synthetic pole of these cells, whereas the aminergic type B fibres possibly arise from the nucleus lateralis tuberis and terminate on the release pole of the same cells. Knowles interpreted these observations as indicating that synthesis and release are under independent control. More recently, however, evidence has been obtained that the synthetic pole is innervated by both type A and B fibres (*Raja,* Meurling and Björklund 1970; *Scyliorhinus;* Knowles et al. 1970).

Mellinger (1962b, 1963, 1964) showed that lesioning of the tractus preopticohypophyseus in *Scyliorhinus* lead to ultrastructural signs of hyperactivity in the intermedia cells and to continuous release of MSH, a result that suggests that there is an inhibitory hypothalamic control of the secretion of MSH. This hypothesis has been additionally supported in studies by Meurling (1972, *Squalus*), Chevins and Dodd (1970, several Rajidae), Meurling et al. (1969, *Raja*), and Wilson et al. (1974, *Scyliorhinus*).

Chevins and Dodd (1970) found that transection of the tractus preopticohypophyseus, ectopic transplantation of the neuro-intermediate lobe, and lesioning of the preoptic nucleus leads to an increase in MSH secretion in Rajidae. This result suggests that there is either a single inhibitory control of both synthesis and release (contrary to Knowles' hypothesis) or a common origin for both type A and type B fibres in the preoptic nucleus. But as selective sectioning of the type A neurosecretory fibres has little effect on MSH-secretion, whereas treatment with reserpine or the sectioning of type B tracts results in sustained MSH secretion (Meurling et al. 1969), it appears that type B, aminergic, fibres inhibit both synthesis and release of MSH in *Raja*. Type A neurosecretory fibres do not seem to be involved in the control of the pars intermedia. The histochemical study by Wilson and Dodd (1973b) combined with tract lesioning and drug treatment (nialamide and L-dopa) has shown that the aminergic type B fibres originate from the nucleus medius hypothalami. The same authors have also studied the effects of a series of pharmacological agents on the in vivo release of MSH in *Scyliorhinus*. They found that injections of adrenaline, noradrenaline, dopamine, and isoprenaline gave paling responses whereas blackening responses were obtained to injections of reserpine and 6-hydroxydopamine. These results are consistent with the view that amines are involved in the inhibitory control of MSH release. The control of MSH release is also influenced by lesioning of the optic system, including the optic nerves, tractus opticus, or tectum (Wilson et al. 1974).

It has been suggested that the control of the other pituitary lobes is regulated by the hypothalamus via the hypophysial portal system (Fig. 9-8). The presence in the dogfish of peptidergic neurosecretory regulatory mechanisms has been indicated by Deery (1974) and Deery and Jones (1975), who found that median eminence extracts cause a significant activation of the adenyl cyclase of each lobe of the dogfish pituitary. However, the (putative) neurotransmitters melatonin, serotonin, adrenaline, noradrenaline, dopamine and acetylcholine had no effect. Two synthetic mammalian hypothalamic hormones—thyrotrophin releasing hormone and gonadotrophin releasing hormone—activated the ventral lobe enzyme, but had no effect on the adenyl cyclase of the other three lobes. Since the ventral lobe of the dogfish receives no portal blood from the hypothalamus and has no direct innervation, it seems likely that hypothalamic regulation of the cells in this lobe may be mediated via the systemic circulation (Fig. 9-8).

Recently Jenkins et al. (1980) have investigated a possible oestradiol feedback system in female *Scyliorhinus* where marked annual cycles have been observed in plasma oestradiol and testosterone levels, and in the gonadotrophic content of the ventral lobe (Sumpter 1976), synchronized with changes in the external environment. Since preliminary evidence indicates that oestradiol has an antigonadotrophic effect in the female dogfish (Jenkins 1978), it is of interest to determine which brain and pituitary regions may be involved in a oestradiol feedback mechanism in this species. Jenkins et al. (1980) found that the main areas of label concentrations, following injections of (^{3}H) oestradiol, were the nucleus preopticus, the ganglia habenulae, the nucleus medius hypothalami, the nucleus lateralis tuberis, the nucleus lobi lateralis, and the ependyma of the third ventricle, with the labelling in the preoptic nucleus being the most intense. Most of the hypothalamic areas found to accumulate oestradiol in the dogfish have also been shown to contain catecholamines (Wilson and Dodd 1973c) and exhibit monoamine oxidase activity, thus leaving open the possibility that the aminergic system may be involved in steroid feedback in these fishes.

9.8 Functional Correlations

The regions of the central nervous system considered in previous chapters are associated for the most part with the control of movement and equilibrium, and with the initial stages in the analysis of sensory information. The regions contained within the diencephalon are associated with more complex functions and are involved in various behavioural activities. In mammals the thalamus is a major subcortical centre that relays sensory information from all over the body to the telencephalon. The hypothalamus, in association with the limbic centres of the telencephalon, maintains homeostatic control of body functions and mediates "motivated" behaviours such as feeding, escape, attack, aggression and sex. Behaviour of this kind is modified when lesions or stimulations are made in the hypothalamus.

Our knowledge of the functions of these parts of the brain in cartilaginous fishes is very limited at present. The hypothalamus is well developed, however, and it is conceivable that, as in other vertebrates, in association with the pituitary gland it is involved in the regulation of skin colour, growth, salt and water balance and sexual development. Stimulation of the inferior lobe of the hypothalamus in a free-swimming nurse shark, *Ginglymostoma,* via implanted electrodes elicited feeding responses, a result suggesting that this zone may regulate feeding behaviour (Demski 1977).

For cartilaginous fishes the traditional view has been that the telencephalon and diencephalon are dominated by olfactory input and that relationships are established in subsequent evolution with other sensory systems. In the elasmobranchs olfactory input does make important connections with the diencephalon but recent work has revealed that there is a considerable visual input to this region. Our knowledge of the higher projections of other sensory systems is presently limited but it is likely that they, too, make thalamic connections. It is clear that the diencephalon of cartilaginous fish forms a significant relay centre. Hardly any ascending fibres appear to reach the telencephalon without synaptic interruption in the diencephalon, while conversely the majority of telencephalic descending fibres project only to the diencephalon. Clearly, the diencephalon of these fishes is far more than an olfactory relay.

10 Telencephalon

10.1 Morphogenesis and Overall Structure

Although evagination is the major morphogenetic process in the telencephalon of all cartilaginous fishes, the adult forebrains of the main groups of this class show considerable differences.

The telencephalon of the Holocephali is relatively simple (Fig. 3-5). Two hollow hemispheres arise as a rostrally directed evagination, the lateral walls of which are considerably thicker than are the medial walls. The olfactory bulbs are inserted at the rostral end of each hemisphere. These structures have no separate ventricular cavities. It is important to note that the hemispheres, though relatively large in size, are smaller than the unevaginated telencephalon impar. The latter can be subdivided into a rather massive rostral portion and a long, mostly fibrous, caudal portion (Fig. 3-3). This caudal portion, which essentially comprises the extremely elongated regio preoptica, connects the telencephalon sensu strictiori with the diencephalon. The formation of this remarkable pedunculus telencephali has doubtless to do with the fact that the orbitae of the very large eyes meet each other in the median plane, leaving room for the passage of only a thin and elongated stalk. The most dorsal parts of the lateral walls of these stalks have recurved somewhat, but this eversion is less pronounced than in the telencephalon of actinopterygian fishes.

The studies of Sterzi (1912) have shown that in sharks the rostral parts of the prosencephalic side plates bulge out laterally and that later the hollow anlagen of the olfactory bulbs are formed as an outgrowth of the lateral walls of the expanded portions. During the formation of the olfactory bulbs another rostrally and laterally directed evagination occurs which provides the lateral lobes or hemispheres (Fig. 3-5). Late in development the olfactory bulbs separate more and more from the hemispheres, interconnected by stalks of varying lengths, the pedunculi olfactorii. Thus in *Scyliorhinus* these stalks are short but in *Squalus* they are of a considerable length (Fig. 3-3).

A remarkable feature of the selachian forebrain is the connection of the right and left hemispheres. As a part of the evagination process the dorsal parts of the hemispheric walls undergo inversion and meet each other in the median plane. Johnston (1911) supposed that the connection of both hemispheres is the result of a secondary fusion of the meningeal surfaces of the medial walls of the hemispheres. However, Bäckström (1924) concluded from his embryological studies that the union of the hemispheres results largely from a migration of nervous elements into the initially membranous structures which interconnect the two telencephalic halves. The degree of fusion varies from species to species (Figs. 1, 34, 66, 100). In primitive forms such as *Chlamydoselachus* and *Hexanchus* it is only very restricted, but in *Mustelus* and *Scyliorhinus* the medial hemispheric walls are completely united. *Squalus* occupies an intermediate position in this respect. It is noteworthy that the most caudal portion of the selachian telencephalon is formed by the unevaginated telencephalon impar. This structure is dorsally covered by a membranous roof, the tela telencephali, and by the most caudal part of the pallium.

Ventrally the two halves of the telencephalon are connected by a generally well-developed anterior commissure. In most species this commissure is located in the transition area between the hemispheres and the telencephalon impar.

The early development of the telencephalon of batoids closely resembles that of the selachians. However, later in development the two groups diverge with regard to the extent of the ventricular cavities. Whereas in most sharks the lateral ventricles remain wide, in rays they become gradually reduced to narrow slitlike spaces (Fig. 3-5). This reduction of the ventricles is due to a considerable increase in thickness of the hemispheric walls (Bäckström 1924). Finally, the bulbar ventricles and the lateral parts of the ventricles obliterate completely, and in full-grown rays the lateral ventricles are merely represented by two short, horn-shaped rudiments. As in certain sharks, the medial walls of the batoidean hemispheres are completely united in the median plane (Fig. 66). It should be noted that the pattern of development just outlined is not strictly confined to rays. Solid hemispheres are also found in certain selachian species, such as the nurse shark *Ginglymostoma.* The olfactory peduncles in rays attain a considerable length and it is also remarkable that in these forms the olfactory bulbs are of a flattened, saucer-shaped appearance.

10.2 Survey of Literature on the Histological Structure and Fibre Connections

Studies on the structure of the telencephalon of cartilaginous fishes, based on normal material, have been made by Johnston (1911, *Squalus, Scyliorhinus*), Ariëns Kappers and Carpenter (1911, *Chimaera*), Holmgren (1922, *Squalus, Chimaera*), Bäckström (1924, *Squalus, Scyliorhinus, Raja, Chimaera*), Faucette (1969a, b, *Chimaera, Hydrolagus*), Kuhlenbeck (1973, 1977, various species), Williams (1973, *Squalus*), Northcutt (1978, 1981, various species), and Ebbesson (1980b, *Ginglymostoma*). However, experimental work on the telencephalon is more limited (Ebbesson and Heimer 1970; Ebbesson and Schroeder 1971; Graeber et al. 1978; Bodznick and Northcutt 1979; Northcutt and Wathey 1979; Luiten 1981b; Smeets 1983).

A survey of this literature indicates that a detailed and systematic study of the chondrichthyan telencephalon is lacking. In the studies of Johnston (1911), Ariëns Kappers and Carpenter (1911), Holmgren (1922), Bäckström (1924) and Northcutt (1978, 1981) only rough subdivisions of the telencephalic hemisphere are presented whereas the more detailed papers of Faucette (1969a, b) and Williams (1973) contain several inaccuracies which render the value of these studies questionable. Therefore, a systematic analysis of the telencephalon of the four species considered in the present work seemed to be warranted.

In the following sections the main parts of the chondrichthyan telencephalon will be considered first (10.3), then the ventricular sulcal pattern (10.4), the cell masses (10.6) and the fibre connections (10.7). Because the interpretation of neural grisea is partly based on fibre relations it appeared to be necessary to preface the descriptions of the cell masses with a brief survey of the main fibre paths (10.5). Next, an attempt will be made to consider the cell masses and fibre connections of the chondrichthyan telencephalon in the light of what is known of the morphogenesis of this structure (10.8). Finally, a synopsis will be presented (10.9) of what is known of the functions of the chondrichthyan telencephalon.

10.3 Subdivision of the Telencephalon

The telencephalon of the four species studied can be readily divided into three parts. These are from caudal to rostral: (1) the telencephalon impar; (2) the lobus hemisphericus telencephali, and (3) the bulbus olfactorius.

The telencephalon impar is that portion of the telencephalon which encloses, as its name implies, an unpaired ventricle. Its caudal boundary can be defined as a plane passing from the velum transversum to the caudal pole of the optic chiasm (Johnston 1911). Some authors (Kuhlenbeck 1929a, b; Gerlach 1947; Kuhlenbeck and Niimi 1969) defined the tel-di-encephalic boundary as a plane passing from the velum transversum to the anterior commissure, an interpretation that would include in the diencephalon a part of what is here called telencephalon impar.

The lobus hemisphericus has been subdivided by all students of the chondrichthyan forebrain into a dorsal pallium and a ventral subpallium. However, differences of opinion exist with regard to the boundaries of these two principal parts. According to Johnston (1911) and Holmgren (1922) medial and lateral cell-free zonae limitantes indicate the pallio-subpallial boundaries, whereas Kuhlenbeck (1929b) and Gerlach (1947) considered ventricular sulci together with zonae limitantes as important landmarks. Northcutt (1978, 1981) on the other hand left both sulci and zonae limitantes out of consideration and paid more attention to fibre connections.

In our material we have looked for zonae limitantes and ventricular grooves which could serve as boundary sulci. We have found more or less distinct cell-free zones in both sharks studied. Both a zona limitans lateralis in the lateral wall and a zona limitans medialis in the medial wall could be recognized in the caudal part of the telencephalic hemisphere. The positions of these zonae are marked on the ventricular surface by two distinct sulci, the sulcus limitans lateralis and medialis. In *Scyliorhinus* the position of the zona limitans lateralis is also marked by an external groove at caudal levels (Figs. 59, 60). At rostral telencephalic levels we were unable to distinguish zonae limitantes, though the sulci limitantes could be traced farther rostralward than the corresponding zonae, especially in *Squalus* (Figs.29-32).

In *Raja,* the subdivision of the telencephalic hemisphere into a pallium and a subpallium is rather difficult, since distinct zonae limitantes and sulci limitantes cannot be recognized. A zona limitans lateralis and corresponding ventricular and external sulcus are present only in the most rostral part of the telencephalon impar (Fig. 92).

In *Hydrolagus* the pallio-subpallial boundary is marked medially and laterally by ventricular sulci; distinct zonae limitantes are absent, though at some levels a trace of a zona limitans can be discerned (Figs. 134, 135). As in *Scyliorhinus* and *Raja,* a sulcus limitans lateralis externus can be identified in *Hydrolagus* (Figs. 130–134).

The third telencephalic subdivision, the bulbus olfactorius, can be clearly delimited from the telencephalic hemisphere in both *Squalus* and *Raja,* because in these species the bulbus is separated from the hemisphere by a long olfactory peduncle. In *Scyliorhinus* and *Hydrolagus,* on the contrary, the bulbar structures are closely apposed to the hemispheres.

The receptive element of the olfactory apparatus is represented by slender bipolar cells situated in the mucosa of the nasal cavity. These peripheral elements have, in addition to their receptive function, also a conductive function: they give rise to axonal processes which relay directly towards the brain. These primary olfactory fibres or fila olfactoria enter into a separate extension of the telencephalon, the bulbus olfactorius.

Though the olfactory bulbs of the cartilaginous fish studied vary considerably in shape, histologically four layers can always be distinguished (Nieuwenhuys 1967b): (1) a stratum nervosum, as in all vertebrates made up of intertwined primary olfactory fibres; (2) a stratum glomerulosum, consisting of several layers of well defined glomeruli; (3) a stratum mitrale, built up of large polymorph elements, and (4) a wide stratum granulare. Golgi studies (Catois 1901; Sterzi 1909) and electron microscopic studies (Andres 1970, 1975) have revealed that the

Table 10-1. Comparison of the nomenclature of telencephalic ventricular grooves in cartilaginous fishes

Johnston (1911)	Williams (1973)	Kuhlenbeck (1973, 1977)	Kuhlenbeck and Niimi (1969)	Present study
Sharks			Holocephali	
Sulcus limitans hippocampi lateralis	–	fd1	fd1	sll
Sulcus limitans hippocampi medialis	Sulcus septo-hippocampalis	fd4	fb3	slm
–	–	fb2	–	stv
–	–	fd3	–	stdm[a]
–	Sulcus septo hippocampalis	y	–	stvm[a]
–	–	x	–	stdl[a]
–	–	–	fb2	stvl
–	–	fd2	–	sulcus g[a]
–	–	fb1	–	sulcus h[a]
–	–	–	fd1	sulcus j
–	–	–	fb1	sulcus k
–	–	–	fd3	sulcus l

[a] Not found in *Hydrolagus*

dendrites of the mitral cells generally enter into the formation of several glomeruli. Besides glomerular branches, these cells often also possess dendrites which ramify and terminate outside the glomeruli (Sterzi 1909). The axons of the mitral cells traverse the granular layer obliquely and assemble in the periventricular zone of the bulb forming the secondary olfactory tracts. The granule cells in chondrichthyans are, contrary to those in the tetrapods, provided with distinct axons. It is thought that the granular cells have an inhibitory action on the mitral cells (for details see Andres 1970, 1975).

10.4 Sulcal Pattern

The telencephalic sulcal pattern has received little attention in the older literature and apart from Kuhlenbeck (1929b, 1973, 1977) and his students (Gerlach 1947; Kuhlenbeck and Niimi 1969), workers on the chondrichthyan telencephalon have only incidentally labelled the ventricular sulci. A list of synonyms is provided in Table 10-1. Our analysis revealed the following distinct grooves.

A *sulcus limitans lateralis* is very distinct in both shark species, as well as in *Hydrolagus,* and can be recognized throughout the whole extent of the telencephalon. This sulcus marks the position of a more or less distinct zona limitans lateralis on the ventricular surface. The sulcus limitans lateralis can, therefore, be considered as the ventricular landmark of the pallio-subpallial boundary in the lateral neural wall. In *Raja,* where a sulcal pattern is almost completely absent due to the extreme reduction of the ventricular cavity, only a single sulcus is present which is probably equivalent to the sulcus limitans lateralis of the other three species studied (Fig. 92).

A *sulcus limitans medialis* is present in the medial hemispheric wall. It is a distinct groove in *Squalus* and *Scyliorhinus* as well as in *Hydrolagus.* In both sharks this sulcus, just like the sulcus limitans lateralis, marks at caudal telencephalic levels the pallio-subpallial boundary and the ventricular starting-point of a more or less distinct zona limitans medialis. However, at rostral telencephalic levels this sulcus shows no clear relationship with the subdivision into pallium and subpallium. In *Hydrolagus* the suclus limitans medialis marks throughout its whole caudorostral extent the pallio-subpallial boundary in the median hemispheric wall.

In both sharks studied a very distinct *sulcus telencephalicus ventralis* can be recognized, extending thoughout the rostrocaudal extent of the hemispheric wall. The same sulcus could also be identified in *Hydrolagus* where it extends from the telencephalon impar into the caudal part of the hemisphere.

A *sulcus telencephalicus dorsomedialis* and a *sulcus telencephalicus ventromedialis* could only be identified in the two sharks studied. Of these, the former is a rather distinct groove forming the dorsal boundary of the medial pallium, whereas the latter is, at least in *Squalus*, a very distinct sulcus extending throughout the rostral half of the medial hemispheric wall. The sulcus telencephalicus ventromedialis is less well developed in *Scyliorhinus* and limited in its rostrocaudal extent.

A *sulcus telencephalicus dorsolateralis* could be recognized in the rostral part of the telencephalon impar and in the caudal part of the lateral hemispheric wall of *Squalus*. This sulcus, which is quite distinct, forms the dorsal boundary of an area which we have labelled area periventricularis dorsolateralis (pallii). A comparable sulcus could not be identified with certainty in *Scyliorhinus*, but it is possible that the sulcus indicated as such (Fig. 61) is equivalent to that of *Squalus*.

In *Hydrolagus* a distinct sulcus extends throughout almost the whole lateral hemispheric wall as far as the olfactory bulb (Figs. 135–139) which we have labelled, on account of its position, as the *sulcus telencephalicus ventrolateralis*. This sulcus does not seem to be equivalent to any ventricular groove in the telencephalon of the sharks studied.

In addition to the main grooves just described some accessory sulci occur in *Squalus* (g, h) and in *Hydrolagus* (j, k, l). Sulci g and h can be recognized throughout the whole rostrocaudal extent of the hemispheric wall in *Squalus* (Figs. 26–31). The former lies in the dorsal ventricular wall and is, at some levels, very distinct, whereas sulcus h is less distinct and takes a ventrolateral position. The accessory sulci j (Figs. 129–131), k (Figs. 132–133) and l (Figs. 136–138) although distinct in *Hydrolagus* are limited in extent.

From the survey presented above it can be concluded that a distinct sulcal pattern can be recognized in *Squalus* and *Scyliorhinus* as well as in *Hydrolagus,* whereas sulci are almost completely absent from *Raja*. If we compare our results with those of Kuhlenbeck (1929b, 1973, 1977) for sharks and with those of Kuhlenbeck and Niimi (1969) for Holocephali, considerable differences appear to exist (Table 10-1). These may be due partly to the poor histological condition of the material available to these authors and partly to a different interpretation of some telencephalic areas (see below). Furthermore, just as in the brain stem, the sulcal pattern can only be revealed by a systematic analysis of the sulci. For example, the sulcus fd1 of Kuhlenbeck and Niimi (1969, *Hydrolagus*) is represented in the caudal part of the telencephalon impar by sulcus j of the present study and, at more rostral levels, by the sulcus limitans lateralis. However, close examination reveals that these two sulci are coexisting over some distance and thus represent two separate entities.

10.5 Main Fibre Systems

Because fibre relations have to be taken into account repeatedly in the ensuing description and interpretation of the telencephalic cell masses, a brief overview of the main fibre systems will be provided in the present section. For a more detailed description the reader is referred to Sect. 10.7.

1. *Secondary olfactory fibres*. These are formed by the olfactory tracts which originate in the olfactory bulb. In all four species studied a lateral and a medial olfactory tract could be distinguished. The former is the more developed of the two tracts and can be traced in normal material over a considerable distance. Its fibres mainly pass to the lateral pallium. The smaller medial olfactory tract probably distributes its fibres to the rostrodorsal region of the telencephalic hemisphere. The fibres of the medial olfactory tract, which are mostly unmyelinated, lose their identity by becoming mixed with other fibre systems.

2. *Commissural and Decussating fibres*. Both pallium and subpallium of elasmobranch fish contain well-developed commissural and decussating fibre systems. The decussatio dorsalis pallii consists, in our opinion, mainly of tertiary olfactory fibres and intrinsic telencephalic fibre systems. The large subpallial commissura anterior is formed by numerous fibre components, of which those originating from or projecting to the striatum and the area superficialis basalis are the most important. In *Hydrolagus* only ventral commissures are present.

3. The *fasciculus basalis telencephali* or *basal forebrain bundle* consists mainly of fibres which originate from or project to the striatum and area superficialis basalis. This fibre system can be considered as the main communication channel between the subpallium and the more caudal brain regions.

4. The *stria medullaris* consists of various fibre components which collect at caudal telencephalic levels. This fibre system passes in close relationship with the tractus pallii through the telencephalon impar. Upon reaching the diencephalon the stria medullaris courses dorsomedialward and terminates in both the ipsi- and contra-lateral habenular ganglion. Its constituent fibres are believed to arise mainly from subpallial regions (Johnston 1911).

5. The *tractus pallii* is strongly developed in the four species and consists of both myelinated and unmyelinated fibres. It connects the pallium with the rostral parts of the ipsilateral as well as the contralateral hypothalamus. The decussating fibres form a decussation immediately ventrocaudal to the optic chiasm. The tractus pallii of *Squalus* can be subdivided into a pars superficialis and a pars periventricularis. In *Hydrolagus* this tract also consists of two separate bundles, which have been labelled pars medialis and pars lateralis on account of their position. Only a single tractus pallii could be identified in *Scyliorhinus* and *Raja*.

10.6 Cell Masses

10.6.1 Cell Masses in the Telencephalon Impar and Subpallium

Because some subpallial cell masses extend beyond the caudal boundary of the lobus hemisphericus, i.e., the foramen of Monroi, we shall include in this section the cell masses of the telencephalon impar as well as those of the subpallial division. These cell masses are represented together in the topographical reconstructions (Figs. 6, 39, 71, 105).

The most caudal telencephalic cell mass is the *nucleus preopticus* which could be recognized in all four species studied. This cell mass has already been described in the previous chapter (Sect. 9.4.4) and will, therefore, be omitted here. The nucleus preopticus merges rostrally into the nucleus *interstitialis commissurae anterioris,* a nucleus which could be identified as a cluster of rather densely packed, small, round to ellipsoidal cells which lie dorsal to the anterior commissure. In *Hydrolagus* the nucleus interstitialis of the anterior commissure is separated from the preoptic nucleus by the long, fibrous pedunculus telencephali.

The fasciculus basalis telencephali contains throughout its whole extent numerous cells which accumulate in the rostral part of the telencephalon impar. This accumulation, labelled by us in the transverse sections as the *nucleus interstitialis fasciculi basalis telencephali,* consists of small, ellipsoidal cells that intermingle with the fibres of the basal forebrain bundle. The interstitial nucleus of the basal forebrain bundle is less well developed in *Hydrolagus* than in elasmobranchs.

In the caudal part of the telencephalon impar a cell mass can be recognized which was termed previously the area somatica by Johnston (1911). He thought that the telencephalon of cartilaginous fishes is almost completely dominated by the olfactory system and that only its most caudal part is reached by non-olfactory somatosensory stimuli. Consequently, the name area somatica reflects this interpretation. However, recent experimental neuroanatomical findings have shown this concept to be incorrect and hence we prefer the name *nucleus entopeduncularis* for the cell mass in question. We should like to emphasize that this term is used here in the strictly descriptive sense of "lying within the peduncle", for it is not our intention to imply any homology of the chondrichthyan entopeduncular nucleus with the nucleus of the same name in tetrapods.

An entopeduncular nucleus consisting of small, round to ellipsoid cells, could be recognized in all species studied. It lies largely lateral to the basal forebrain bundle, but has a medial extension which lies ventral to that bundle. Rostrally the nucleus entopeduncularis is replaced by the area superficialis basalis. The main difference between these two cell masses is the distinct laminar organization of the latter.

The *area superficialis basalis* consists of small cells which are arranged in a u-shaped layer. This cell mass is very large in its rostrocaudal extent and occupies the whole ventral outer zone of the subpallium in both *Scyliorhinus* and *Squalus* but in *Raja* and *Hydrolagus* it has a restricted rostrocaudal extent and is confined to the area between the nucleus entopeduncularis and the lateral pallium. Moreover, the lamina of cells constituting the area superficialis basalis of *Raja* is considerably thicker than that found in the other species studied (Figs. 10-1 C, E, 10-2 E), while the same cell mass in *Hydrolagus* is distinguished by its lateral position (Figs. 130–131).

The area superficialis has been recognized by all students of the chondrichthyan telencephalon although frequently labelled with different names. It has been called nucleus postolfactorius (Houser 1901), hypostriatum (Catois 1901), nucleus taeniae (Ariëns Kappers 1906), cortex olfactoria (Edinger 1908), and tuberculum olfactorium (Dart 1920; Holmgren 1922; Bäckström 1924; Faucette 1969a, b). Because recent experimental studies (Ebbesson and Heimer 1970; Smeets 1983) have revealed that only a part of this cell layer receives secondary olfactory projections we have called this cell mass the area superficialis basalis, following Johnston (1911), Northcutt (1977, 1978, 1981) and Ebbesson (1981 b).

At rostral telencephalic levels the area superficialis basalis is replaced by another superficial cell layer. Many authors (e.g. Faucette 1969a, b; Williams 1973; Northcutt 1978) considered this layer to be the rostral continuation of the area superficialis basalis. However, we find this designation inappropriate because its cells are larger and more diffusely arranged than those of the area superficialis basalis (cf. Fig. 10-1 C, D) and moreover, the two cell masses differ with respect to their fibre connections. We have designated this cell layer, which could not be exactly delineated, as *nucleus N* in the transverse sections.

Dorsal to both the rostral pole of the nucleus entopeduncularis and the caudal part

Fig. 10-1. Photomicrographs from transverse, Nissl stained sections through the telencephalon of *Squalus* (**A–D**), and *Scyliorhinus* (**E–F**). High power photomicrographs to show cells of some nuclei. **A** nucleus M; **B** nucleus septi caudodorsalis; **C** area superficialis basalis; **D** nucleus N; **E** area superficialis basalis; **F** pallium dorsale pars superficialis posterior. *Bar scale* = 100 μm

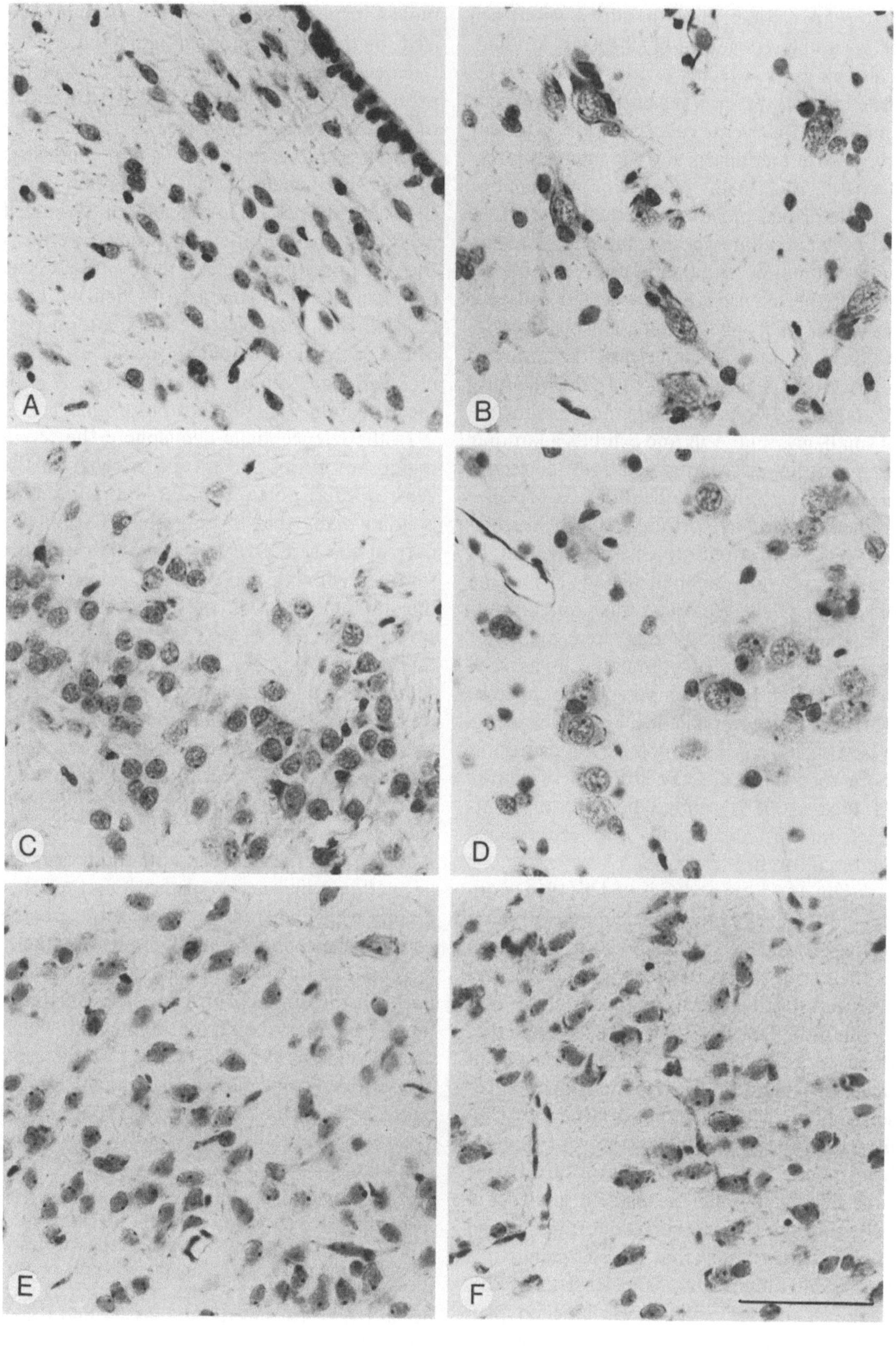
A
B
C
D
E
F

of the area superficialis basalis a small cell mass can be recognized in all species studied. This cell mass, which we have tentatively labelled *nucleus M*, consists of rather small cells and has a periventricular position adjacent to the stria medullaris and the taenia choroidea (Fig. 10-1 A). As far we know it has not been described previously in the literature.

In the two shark species studied a conspicuous cell mass is situated in the ventrolateral wall of the telencephalon, where it causes a rather distinct protrusion. This cell mass, which is situated at the level of the foramen of Monroi, is labelled in the present study as *striatum*. It consists of numerous clusters of small, granular cells and exhibits a laminar organization. A striatum could not be recognized with certainty in the other two species studied. In *Raja*, Northcutt (1978) labelled an area as striatum which is equivalent to the u-shaped area subpallialis 3 (SP3) of the present analysis. We prefer this non-committal label because: (1) this area does not consist of clusters of granular cells as seen in the sharks, but is composed of large cells, which are accompanied by glial elements (Fig. 10-2F), (2) its topological position is more rostral than is the striatum of sharks (cf. Figs. 6, 39, 71), and (3) no fibres originating from this area could be traced toward the basal forebrain bundle.

Ariëns Kappers and Carpenter (1911) considered the large intraventricular protrusion of the lateral telencephalic wall in *Chimaera* as the corpus striatum. Holmgren (1922) recognized both rostral and caudal striatal swellings in *Chimaera* and held the view that these were homologous with the striatum of *Squalus*. Faucette (1969a, b) recognized three separate divisions of the striatum in *Chimaera*, in the lower half of the ventrolateral hemispheric wall (with exception of the area superficialis basalis) extending rostrocaudally through the telencephalon. It was found in the present study that in *Hydrolagus* this prominent bulge contains at least three distinct cell masses, which were labelled provisionally as area subpallialis 6, 7 and 8. The most conspicuous of these, SP6, is characterized by fusiform and triangular, medium-sized cells, and lies in the caudal part of the protrusion (Figs. 10-3C, 130). Rostralward, the lateral part of this area is replaced by a cell mass indicated as SP 7, which consists of densely packed, small, bipolar cells (Figs. 10-3D, 131). Dorsal to area SP7 a cell mass could be identified which has a periventricular position lying just below the sulcus limitans lateralis. This area, which we have labelled SP8, occupies the dorsolateral part of the protrusion just mentioned. Our results in *Hydrolagus* are difficult to reconcile with those of Faucette (1969a, b) in *Chimaera*.

In the telencephalic hemisphere the septal region comprises the medial as well as the ventromedial portion of the subpallium. It is noteworthy that in both sharks studied a part of the septal region arches around the foramen Monroi and occupies a position in the dorsomedial wall of the cerebral hemisphere (Figs. 10-1 B, 26, 27, 60, 61). Although this supraforaminal part of the caudal septum occupies a seemingly pallial position we consider it, with Johnston (1911) as a subpallial structure and termed it *nucleus septi caudodorsalis*. The ventral part of the caudal septum has been labelled accordingly as *nucleus septi caudoventralis*.

In *Squalus* the nucleus septi caudodorsalis lies within a protrusion of the dorsal telencephalic wall, whereas a protrusion of the ventral telencephalic wall contains the nucleus septi caudoventralis (Figs. 26, 27). In *Scyliorhinus* the dorsal protrusion is less distinct, but a cell mass could be delineated which corresponds in position to the nucleus septi caudodorsalis of *Squalus*. In *Raja* a cell mass could be delineated in the caudal part of the

Fig. 10-2. Photomicrographs from transverse, Nissl stained sections through the telencephalon of *Raja*. High power photomicrographs to show cells of some nuclei. **A** pallium dorsale pars centralis; **B** area pallialis 1; **C** area pallialis 2; **D** area pallialis 4; **E** area superficialis basalis, and **F** area subpallialis 3. *Bar scale* = 100 μm

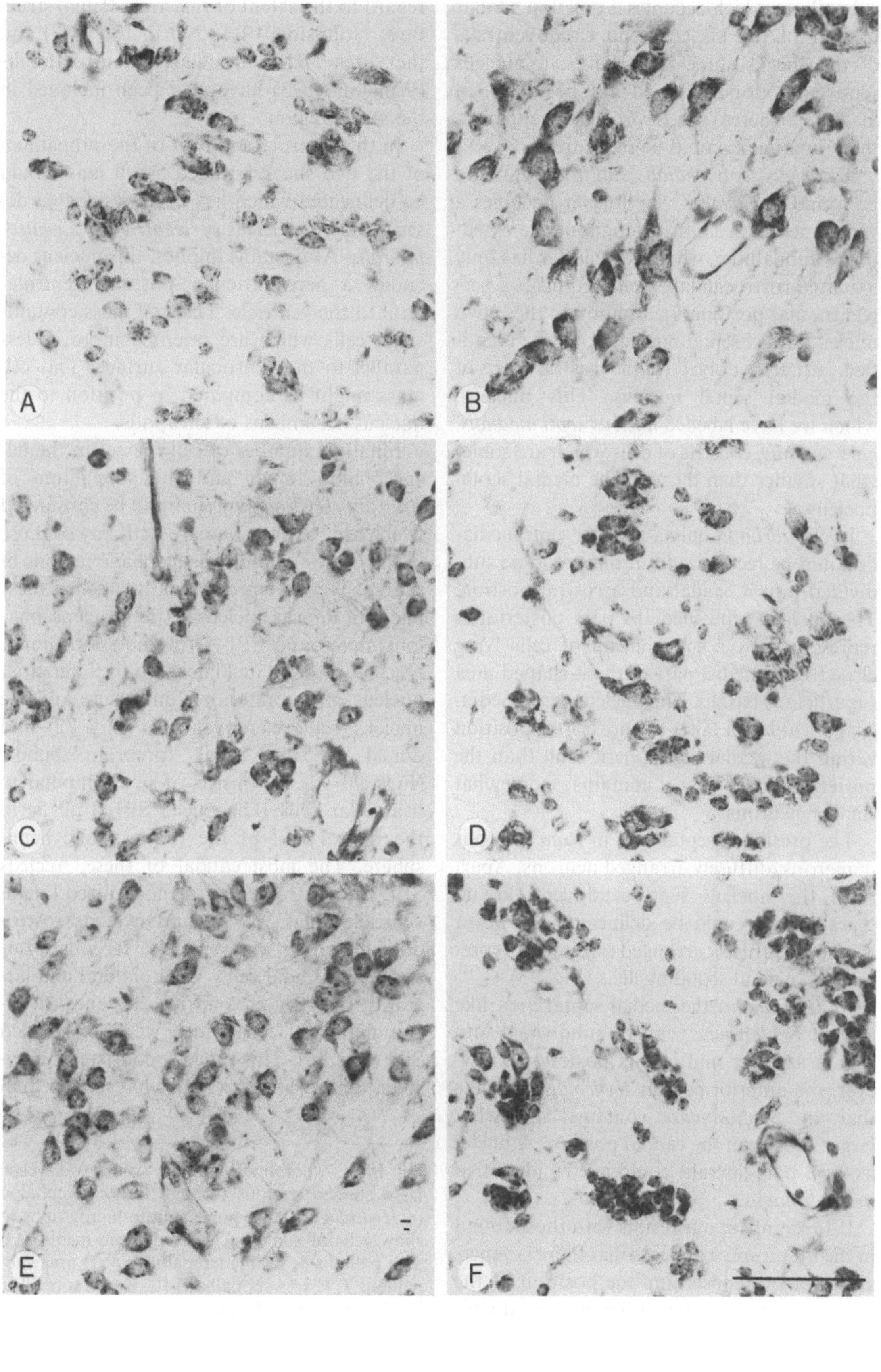
A
B
C
D
E
F

subpallium which occupies a position similar to that of the nucleus septi caudoventralis of the sharks studied. However, a nucleus septi caudodorsalis could not be identified in *Raja,* whereas in *Hydrolagus* neither of these two nuclei could be recognized.

A *nucleus septi medialis* and *lateralis* could be found in *Squalus.* The former occupies a ventromedial position throughout the extent of the subpallium, whereas the latter has only a limited rostrocaudal extent and takes a periventricular position. In addition to these two nuclei, a third septal nucleus could be identified, situated dorsal to the rostral part of the medial septal nucleus. This nucleus, which we have labelled *nucleus septi medialis, pars dorsalis,* consists of cells which are somewhat smaller than those of the medial septal nucleus.

In *Scyliorhinus* only a nucleus septi medialis could be recognized but this could be subdivided into a caudal and a rostral portion. The nucleus septi medialis pars posterior is represented by a small group of cells lying close to the medial part of the u-shaped area superficialis basalis. The nucleus septi medialis, pars anterior takes a more dorsal position within the medial hemispheric wall than the posterior part and contains somewhat smaller neurons.

The presumed septal area in *Raja* contains numerous, diffusely arranged neurons. Apart from the nucleus septi caudoventralis no septal nuclei could be delineated. However a zone of diffusely arranged cells might represent the medial septal nucleus (Fig. 93).

In *Hydrolagus* the medial septal area like that in *Scyliorhinus,* can be subdivided into a pars anterior and a pars posterior. However, the anterior part in *Hydrolagus,* unlike that in *Scyliorhinus,* contains somewhat larger cells than the caudal portion. A nucleus septi pars lateralis could not be identified in *Hydrolagus.*

If we compare our results with those found in the literature, it appears that there is general agreement concerning the position of the septal region, but some differences exist with regard to the extent of this area. Pallial structures (Johnston 1911; Northcutt 1977) and the area periventricularis ventrolateralis (Williams 1973) have also been included in the septal region.

In the ventrolateral part of the subpallium of the two sharks studied a cell mass could be delineated which has been designated descriptively as *nucleus periventricularis ventrolateralis.* As its name implies, this nucleus occupies a periventricular position ventrolateral to the ventricle. This cell mass contains small cells which are oriented more or less parallel to the ventricular surface. This cell mass might be compared in position to the nucleus accumbens of tetrapods.

Finally a number of cell masses in the telencephalon impar and the subpallium of *Raja* and *Hydrolagus* remain to be considered which are hard to compare with any of those in the corresponding telencephalic regions of sharks. We have provisionally labelled these nuclei in *Raja* as nucleus O (in the telencephalon impar) and SP1–4 (in the subpallium). Nucleus O is situated lateral to the interstitial nucleus of the anterior commissure and the nucleus septi caudoventralis (Fig. 72) and dorsal to the basal forebrain bundle (Figs. 90–92). It consists of small, bipolar to triangular cells. The nuclei SP1–4 all lie in the rostral half of the telencephalic hemisphere. The most caudal of these, nucleus SP1, lies ventromedial to the reduced lateral ventricle (Figs. 95, 96) and extends rostrodorsally along the ventricle. It consists of small, ellipsoidal cells. Ventrolateral and lateral to the reduced ventricle lies another accumulation of cells which we have labelled SP2 (Fig. 96). This nucleus consists of ellipsoidal cells, which are somewhat smaller than

Fig. 10-3. Photomicrographs from transverse, Nissl stained sections through the telencephalon of *Hydrolagus.* High power photomicrographs to show cells of some nuclei. **A** pallium mediale; **B** area pallialis 8; **C** area subpallialis 6; **D** area subpallialis 7; **E** area subpallialis 10; **F** area subpallialis 11. *Bar scale* = 100 μm

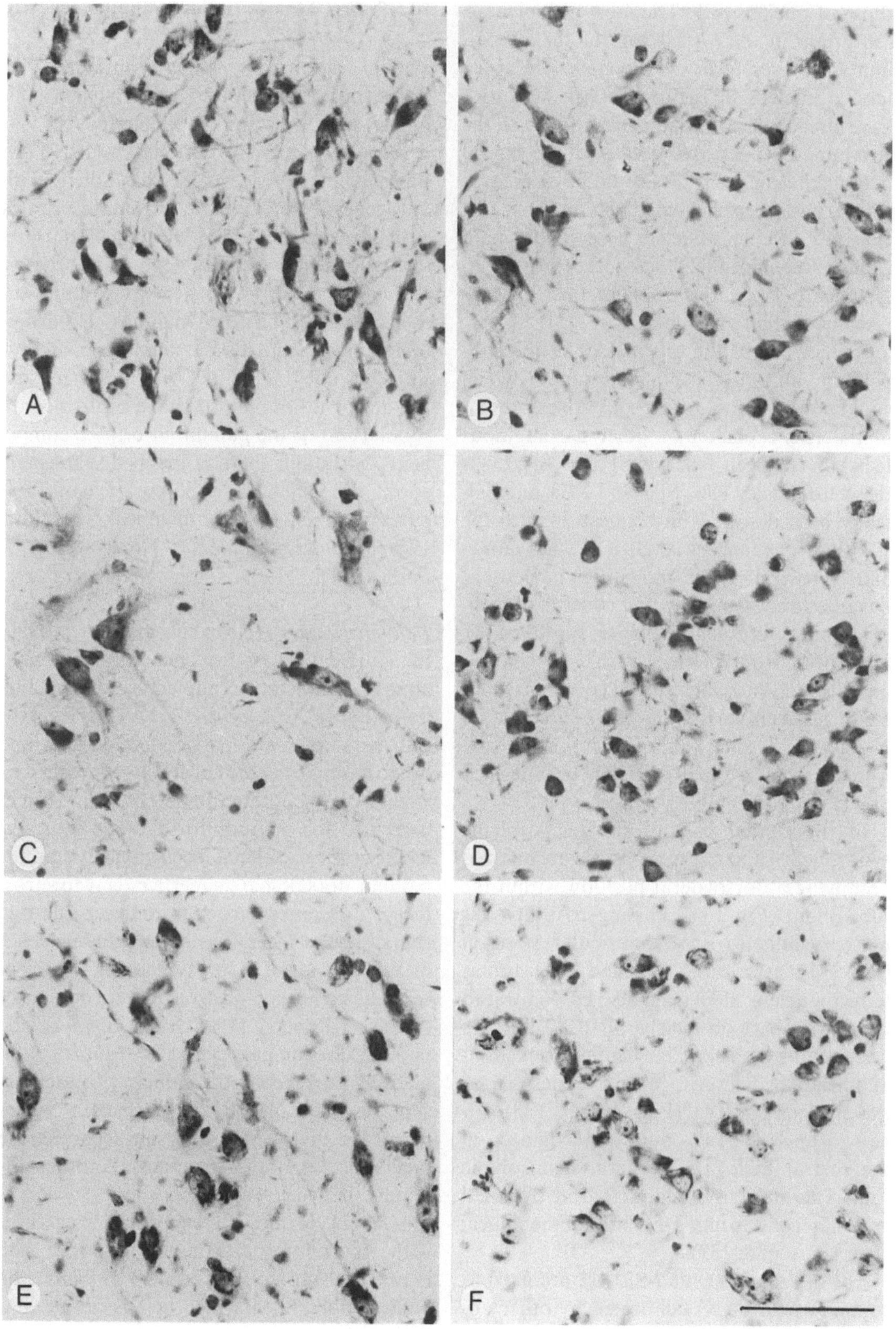
A
B
C
D
E
F

those of nucleus SP1. The most conspicuous cell group in the rostral part of the subpallium is represented by the u-shaped area, termed nucleus SP3 (Figs. 97–99), which we have already described in connection with the striatum. Between this area and the lateral pallium another cell mass could be delineated which has been indicated as nucleus SP4 (Figs. 97, 98). This nucleus consists of small, round to ellipsoidal cells. According to Northcutt (1978), this cell mass possibly represents a pallial amygdala.

In *Hydrolagus* the cell masses in the subpallium and telencephalon impar which could not be homologized with grisea present in the telencephalon of other cartilaginous fishes, have been labelled SP5 to SP12. Of these the nuclei SP6, SP7 and SP8 have already been described in connection with the striatum. The nucleus labelled SP5 lies immediately rostral to the interstitial nucleus of the anterior commissure and consists of rather small, ellipsoidal cells (Figs. 129–131). Slightly in front of the rostral pole of this nucleus the ventrolateral wall contains a submeningeal cell mass lying between cell layer N and the lateral pallium. This cell mass, which consists of small cells, has been termed SP9 (Figs. 132, 133).

In the caudal part of the telencephalic hemispheres a nucleus could be recognized occupying a dorsolateral position within the subpallium (Fig. 134). Passing rostrally this nucleus shifts to a periventricular position (Fig. 135). It consists of small cells and is termed nucleus SP10 (Fig. 10-3 E). Ventral to the rostral part of nucleus SP10 a cell mass was found which contains densely packed bi- and tripolar neurons. We have designated this nucleus as SP11 (Figs. 10-3 F, 135). Finally a nucleus SP12 could be recognized in the rostral half of the lateral hemispheric wall. This nucleus is characterized by small granular cells forming a thin submeningeal layer (Figs. 136, 137).

The cell masses O and SP1–12 are hard to compare with nuclei delineated by other authors.

10.6.2 Cell Masses in the Pallium

Though Johnston (1911) considered the entire roof of the chondrichthyan telencephalon as a primordium hippocampi, the embryological studies of Holmgren (1922) and Bäckström (1924) revealed that during early ontogenesis three pallial formations can be clearly recognized. They termed these three formations the medial, or hippocampal pallium, the dorsal, or general pallium and the lateral, or piriform pallium. According to Holmgren and Bäckström, these subdivisions become less distinct with increasing age. Studies on adult material have confirmed the subdivision of the pallium into three components, although there is much disagreement about the extent and boundaries of these parts (Gerlach 1947; Kuhlenbeck and Niimi 1969; Kuhlenbeck 1977; Northcutt 1977, 1978, 1981).

Pallium Dorsale. In *Squalus* and *Scyliorhinus* the caudal part of the telencephalic roof is almost completely occupied by the pallium dorsale (Figs. 26, 59, 10-1 F). At this level the cell mass can be subdivided into inner and outer laminae separated from each other by a zone that contains few cells. The former is termed the pallium dorsale pars centralis, whereas the latter has been indicated as the pallium dorsale pars superficialis. More rostrally, differences are seen between the two shark species which result from the presence in *Scyliorhinus* of a dorsal recess of the telencephalic ventricle, the diverticulum dorsale of Ariëns Kappers (1906). In *Squalus* the pallium dorsale pars centralis is dorsally and laterally covered by the pallium dorsale pars superficialis (Figs. 26, 27) whereas in *Scyliorhinus* the diverticulum dorsale separates the medial parts of both structures. Another difference between *Scyliorhinus* and *Squalus* is that in the former the superficial part fuses with its counterpart on the opposite side (Fig. 61). Moreover, in *Scyliorhinus* the pallium dorsale pars superficialis can be subdivided into posterior and anterior portions, the former lying caudal, the latter rostral to the recessus neuroporicus externus. A remarkable feature of the pallium dorsale pars superficialis is the presence of numerous fibres which traverse the lateral portion of this cell mass to join the tractus pallii. Characteristically, these fibres are oriented perpendicular to the main fibre stream of the pallial tract (Fig. 10-6).

In *Scyliorhinus* the pars centralis of the dorsal pallium consists of two parts, a pars medialis and a pars lateralis (Fig. 60). This subdivision is based on the somewhat smaller size and the higher cell-density of the medial part. In *Squalus* only a single pars centralis pallii dorsalis could be recognized (Figs. 25, 26).

In the dorsal pallium of *Raja* several more or less distinct cell masses could be delineated. The most caudal portion of the pallium in *Raja* is almost completely occupied by a cell mass consisting of small, round to ellipsoidal cells, which is here labelled as pallium dorsale pars centralis because of its position and connections (Figs. 91, 92, 10-2A). A corresponding cell mass in *Raja eglanteria* was called the central nucleus by Northcutt (1978) who recognized a separate layer of cells peripheral to this cell mass which he termed the pallium dorsale. We were unable to delineate separate inner and outer pallial layers, at least at caudal pallial levels. However, going rostralward, the pallium thickens considerably and a distinct subdivision into two zones becomes possible (Fig. 93). The *outer zone* consists of at least three distinct cell masses constituting median, intermediate and lateral nuclei. The latter forms the rostral continuation of the pallium dorsale pars centralis, whereas the median cell group, labelled P4, and consisting of small, round neurons (Fig. 10-2D), might be a part of the medial pallium (see below). The intermediate part of the outer zone differs from the median and lateral parts in containing medium-sized, bi- and tripolar cells (Fig. 10-2B). This cell mass, which has only a short rostrocaudal extent, has been labelled *P1*.

In *Raja* the *inner zone* of the pallium contains small clusters of neurons and glial elements that form a flattened plate above the reduced ventricles. This configuration has been probably imposed upon this zone by the caudal part of the decussatio dorsalis pallii.

Rostralwards, the nucleus labelled P1 becomes replaced by another pallial structure, which is termed *P2* and consists of considerably smaller neurons (Fig. 10-2C). This cell mass lies within a dorsal eminence of the pallial wall. However, it is noteworthy that both, P1, P2 and, to a lesser extent, P3 (see below) are characterized by fibres which, in their peculiar way of coursing to the pallial tract, resemble those found in both sharks studied. Because of this similarity, P1, P2 and P3 of *Raja* may be considered to be equivalent to the caudal part of the pallium dorsale pars superficialis of sharks.

The cell mass *P3* lies ventrolateral to the rostral portion of P2 (Figs. 96, 97) and consists of rather small cells. Only its extreme caudal portion is traversed by fibres joining the tractus pallii. The rostral extent of this cell mass is accompanied by a nucleus labelled *P7*, which differs from it by a more dense packing of the constituent neurons. On the basis of their position and lack of traversing fibres, P7 and the rostral part of P3 may be compared with the rostral part of the pallium dorsale pars superficialis of sharks.

At least three cell masses can be recognized in the pallium of *Hydrolagus* which, with respect to the peculiar course of fibres traversing their territories and joining the pallial tract, are comparable to the caudal part of the pallium dorsale pars superficialis of the other three species studied. These cell masses are indicated by the labels *P8*, *P9m* and *P9l*. The nucleus P8 extends caudally into the somewhat everted dorsal part of the telencephalic wall (Figs. 130, 131). Rostrally the dorsal part of the telencephalic wall, including cell mass P8, bends medialward (Fig. 132) and subsequently ventralward (Fig. 133) to touch finally the septal region (Fig. 134). The cell mass P8 consists of rather small, ellipsoidal and bipolar cells (Fig. 10-3 B) and constitutes the deeper part at levels where the pallium thickens (Figs. 134, 135). The outer zone of the dorsal pallium is formed by a cell layer consisting of fusiform and polygonal cells which are oriented in a more or less tangential direction. On the basis of small differences in cell density this area has been subdivided into a medial and a lateral part, labelled P9m and P9l respectively. Of these nuclei, P9m extends more rostralward (Fig. 136).

Rostrally, cell group P9m is replaced by another griseum, which has been labelled *P10* and which, because of its position and lack of traversing fibres, might be considered as the rostral part of the pallium dorsale pars superficialis.

Finally, two cell masses in the dorsal pallium can be recognized only in *Squalus* and *Scyliorhinus*. In these animals the rostral part of the pallium contains a very conspicuous laminar structure which we have termed the *area periventricularis pallialis*. This cell mass covers the ventricle on the medial and dorsal side and contains small granular cells, which form clusters like those of the striatum. Caudally this area divides into two parts, one medial, the other dorsal to the ventricle (Figs. 28–33, 62–65). The medial component which extends farther caudalward, was interpreted by Johnston (1911, Fig. 19) and Northcutt (1981, Fig. 5D) as a distinct septal nucleus. In *Scyliorhinus*, where the two hemispheres are almost completely fused, the medial part of this area passes to the median plane to join its contralateral equivalent (Figs. 64, 65) but such fusion could not be recognized in *Squalus*. As shown by the reconstructions (Figs. 7, 40), the area periventricularis pallialis occupies the anterior part of the telencephalic roof, rostral to the recessus neuroporicus externus.

Another periventricular cell zone could be recognized in the caudal half of the pallium of the sharks studied. This cell zone, termed *area periventricularis dorsolateralis* by the present authors, is particularly well developed in *Squalus* (Figs. 26, 27) and contains medium-sized cells. It causes a slight protuberance into the ventricle of the lateral pallial wall. In *Scyliorhinus* this cell mass is less well developed and consists of small cells (Figs. 61, 62).

Pallium Mediale. The pallium mediale or hippocampale in both sharks consists of rather small, ellipsoidal cells and fuses with its counterpart across the midline. It is relatively small, when compared with the pallium dorsale and it causes a ventricular protrusion of the dorsomedial wall into the telencephalic ventricle, which is larger in *Squalus* (Figs. 27–29) than in *Scyliorhinus* (Figs. 62, 63).

In *Raja* a medial pallium comparable to that of the sharks could barely be recognized. However, the cell areas labelled P4 and P5 may represent parts of the pallium mediale (cf. Bäckström 1924, Figs. 10, 11). Both cell masses consist of small cells, which show a fairly high cell density.

The only nucleus in *Hydrolagus* which could possibly represent the pallium mediale lies at the level where the pallium meets the subpallium (Figs. 134, 135). It consists of small to medium-sized cells (Fig. 10-3 A) and lies in the ventromedial part of the pallium. Some previous students of the holocephalian telencephalon have considered almost the entire pallium of this group as belonging to the hippocampal pallium.

Pallium Laterale. In *Squalus* and *Scyliorhinus* a nucleus which has a triangular form (Figs. 25–27, 59–61) can be recognized lying at the same level as, but dorsolateral to, the area superficialis basalis. Although this cell mass lies ventral to the zona limitans lateralis, we consider it as the *pallium laterale* as did Northcutt (1977). The main reason for doing so is that this cell mass receives the main olfactory input, just like the lateral pallium of tetrapods. In fact, in older studies (Johnston 1911; Holmgren 1922) this cell mass was labelled as the nucleus olfactorius lateralis. The region labelled as the pallium laterale by Holmgren (1922) is represented in our study by the lateral part of the pallium dorsale.

In *Scyliorhinus* the lateral pallium is divided into an inner and outer layer by the tractus olfactorius lateralis and at some levels causes a slight protuberance of the meningeal surface (Fig. 59).

In *Raja* the lateral pallium is more extensive than in the two sharks studied. However, we have been unable to distinguish a striatum in the ray and so the possibility cannot be excluded that the lateral pallium of this species as delineated by us also contains striatal neurons. It is noteworthy that numerous large neurons of this cell mass are surrounded by satellite glial cells. An experimental study (Smeets 1983) has shown that this area is the main target area of the secondary olfactory fibres.

The lateral pallium of *Hydrolagus* is formed by a large cell mass which lies in the lateral hemispheric wall and extends caudally into the lateral hemispheric wall and into the telencephalon impar. On the basis of the size of its constituent neurons the lateral pallium can be subdivided in this species into a rostral and a caudal part; the rostral part contains generally somewhat larger neurons.

Remaining Pallial Nuclei. Under this heading the following cell masses will be described: nucleus P (*Squalus*), area P6 (*Raja*) and P11 and P12 (*Hydrolagus*).

In *Squalus* a cell mass labelled P could be recognized at about the same level as the nucleus septi pars caudodorsalis (Fig. 27) lying lateral to the recessus neuroporicus. This cell mass is very distinct and consists of rather densely packed small cells.

The cell mass indicated in *Raja* as nucleus *P6* lies dorsal to the reduced ventricle between the dorsal pallial complex and the pallium laterale (Fig. 95). It comprises relatively densely packed, small ellipsoidal cells.

The two remaining cell masses in *Hydrolagus* lie close to the dorsolateral corner of the ventricle (Figs. 133–137). The more caudal nucleus, *P11*, consists of rather small, ellipsoidal and bipolar cells, and shifts in relation to the ventricle from a dorsolateral position at its caudal limit to a dorsomedial position rostrally. Rostrally this cell mass is replaced by nucleus *P12*, which although consisting of similar neurons to P11, lies dorsal to the dorsolateral corner of the ventricle.

10.7 Fibre Tracts

A conspicuous feature in the telencephalon of the four species studied is that the majority of the fibres contained within the fibre tracts are unmyelinated. Apart from these main fibre streams the telencephalon also contains a huge number of unmyelinated fibres that intermingle and form an extensive, diffuse meshwork. In normal material it is, therefore, almost impossible to trace precisely a given fibre system from the cells of origin to the sites of termination. In *Scyliorhinus* most of the more compact telencephalic fibre systems contain a considerable number of myelinated fibres, whereas in *Raja* such fibres occur only in a few telencephalic tracts. For this reason we have illustrated the fibre relations in the telencephalon of *Raja* at some levels with photomicrographs from Bodian rather than from Klüver-Barrera sections (Figs. 94–96). In the following account we will describe the telencephalic fibre systems under the following headings: (1) secondary olfactory tracts; (2) intrinsic telencephalic connections; and (3) telencephalic connections with other brain regions.

10.7.1 Secondary Olfactory Tracts

In the classical studies on the chondrichthyan forebrain (Edinger 1908; Johnston 1911; Ariëns Kappers and Carpenter 1911; Bäckström 1924; Ariëns Kappers et al. 1936), but also in some more recent studies (Faucette 1969a, b; Williams 1973; Kuhlenbeck 1977), almost the entire telencephalon has been

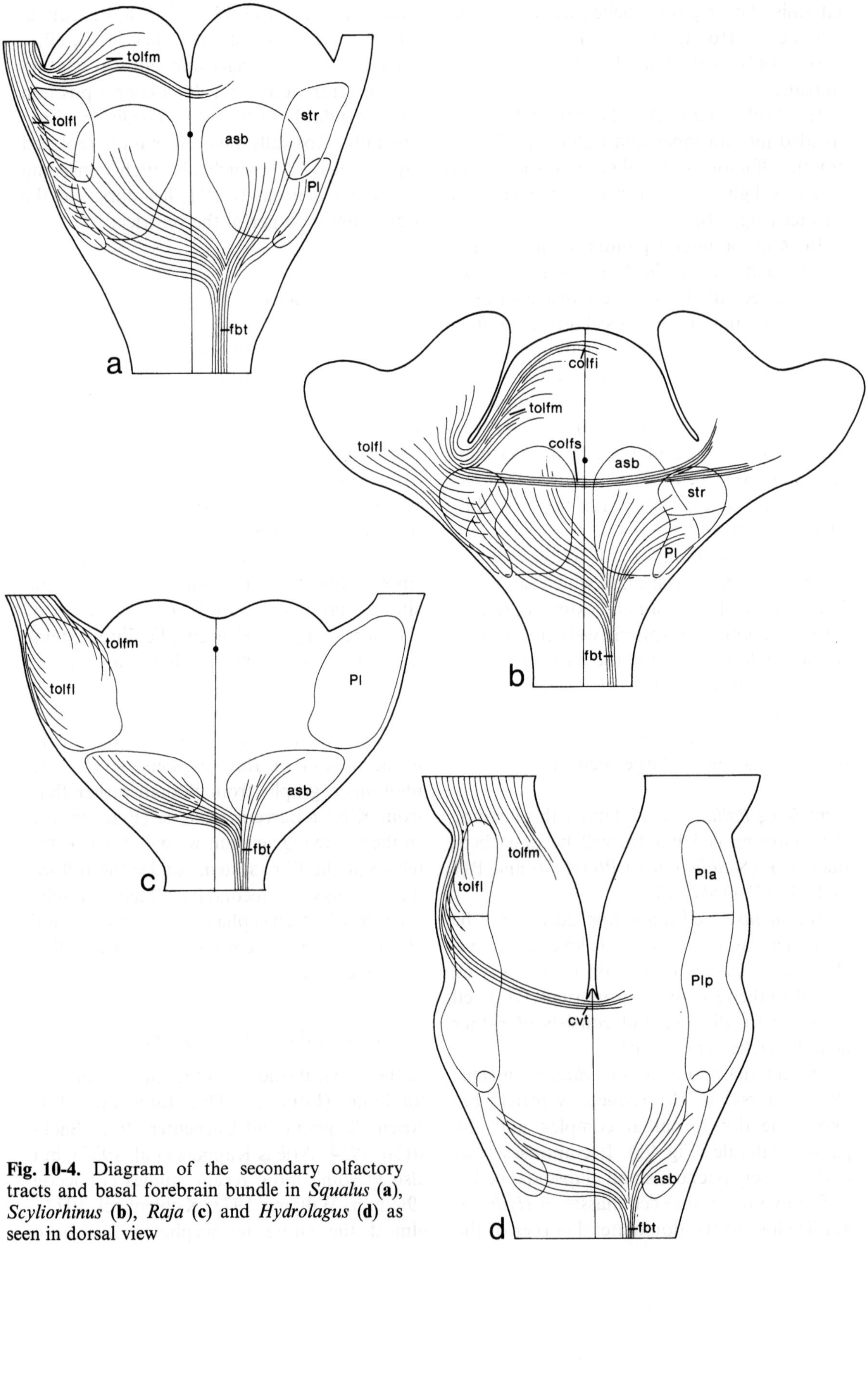

Fig. 10-4. Diagram of the secondary olfactory tracts and basal forebrain bundle in *Squalus* (**a**), *Scyliorhinus* (**b**), *Raja* (**c**) and *Hydrolagus* (**d**) as seen in dorsal view

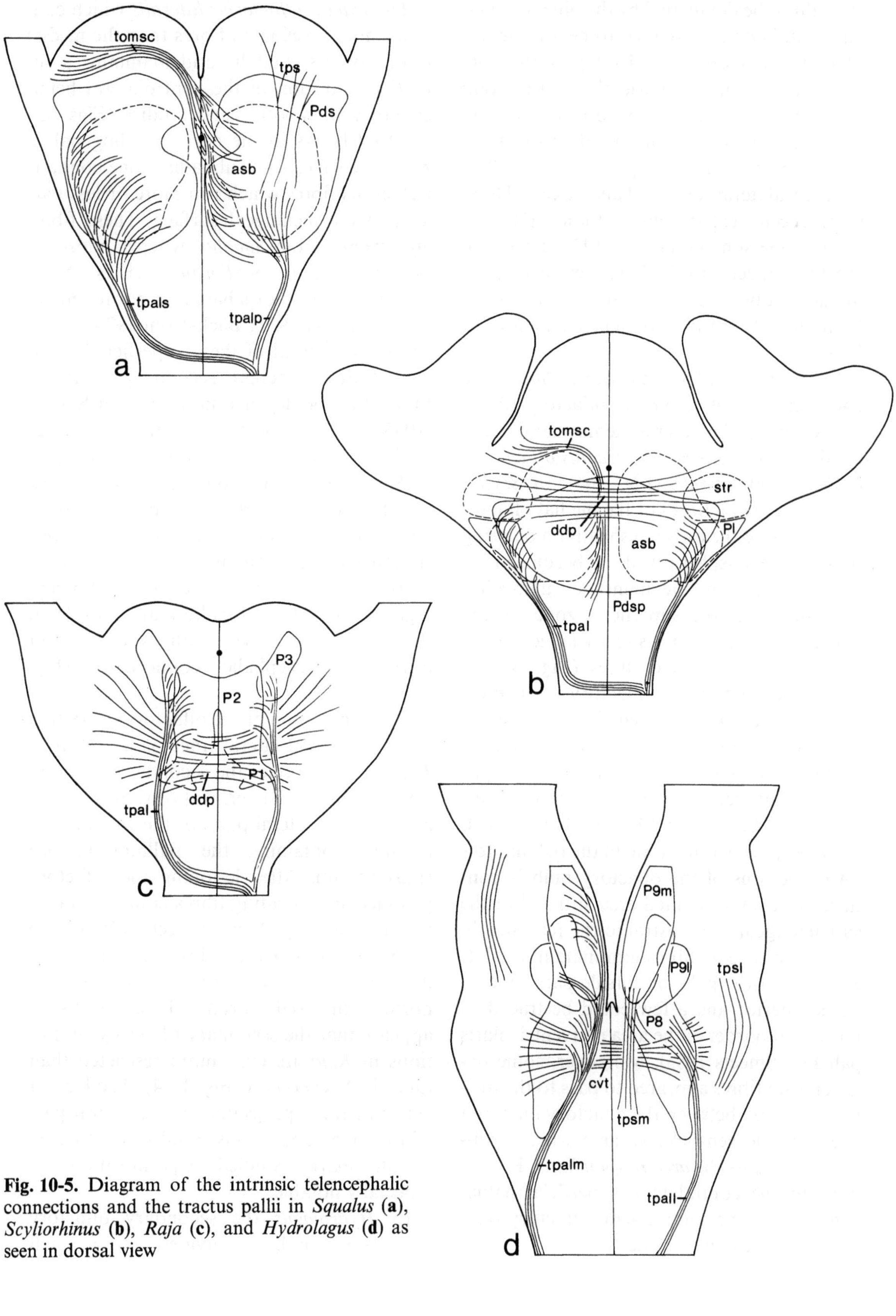

Fig. 10-5. Diagram of the intrinsic telencephalic connections and the tractus pallii in *Squalus* (**a**), *Scyliorhinus* (**b**), *Raja* (**c**), and *Hydrolagus* (**d**) as seen in dorsal view

thought to be dominated by the olfactory system and, hence, considered to be a rhinencephalon. The most detailed study on the fibre anatomy of the chondrichthyan forebrain was done by Bäckström, who recognized no fewer than seven secondary olfactory tracts which he could trace to ipsilateral as well as to contralateral telencephalic areas. However, recent experimental studies (Heimer 1969; Ebbesson and Heimer 1970; Ebbesson 1980b; Smeets 1983) have revealed that, in fact, only a restricted area of the forebrain directly receives secondary olfactory fibres.

In both *Squalus* and *Scyliorhinus* the compact mass of secondary olfactory fibres, on reaching the lateral hemispheric wall, divides into two parts, the *tractus olfactorius medialis* and *lateralis* (Fig. 10-4). In *Scyliorhinus* almost all fibres of the medial olfactory tract loose their myelin sheath on entering the dorsal hemispheric wall and become indistinguishable from the numerous unmyelinated fibres in the superficial zone of the pallium. In *Squalus* this superficial zone contains many myelinated fibres (Figs. 31–33) with which the medial olfactory tract intermingles. The unmyelinated fibres in the superficial zone of *Scyliorhinus* and the myelinated fibres in *Squalus* have both been thought to represent secondary olfactory tracts (Johnston 1911; Bäckström 1924; Williams 1973). However, the degeneration in the pallium following lesions of the olfactory bulb is confined to a rather limited area situated in the submeningeal zone rostral to the recessus. In *Scyliorhinus* some fibres terminate in the lateral part of the pallium dorsale pars superficialis anterior and a few could be traced to the lateral border of the area periventricularis pallialis (Smeets 1983). Moreover, some degenerating fibres appeared to pass to the ipsilateral region between the olfactory bulb and telencephalic ventricle, an area which probably represents the *area retrobulbaris*. Projections to the contralateral superficial pallial zone and to the contralateral olfactory bulb could also be demonstrated.

The *tractus olfactorius lateralis* which contains more myelinated fibres than the medial tract, courses caudally and maintains a superficial position in the ventrolateral hemispheric wall in both sharks studied (Figs. 25–31, 59–64). As in the corresponding pallial zone, the superficial subpallial zone contains numerous fibres which run more or less parallel to the meningeal surface. These fibres are myelinated in *Squalus*, whereas most are unmyelinated in *Scyliorhinus* (Figs. 25–27, 59). In several studies based on normal material (Johnston 1911; Bäckström 1924) it was suggested that all of these superficially situated fibres represent secondary olfactory fibres, but the degeneration study of Smeets (1983) has not confirmed this. The latter study reveals that the lateral olfactory tract of *Scyliorhinus* sends some fibres to the area retrobulbaris and striatum, but the majority terminates in the lateral pallium and in a zone superficial to the lateral part of the area superficialis basalis. The fibres in the submeningeal zone of the subpallium are considered therefore by the present authors to represent intrinsic telencephalic connections (Fig. 10-5).

The subdivision of the olfactory projection into medial and lateral tracts is less clear in *Raja*. In this species the secondary olfactory fibres occupy the superficial zone of the lateral and ventrolateral parts of the winglike expansion containing the pallium laterale (Figs. 95–98). After lesioning the olfactory peduncle degenerating fibres could be traced to the lateral pallium (Smeets 1983). The extent of the olfactory fibres, as seen in the degeneration study, is comparable to that in normal material. From these results it appears that the secondary olfactory projections in *Raja* are even more restricted than those in *Scyliorhinus* (Fig. 10-4). The lack of evidence for a projection to the lateral part of the area superficialis basalis, as reported in all sharks studied experimentally, is especially notable.

In *Hydrolagus* the secondary olfactory projection can be subdivided into a small medial and large lateral tract. In normal material the medial olfactory tract seems to send its fibres to the area which we have labelled P10, whereas the lateral olfactory tract, keeping its superficial position in the lateral hemispheric wall, provides olfactory input to the rostral and caudal parts of the lateral pallium (Fig. 10-4). The existence of a contralateral secondary olfactory projection via the ventral commissure could not be excluded. Though no experimental data are available, it does seem unlikely that the lateral olfactory tract of *Hydrolagus* reaches the area superficialis basalis, and in this respect this tract would resemble the equivalent structure in *Raja*.

10.7.2 Intrinsic Telencephalic Connections

In the previous section we have seen that the secondary olfactory projections in cartilaginous fish are limited in their telencephalic distribution. The same holds true for the projections to or from farther caudal brain regions. There are thus numerous telencephalic fibres that can be classified neither as direct afferent nor efferent fibres. The majority of these are unmyelinated and form a dense meshwork within the telencephalon. However, some collect as rather distinct fibre bundles which can be traced for some distance within the telencephalon, although their exact sites of origin and termination cannot be determined in our material. In the following account of the intrinsic telencephalic connections we will use the terminology of Bäckström (1924) wherever possible. This does not imply that we agree with him fully, but merely wish to avoid a further accumulation of novel names.

A very conspicuous fibre system, at least in *Squalus*, is the tract termed *tractus olfactorius medialis septi cruciatus* by Bäckström. He believed that this tract consisted of secondary olfactory fibres which separate from the tractus olfactorius medialis and curve medially along the anterior surface of the brain until they reached the midplane, where they arched caudoventrally through the septum and cross, ventral to the anterior commissure, to the opposite side. Then the bundle turns dorsolateralward and rostralward and finally distributes its fibres to the striatum, the area superficialis basalis and the lateral pallium. Corresponding tracts were described by Bäckström in *Scyliorhinus* and *Hydrolagus*, but not in *Raja*.

In *Squalus*, *Syliorhinus* and *Hydrolagus* we have found a fibre system which, with regard to its course, resembles the tractus olfactorius medialis septi cruciatus of Bäckström (Fig. 10-5). This tract is particularly distinct and well myelinated in *Squalus* (Figs. 25–33), but in *Scyliorhinus* it consists almost exlusively of unmyelinated fibres which, according to our observations, all remain ipsilateral. Our results differ from those of Bäckström in two respects: (1) experimental studies on *Scyliorhinus* (Smeets 1983) have shown that the tomsc does not consist of secondary olfactory fibres, and (2) careful analysis of the *tractus olfactorius medialis septi cruciatus* in normal material of *Squalus* reveals a decrease in the number of fibres at midhemispheric levels (Figs. 27, 28) which does not tally with the description of Bäckström (1924). The last finding might explain why Johnston (1911) considered the rostral part of the tomsc in *Squalus* as a tractus olfacto-corticalis mediodorsalis and the caudal part as an extension of the tractus olfactorius lateralis.

Another fibre system, recognized in the normal shark material, is the *tractus pallialis septi*. In *Squalus* this tract consists of myelinated fibres and can be traced from caudal hemispheric levels, where its fibres intermingle with those of the tractus pallii pars periventricularis, to the rostral part of the telencephalon, where it distributes to the ventrolateral hemispheric wall (Figs. 28–33). A comparable fibre system could be recognized in *Scyliorhinus* at rostral telencephalic levels (Figs. 63–65) and in *Hydrolagus* at caudal levels (Figs. 132–134). In the latter species this bundle takes a dorsolateral position and consists of two components, a pars lateralis and medialis (Bäckström 1924). In *Raja* we could not identify a distinct tractus pallialis septi, but the tractus pallialis septi in *Raja* according to Bäckström (1924, Fig. 28) is constituted by the rostral continuation of the tractus pallii.

In the pallium of *Scyliorhinus* several decussating fibre systems can be identified in normal material. The most rostral, the *commissura olfactoria inferior*, consists mainly of unmyelinated fibres which interconnect the pallial parts of both hemispheres rostral to the neuroporus externus. Some of these fibres may not be true commissural fibres, but rather decussating secondary olfactory fibres (Smeets 1983). A comparable commissure could not be identified in the other three species studied, a result in harmony with the findings of Bäckström (1924).

A *commissura olfactoria superior* could be identified only in *Scyliorhinus* (Fig. 10-4). This commissure, which is very distinct in normal material (Figs. 62, 63), contains myelinated fibres which connect both olfactory bulbs, as has been experimentally confirmed by Smeets (1983). The commissura olfactoria superior lies immediately caudal to the recessus neuroporicus externus and superficial to the decussatio dorsalis pallii (see below).

In *Raja,* Bäckström (1924) identified a commissura olfactoria superior whose constituent fibres cross within the territory of the large decussatio dorsalis pallii. We were unable, however, to trace such a pathway in our normal material, and a recent degeneration study (Smeets 1983) also fails to reveal the presence of crossing secondary olfactory tracts in *Raja*.

In *Scyliorhinus* as well as in *Raja* a strongly developed *decussatio dorsalis pallii* could be recognized in the present study (Figs. 10-5, 62, 63, 93, 94). Most of the fibres constituting this decussation are myelinated in *Scyliorhinus,* but unmyelinated in *Raja.* In both, the decussation consists of several components, which are hard to trace to their sites of origin or termination. However, it may well be that this complex consists of both decussating and true commissural fibres which connect lateral pallium, striatum, area superficialis basalis and dorsal pallium. In *Squalus* a small pallial commissure has been described by Johnston (1911), but the present authors were unable to confirm the existence of such a commissure.

10.7.3 Telencephalic Connections with More Caudally Situated Brain Regions

Three distinct fibre tracts, connecting the telencephalon with the diencephalon, can be recognized in normal material of the four species studied: the fasciculus basalis telencephali, the tractus pallii, and the stria medullaris. As the diencephalic course of these fibre tracts has already been discussed in the previous chapter we shall confine ourselves here to their telencephalic parts.

The *fasciculus basalis telencephali* or *basal forebrain bundle* is well developed in the four species studied and consists of both myelinated and unmyelinated fibres. In normal material it is obvious that the fibres form a well-defined tract at the level of the caudal pole of the area superficialis basalis (Fig. 10-4). Before passing caudalward in the basal forebrain bundle, the majority of these fibres decussate in the anterior commissure. The rostrocaudal extent of this commissure is large in *Squalus, Scyliorhinus* and *Hydrolagus,* but small in *Raja.* However, normal material indicates that the number of decussating fibres is larger in elasmobranchs than in *Hydrolagus,* a finding which has also been reported by Bäckström (1924). This author described 11 components within the basal forebrain bundle originating from both pallial and subpallial regions. However, we can recognize only the striatum and the area superficialis

Fig. 10-6. Photomicrographs of transverse Bodian stained sections to show the peculiar course of the fibres joining the tractus pallii. **A** *Squalus;* **B** *Scyliorhinus;* **C** *Raja,* and **D** *Hydrolagus. Bar scale* = 100 μm

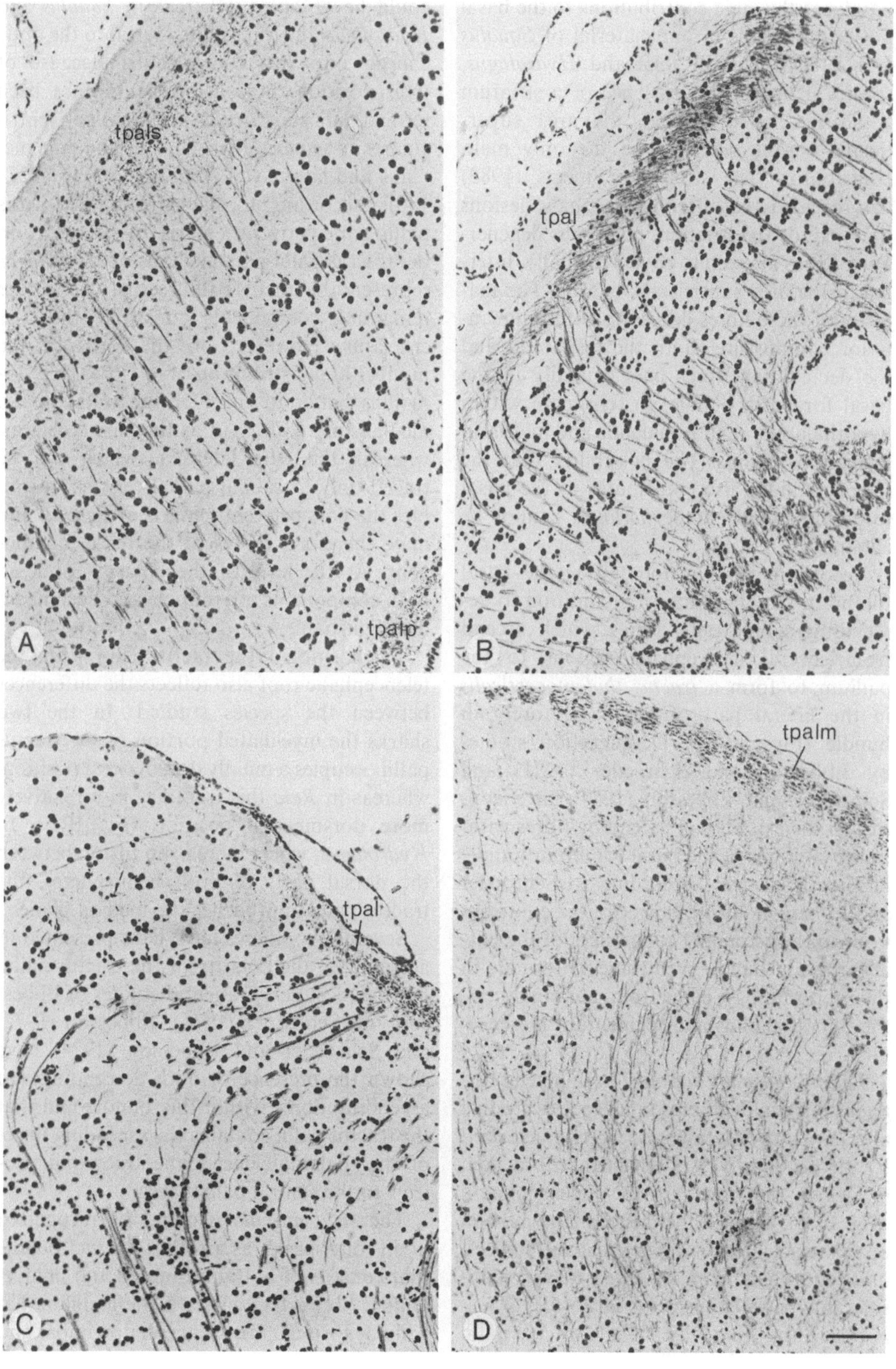
tpals
tpalp
A
tpal
B
tpal
C
tpalm
D

basalis as the main contributions to the basal forebrain bundle in our material of *Squalus* and *Scyliorhinus*. In *Raja* and *Hydrolagus*, species in which the existence of a striatum could not be confirmed, the area superficialis basalis seems to be the only main contributor. The study by Smeets (1983) has revealed that after telencephalic lesions that include the striatum, numerous degenerating fibres are seen to course medially, internal to the area superficialis basalis. The majority of these fibres decussate within the anterior commissure. Both the ipsilateral and the decussating fibres pass caudally in the basal forebrain bundle, to terminate within hypothalamic, ventral thalamic and midbrain tegmental regions. The possibility that the lateral pallium also contributes to the basal forebrain bundle in *Scyliorhinus* cannot be eliminated.

According to Johnston (1911) the basal forebrain bundle contains, in addition to descending components, fibres which ascend from several diencephalic regions to the pallium to form a *tractus thalamocorticalis* in the lateral part of the basal forebrain bundle (Figs. 56–58). Degeneration studies by Ebbesson and Schroeder (1971) and Schroeder and Ebbesson (1974) have confirmed the existence of ascending fibres within the confines of the basal forebrain bundle in *Ginglymostoma*. According to Schroeder and Ebbesson (1974), however, the contributions to the ascending component of the basal forebrain bundle (ventromedial bundle in their terminology) come from, among others, the hypothalamus and tegmentum mesencephali.

The *tractus pallii* contains both myelinated and unmyelinated fibres in all species studied and is, as already mentioned, characterized by the peculiar way in which the fibres join or leave the tract in the telencephalon (Fig. 10-6). Though the diencephalic portion of this tract did not show great differences, its telencephalic part appeared to vary considerably between species (Fig. 10-5). In *Scyliorhinus* and *Raja* only a single tractus pallii could be observed, whereas in *Squalus* and *Hydrolagus* the tract splits rostral to the optic chiasm into two components. Because of their position these components have been termed pars superficialis and pars periventricularis in *Squalus* (Figs. 25–30), and pars medialis and lateralis in *Hydrolagus* (Figs. 127–136). According to Bäckström the tractus pallii pars periventricularis of *Squalus* and the tractus pallii pars medialis of *Hydrolagus* represent the tractus pallii rectus, whereas the remaining components constitute the pars cruciatus. However, on the basis of the number and distribution of its fibres (Fig. 10-5) it is better to consider the tractus pallii medialis of *Hydrolagus* as the pars cruciatus, whereas the pars lateralis, which can be traced only to caudal pallial regions, resembles the pars periventricularis of *Squalus* and may, therefore, represent the rectus component. At the level of the chiasma opticum both components intermingle and become indistinguishable.

The position of the tractus pallii within the telencephalic roof also reflects the differences between the species studied. In the two sharks the myelinated portion of the tractus pallii occupies a mainly dorsolateral position, whereas in *Raja* this tract lies in a relatively more dorsomedial position (Fig. 10-5). In *Hydrolagus*, where there is no fusion between the dorsal parts of both hemispheres, the tractus pallii approaches the median plane.

Several hypotheses have been presented to account for the polarity of the tractus pallii, but the recent experimental studies of Ebbesson and Schroeder (1971), Ebbesson (1972) and Schroeder and Ebbesson (1974) have shown the presence of both descending and ascending fibres within this bundle and that the tractus pallii contains the ascending fibres of the thalamotelencephalic tract that projects to the central pallial area.

The *stria medullaris* could be recognized in all four species as a rather distinct bundle that consists of both myelinated and unmyelinated fibres. Its course to the ganglion habenulae through the telencephalon impar and diencephalon is readily traced and has been described in the previous chapter. According to our observations, the stria medullaris collects fibres from the caudal parts of both the pallium and the subpallium. The exact site of origin of these fibres could not be determined. The stria medullaris, in passing to the diencephalon, lies in close relationship to the tractus pallii and we cannot exclude the possibility that these two systems interchange fibres, as reported by Kuhlenbeck (1977).

A detailed study of the stria medullaris complex was made by Johnston (1911, *Squalus, Scyliorhinus*) who recognized the following components:

1. Tractus septohabenularis. The fibres constituting this tract originate from the septal area and run at first within the basal forebrain bundle. Upon entering the diencephalon, these fibres ascend directly to the ganglion habenulae.
2. Tractus olfactohabenularis lateralis, the fibres of which arise from the area superficialis basalis and the lateral pallium.
3. Tractus corticohabenularis. Originating from the pallium, its fibres join those of (2) behind the foramen interventriculare.
4. Tractus olfactohabenularis posterior. This fibre tract consists of a diffuse bundle of fibres arising from the preoptic region interlaced with the outermost fibres of the basal forebrain bundle.
5. Tractus taeniae consisting of fibres which have their origin in the area somatica (nucleus entopeduncularis of present study).
6. Tractus habenulothalamicus. Johnston (1911) did not know the origin of these fibres, which enter the dorsal thalamus and, for a small part, the lateral geniculate body.
7. Commissura pallii posterior (commissura superior telencephali of present study), a true pallial commissure consisting of many fine, unmyelinated fibres.

Apart from the commissura superior telencephali none of these components could be traced with certainty to their place of origin by the present authors.

Finally it is noteworthy that several authors, such as Edinger (1888), Ariëns Kappers (1906), Johnston (1910) and Kuhlenbeck (1977, termed the stria medullaris the tractus taeniae.

10.7.4 Remaining Fibre Systems

Under this heading we will deal with the fornix and the nervus terminalis.

A presumed *fornix* has been described by Johnston (1911) and Bäckström (1924) in several cartilaginous fishes. According to these authors fibres collect from the medial pallium of either side and pass medioventrally to form two compact bundles lateral to the recessus neuroporicus externus. As they descend caudalward they gradually take a position ventral to this recess and finally join the fasciculus basalis telencephali. They pass with this bundle to a hypothalamic centre which Johnston and Bäckström considered as a primordial corpus mamillare. In *Squalus* and *Scyliorhinus,* as well as in *Raja,* fibres could be traced from the pallium to caudal telencephalic levels which resemble the fornix as described by Johnston and Bäckström. However, we were unable to trace these fibres to the hypothalamus. Faucette (1969a) labelled a distinct fibre system, that lies adjacent to the sulcus telencephalicus ventralis at mid-telencephalic levels in *Hydrolagus* (Figs. 133, 134), as both medial forebrain bundle and fornix. We consider it more likely that this fibre system is an intrinsic telencephalic pathway, since caudalward the number of its constituent fibres decreases gradually.

A *nervus terminalis* could be recognized in both sharks, but not in *Raja,* and *Hydrolagus.* In *Squalus* the nervus terminalis, which consists of unmyelinated fibres, passes from the olfactory bulb over the dorsal surface of the lobus hemisphericus to the cleft between the two hemispheres (Figs. 31–33). Then it enters the medial hemispheric wall, and becomes unrecognizable as it intermingles with other unmyelinated fibres. In *Scyliorhin-*

us this nerve takes a different course, entering the brain ventrally between the subpallial parts of the hemispheres, before passing caudally for some distance (Figs. 61, 62).

Extensive studies on the nervus terminalis made by Locy (1905), Johnston (1911) and Bäckström (1924) yielded similar results. Bäckstöm identified a nervus terminalis in several species of *Raja* where he reported considerable variability in the position of its entry.

10.8 The Telencephalon and the Problem of Homology

In this section we will make an attempt to homologize the telencephalic cell masses of the four species studied, taking their relative positions and fibre connections into consideration.

The telencephalon impar of the four species studied can be easily compared, except for the elongated preoptic area in *Hydrolagus*. In every species this part of the telencephalon contains a nucleus preopticus, a nucleus interstitialis fasciculi basalis telencephali, a nucleus entopeduncularis, a nucleus M and an area superficialis basalis. Moreover, the main fibre tracts which pass through the telencephalon impar occupy corresponding positions. So, we can state that there is no problem in homologizing structures in the telencephalon impar.

The bulbus olfactorius varies considerably with regard to shape and mode of connection with the telencephalic hemispheres in the four species studied. The whole range from a rostral (*Hydrolagus*) to a lateral insertion (*Scyliorhinus*) is found. Embryological studies (Holmgren 1922; Bäckström 1924) have shown that the primary position of the bulbus is a lateral one and that, therefore, the frontal position of the olfactory bulb in *Hydrolagus* is to be considered as secondary. Holmgren (1922) has pointed out that the olfactory peduncle is a subpallial structure and that its shift is generally effected without changes in the configuration of the pallium. However, in our opinion the spatial relations of the olfactory bulb, the secondary olfactory connections and their terminal nuclei strongly suggest that the point of attachment of the olfactory peduncle is not a fixed point (Fig. 10-4).

The telencephalic hemispheres vary considerably in both macroscopical and microscopical structure. However, the following relatively constant relationships can be recognized:

1. The relationship between secondary olfactory fibres and the lateral pallium, the latter receiving most of the olfactory input (Fig. 10-4).
2. The relation of the area superficialis basalis and, if present, striatum to the basal forebrain bundle (Fig. 10-4).
3. The relation of the tractus pallii to the dorsal pallium, or its subdivisions, and to the lateral pallium (Fig. 10-5).

The present analysis indicates that the position and course of the tractus pallii is strongly dependent upon the shape and position of the pallium. The main portion of this tract, which decussates behind the optic chiasm, shows a constant relationship with the morphologically lateral portion of the dorsal pallium (Fig. 10-5). The topographically medial position of the main pallial tract in *Hydrolagus* can be explained by the strong inversion of the pallium dorsale, moving the lateral part of the dorsal pallium to a medial position.

Another constant feature of the tractus pallii is its position medial to the lateral pallium. This explains why in *Raja,* with its particularly strongly developed lateral pallium, the tractus pallii occupies a relatively far medial position.

The following constant relationships can be recognized for the commissural and decussating fibre systems:

1. the decussatio tractus pallii is always found to be located caudoventral to the optic chiasm, and

2. the commissura anterior, which contains decussating fibres of the basal forebrain bundle, is always situated in the rostral part of the telencephalon impar.

Apart from these two constant structures, the hemispheres contain several other commissures or decussations, but these show considerable variations in both position and extent (Figs. 10-4, 10-5). These differences are due to the degree in which both hemispheres are fused. For example, in *Scyliorhinus* and *Raja,* where fusion between the two hemispheres is almost complete, a strongly developed pallial decussation is present, whereas in *Squalus,* with only moderate fusion, such a decussation cannot be recognized. However, in *Squalus* decussating fibres appear to pass through the most rostral part of the telencephalon to the contralateral hemisphere. In *Hydrolagus,* which lacks any fusion of the dorsal parts of the hemispheres, only a subpallial commissure, designated here as commissura ventralis telencephali, is present. The latter commissure, which is situated rostral to the commissura anterior probably contains fibres which connect the pallia of both hemispheres.

10.9 Functional Correlations

Until recently it was thought that the secondary olfactory tracts distribute their fibres to most regions of the telencephalon which, therefore, was considered to be mainly concerned with olfaction (Johnston 1911; Bäckström 1924; Ariëns Kappers et al. 1936; Faucette 1969a, b; Williams 1973). However, degeneration studies on olfactory bulb projections in *Ginglymostoma* (Ebbesson and Heimer 1970) revealed that the secondary olfactory projection appears to be entirely ipsilateral and restricted to the lateral pallium and to a region superficial to the lateral part of the area superficialis basalis. Subsequently this restricted distribution of secondary olfactory fibres in chondrichthyans was confirmed by the electrophysiological studies of Bruckmoser (1973), Bruckmoser and Dieringer (1973) and Vesselkin and Kovačević (1973). Northcutt (1978) presented experimental anatomical evidence for *Raja eglanteria* that showed the secondary projections to be also restricted to the lateral pallium and an area which he labelled *a* (SP4 of the present study). In a preliminary report on *Heterodontus francisci* Bodznick and Northcutt (1979), using degeneration, autoradiographic and HRP techniques, demonstrated that the bulk of the secondary olfactory fibres projects to the ipsilateral pallium. They reported that these fibres arise from both mitral cells and smaller neurons located within the internal granular cell layer. In addition to the olfactory input, the lateral pallium receives afferents from the ipsilateral and contralateral dorsal and medial pallium and from the contralateral lateral pallium. Bodznick and Northcutt (1979) have also found electrophysiological evidence for the existence of these connections. Recently the degeneration study by Smeets (1983) confirmed Northcutt's observation that there is a restricted olfactory input to the lateral pallium in *Raja,* but in *Scyliorhinus* an additional projection to the contralateral olfactory bulb and to the zone superficial to the lateral part of the area superficialis basalis could be recognized. It might be supposed, therefore, that most telencephalic regions receive tertiary, if not secondary, olfactory fibres, but an experimental study on higher order olfactory areas failed to reveal extensive projections to the hemispheres (Ebbesson 1972).

These studies highlight the important question of what could be the function of those large telencephalic regions that apparently do not receive olfactory information. Classical studies on the function of the telencephalon of elasmobranchs have revealed that partial or complete removal of the telencephalon produces no deficits in motility, equilibrium, or vision (Bethe 1899; Polimanti 1911, 1913; Rizzolo 1929). However, it has been recently shown that the telencephalon of *Ginglymostoma* receives visual input from the retina.

This visual projection relays in the dorsal thalamus and terminates in the caudal part of the pallium, i.e., pallium dorsale pars centralis (Ebbesson and Schroeder 1971; Schroeder and Ebbesson 1974; Luiten 1981b).

Electrophysiological evidence in support of the telencephalic visual function in cartilaginous fish has been presented by Cohen et al. (1973, *Ginglymostoma*) and Vesselkin and Kovačević (1973, *Dasyatis, Raja, Torpedo*). These two groups of authors independently obtained telencephalic evoked potentials in response to electrical stimulation of the optic nerve. The responsive area was found to overlap roughly with that previously shown by Ebbesson and Schroeder (1971) to receive fibres from the thalamic visual region. Vesselkin and Kovačević were, however, unable to obtain evoked potentials in response to electrical stimulation of the optic nerve in *Scyliorhinus*.

Behavioural evidence which supports the telencephalic visual function, has been provided by the studies of Graeber et al. (1973, 1978) and Graeber (1978, 1980). It was found that free swimming nurse sharks (*Ginglymostoma*) with complete tectal ablations could still learn to perform simple visual discriminations such as the distinction between black and white, or horizontal and vertical stripes. This result implies that the tectum mesencephali is not the only brain centre to influence the visually guided behaviour of elasmobranchs. Partial or complete lesioning of the central nucleus led to an inability to perform discriminations between black and white, or horizontal and vertical stripes better than by chance. These results implicate the central pallial nucleus in the control of visually guided behaviour in *Ginglymostoma*.

It has been demonstrated that fibres originating from the cerebellum (Ebbesson and Campbell 1973), tectum (Northcutt and Boord 1981; Smeets 1981b), area octavolateralis (Boord and Northcutt 1979, 1982) and spinal cord (Ebbesson and Hodde 1981) pass to the dorsal thalamus, which may well function as a relay centre for sensory information to the non-olfactory telencephalic areas. The experimental study of Schroeder and Ebbesson (1974) has also revealed direct ascending fibres from levels caudal to the dorsal thalamus to the telencephalon. These fibres terminate in a region which lies internal to the medial part of the area superficialis basalis. The electrophysiological study of Platt et al. (1974) gives some evidence of mechanosensory and electrosensory inputs to the telencephalon, and acoustic centres could be identified in the septal region and several pallial areas of *Platyrhinoidis* (Corwin and Northcutt 1982, Fig. 3G, H) with the use of deoxyglucose autoradiography.

A preliminary experimental study by Ebbesson (1972) has revealed that in *Ginglymostoma* long pathways descend from the telencephalon, distributing their fibres to the epithalamus, the hypothalamus, the tectum mesencephali and the reticular formation of the brain stem.

Though most of the experimental studies on the organization of the telencephalon have been done in a single species, *Ginglymostoma cirratum,* and telencephalic variations among cartilaginous fish are considerable, it may be concluded that the chondrichthyan telencephalon functions not only as a relay for olfactory input, but also as an integrative centre for other sensory modalities.

11 Concluding Remarks

A survey of our present knowledge of the structural organization of the central nervous system of cartilaginous fishes has been presented in the preceding chapters. In this final chapter the main results will be assessed and some of the numerous gaps in our knowledge will be indicated. We have grouped our commentaries under the following five headings: (1) gross morphology, (2) subdivision of grey matter, (3) macrocircuitry, (4) structure of nerve roots and fibre systems, and (5) microcircuitry.

11.1 Gross Morphology

The fact that cartilaginous fishes possess relatively large brains deserves re-emphasis. In his book *The Evolution of the Brain and Intelligence,* Jerison (1973) developed the idea that brain size can be used to estimate the value of several fundamental characteristics of the brain, such as the total number of neurons and the richness of neuronal connections and hence, the amount of information processing and the complexity of overall behaviour. With this idea in mind, Jerison determined the brain and body weights of a large number of vertebrates. When the data thus obtained were plotted on log-log coordinates two distinct, non-overlapping arrays of points appeared, one obtained from anamniotes and reptiles, the other from birds and mammals. Because the ratios derived from the former group were substantially lower than those from the latter group, Jerison felt justified in characterizing anamniotes and reptiles as lower vertebrates and birds and mammals as higher vertebrates. The finding of Ebbesson and Northcutt (1976), Bauchot et al. (1976) and Northcutt (1978) that many cartilaginous fishes possess brain-to-body-weight ratios that fall within the ranges for birds and mammals should be placed against this background. The biological meaning of the relatively large brain size of cartilaginous fishes will not become clear until we have quantitative data on neuronal cell number and volume, and on behaviour for these species.

The brains of cartilaginous fishes show considerable interspecific differences in their overall structure. Figure 3-3 gives an impression of this variability but by no means reflects the full spectrum. For additional information the reader is referred to the descriptions and illustrations in a recent study by Northcutt (1978). This study shows that the cerebellum and the telencephalon are by far the most variable parts of the chondrichthyan brain, both in terms of shape and size.

The study of the ventricular sulci of the brain belongs in the realm of gross morphology, although only in the largest cartilaginous fishes is it possible to study these structures by direct inspection. In all other species the ventricular sulcal pattern can be adequately observed only by carefully preparing three-dimensional or graphical reconstructions, based on continuous histological series. We emphasize this point because (1) as mentioned in previous chapters, several authors consider the ventricular sulci in the chondrichthyan brain as highly important landmarks that indicate the boundaries of periventricular grisea, and (2) investigations concerning the course and extent of these structures have often been based merely on the study of histological material, without making reconstructions, and this has led repeatedly to erroneous results. In the present work the ventricular sulcal pattern has been systematically analyzed in the four species. Our conclusion is that certain sulci are constant in position and mark the boundary of grisea, but that conclusions concerning the homologies between ventricular cell zones certainly cannot be based exclusively on these sulci.

11.2 Subdivision of Grey Matter

It should be appreciated that a study of the grey matter with the aid of the Nissl technique is no more than a first step in the analysis of a central nervous system. Yet, it is a highly essential step, providing an indispensable framework for further studies, both on structure and function. This step has been omitted from quite a number of previous investigations aimed at establishing fibre connections in the chondrichthyan brain.

So far no systematic study of the cytoarchitecture of the spinal cord has yet been made in any cartilaginous fish. The reason for this omission is obvious; the preparation and analysis of histological series or even subseries is a very laborious task. However, there can be no doubt that work of this type is very necessary. It will provide the basis for an understanding of the origin and termination of fibre systems in this part of the central nervous system.

Cytoarchitectonic studies involve two phases, inventory and interpretation. During the first, the various nuclei, areas and cellular zones are delineated; during the second phase the grisea delineated are named. If a particular griseum can be homologized with a griseum present in another species, it is preferable to give both the same name, if possible one that already exists; if not, a new name has to be created. With this in mind we have attempted to subdivide the grey matter in the four chondrichthyan species. We have distinguished a total of 126 grisea: the 123 grisea listed in Tables 4-1, 4-2 and 4-3, plus the cerebellum, tectum mesencephali and bulbus olfactorius. Of these, 64 could be homologized with reasonable certainty with cell masses distinguished in other groups of fishes; 54 have been observed in all four species; 12 have been observed in three species (in different combinations); 11 have been observed in two species (again in different combinations) and, finally, no fewer than 49 have been observed in only one. Needless to say, this first attempt at systematic cytoarchitectonic analysis of the entire brain of some chondrichthyans remains incomplete and should be considered as tentative in several respects. Particularly the diffuse character of the grey matter sometimes made an exact delimitation of cell masses impossible. This difficulty appeared to be greater in the more rostral regions. A fairly complete subdivision could be made in the brain stem and the diencephalon, although in the latter we have been compelled to make arbitrary decisions in several places. We expect that continued studies will lead to a number of changes.

It is no exaggeration to say that the telencephalon is a cytoarchitectonic puzzle. The fact that purely descriptive names or provisional symbols have been given to no fewer than 36 of the 51 grisea illustrates the difficulties we have encountered. Clearly, a considerable amount of work has still to be done on this intriguing and complex part of the brain of cartilaginous fishes.

It is noteworthy that much experimental work on the rostral parts of the chondrichthyan brain has been carried out on the nurse shark, *Ginglymostoma cirratum,* a species for which we have no histological material. Published illustrations of sections through the telencephalon of this species suggest that the grey matter is diffusely arranged. Moreover, orientation is hampered by an almost complete obliteration of the ventricular cavities. Nevertheless, a thorough cytoarchitectonic analysis of the telencephalon (and the thalamus) of the nurse shark is essential for placing the many interesting experimental findings on this species in a comparative perspective.

11.3 Macrocircuitry

The study of fibre connections, as lines of communication within the nervous system, has always occupied a prominent place in neuromorphological research, particularly because it may lead to functional interpretations. Formerly this type of research was mainly carried out on normal material, but during the last decade it has been based almost exclusively on work with a steadily increasing arsenal of experimental techniques. Analysis of the origin and termination of the cranial and spinal nerves and their components often results in elementary interpretations of grisea in terms of sensory and motor functions—in the words of Herrick (1943, p. 44): "a physiological analysis ... by an anatomical method". Experimental anatomical studies on the fibre connections of the grisea thus functionally defined may, in turn, lead to functional interpretations of other parts of the central nervous system, and so on. In this way "wiring diagrams" of "functional systems" can be prepared. Diagrams so derived may considerably gain in strength by combining the experimental neuroanatomical work on which they are based with evoked and unit potential studies, and with analyses based on the recently developed deoxyglucose technique. The approach just indicated is principally aimed at unravelling the intergriseal relations, i.e., the macrocircuitry of the central nervous system, and the bulk of the neuromorphological research carried out on cartilaginous fish during the last 15 years has been of this type. The limitations of this approach are obvious, Functional analysis is not completed merely by labelling centres and pathways as "olfactory", "visual", "somatomotor", and so on. Moreover, we cannot expect, even in the brain of an anamniote, to be able to interpret all centres in terms of purely sensory or motor functions. In fact further research often leads to the withdrawal of previously applied functional labels. The changes of the terms "tectum opticum" into "tectum mesencephali" and "tuberculum olfactorium" into "area superficialis basalis" are illustrative cases.

It should be appreciated that the establishment of all connections of the 80 or so centres in the central nervous system of a chondrichthyan is a formidable task; presently we know some of the connections of only about one-half of the grisea. We will not attempt to enumerate here all of the known connections. The pertinent information is summarized in Figs. 5-6, 5-7, 6-1, 6-7, 6-11, 7-4, 7-5, 8-2, 8-3 and 9-6. However, we shall now indicate some of the significant, outstanding hodological problems.

1. Do cartilaginous fish possess an equivalent, or at least a primordium of a dorsal column-medial lemniscus system? The presence of such a system was denied by Ariëns Kappers et al. (1936); we also remain unable to delineate dorsal column nuclei.

2. Where do the ascending efferents of the principal and descending trigeminal sensory nuclei terminate?

3. What is the destination of the efferent fibres of the large primary viscerosensory zone?

4. It is known that the inner ear of cartilaginous fish contains two separate auditory and five separate vestibular detectors. Do these detectors project entirely or preferentially to particular grisea in the octavus region and, if so, do these primary centres then retain separate projections throughout the mesencephalon, diencephalon and telencephalon?

5. Do the various components of the lateral line system also project to different centres of higher order?

6. What are the efferents of the extrathalamic and extratectal targets of the retinofugal fibres, such as the nucleus suprachiasmaticus and the nucleus opticus basalis?

7. It is now firmly established that the secondary olfactory fibres spread over a limited part of the telencephalon, comprising essentially the lateral pallium and the lateral part of the area superficialis basalis. However, it is not known to which areas in the brain these secondary olfactory centres project.

8. During the last years much work has been devoted to establishing the fibre connections and the overall function significance of the dorsal thalamus and the nonolfactory parts of the telencephalon. Although considerable progress has been made our knowledge concerning these regions is still fragmentary. Thus, it has been shown that the dorsal thalamus receives a direct projection from the spinal cord, but it is unknown whether the information received along this channel is relayed to the telencephalon. It is known that the visual system projects along two separate routes, retino-thalamo-telencephalic and retino-tecto-thalamo-telencephalic, to two separate centres in the dorsal pallium. However, it is uncertain by which fibre systems these two projections reach the telencephalon. Evoked potential recordings indicate that auditory, mechanoreceptive and electroreceptive lateral line information reaches the telencephalon. Nothing is known, however, of the pathways along which these three systems project, beyond the midbrain level. Another, entirely open question concerns the medial reticular formation, i.e., does it subserve only a motor co-ordinating function, or do the axons of (some of) its constituent neurons ascend to thalamic levels, thus forming part of a non-specific sensory projection system? Finally, we should consider the possibility that the dorsal thalamus may serve as a relay for "non-sensory" ascending projection systems. The finding that efferents from the cerebellum reach this centre is noteworthy in this context. However, it is unknown whether the impulses travelling along this cerebellothalamic projection are relayed toward the telencephalon.

9. There are several important outstanding problems concerning the projections descending from higher levels to the brain stem. In all chondrichthyans studied a well-developed stria medullaris-habenulo-interpeduncularis system is present, but detailed information about the input and output of this system is unavailable; that is to say the sites of origin of the stria medullaris have not yet been established with experimental neuroanatomical techniques and the same holds true for the efferents of the interpeduncular nucleus. The two other large projections of the telencephalon, i.e. the basal forebrain bundle and the tractus pallii, both contain several descending components. It will be necessary to know the specific sites of origin and termination of each of these components, before the possible "premotor" roles of these various sites of termination can be adequately studied.

10. Normal material strongly suggests that the hypothalamus and brain stem are interconnected by numerous fibres. Detailed information on the sites of origin and termination of these fibres is needed for, among other things, an understanding of the circuitry involved in such vital acts as the detection and selection of food as well as prey catching.

11. The most medial zone of the brain stem and its caudal continuation as the spinal ventral horn are predominantly concerned in the control of locomotion. Basic swimming movements are generated by intrinsic neuronal networks while fine tuning and modification of the basic locomotor rhythm is provided by the brain. At present, detailed information is not available on the spinal circuitry involved in the generation and maintenance of locomotion. We consider it likely that the following centres, known to project directly to the spinal cord, are involved in some way in the regulation of locomotion: the nucleus of the fasciculus longitudinalis medialis, the nucleus ruber, the rhombencephalic medial reticular formation, the nucleus octavus magnocellularis, the nucleus octavus descendens and the nucleus raphes inferior. However, the routes along which the efferents of most of

these centres reach the cord are unknown, and no information is available on the site and mode of termination of all these encephalospinal projections. It may be safely assumed that the tectum mesencephali and the cerebellum play a rôle in the coordination of spinal somatomotor activity. Both of these centres project to the medial reticular formation, the efferents of which furnish a "premotor" final common path. However, next to nothing is known concerning the pathways along which diencephalic and telencephalic centres influence somatomotor activity.

11.4 Structure of Nerve Roots and Fibre Systems

It is remarkable that in modern comparative neuroanatomy much stress is laid on the study of fibre connections as lines of communication, but that little attention is given to the structure of the fibre systems themselves, i.e. to the numbers and diameters of their constituent fibres and to the spatial arrangement of their components. The morphological studies on the nervous system of cartilaginous fishes are no exception. So far as the structure of central connectivities is concerned about all we know is that the medial longitudinal fasciculus (and its spinal continuation) and the fasciculus medianus of Stieda are composed of particularly coarse, heavily myelinated axons, and that the bundle last mentioned consists of some 25–50 fibres. That studies on the structure of fibre connections may yield highly interesting results is illustrated by the recent work of Coggeshall et al. (1978) and Leonard et al. (1978a) on the spinal nerves of the stingray. These studies revealed, among other things, that the number of fibres increases as the animals grow, and that thin, unmyelinated fibres are virtually absent from these nerves.

Another type of research which so far has found very little application in cartilaginous fishes is the determination with histochemical and immunohistochemical techniques of the transmitters and other neuroactive principles, present in cells, axons and terminals.

11.5 Microcircuitry

Analysis of the intrinsic organization of grisea is an indispensable prerequisite for physiological studies aimed at an understanding of the way in which information is processed within these entities. The Golgi method, electron microscopy, and particularly the technique in which these two approaches are combined, form essential tools in this analysis. Very little work has been done on the chondrichthyan brain at this level of resolution. We have some idea of the organization of the lateral line centres and their overlying cerebellar crest, of the cerebellum and of the olfactory bulb; the rest is *tabula rasa*.

From the foregoing survey it appears that a great number of intriguing problems pertaining to the structural organization of the central nervous system of cartilaginous fishes still awaits solution. This book will have served its purpose if it fosters interest in the neuromorphology of this fascinating group of animals and if it kindles some of its readers with enthusiasm to work on these problems.

ATLAS

Squalus acanthias

(Fig. 1–33)

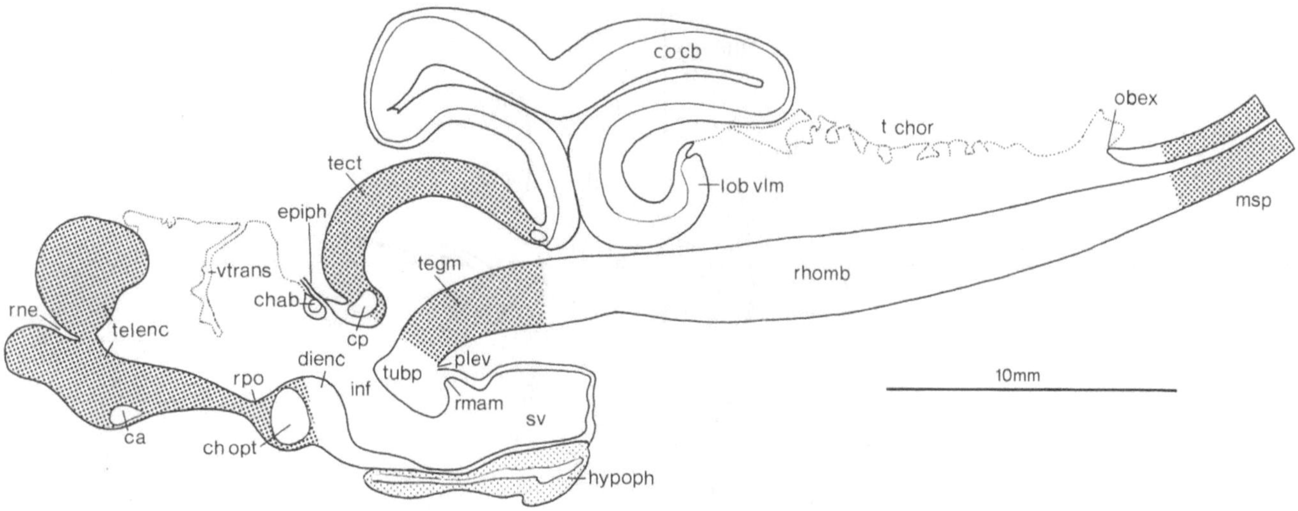

Fig. 1. Midsagittal section of the brain of *Squalus acanthias*

Fig. 2. Topological reconstruction of the brain stem of the shark *Squalus acanthias.* The *heavy line* which constitutes the axis of the figure represents the sulcus medianus inferior. The *curves* which constitute the lateral limits of the figure represent the taenia rhombencephali (*continuous parts*) and the sulcus medianus superior (*dashed parts*). The *interruptions* in these curves, marked by *short horizontal lines,* indicate the areas where parts of the auriculae and of the corpus cerebelli have been omitted from the reconstruction. The *remaining heavy curves* indicate the position of ventricular sulci. The *thin, continuous curves* indicate the boundaries of periventricular cell masses; the outlines of migrated cell masses are indicated by *interrupted curves.* The *black dots* give an impression of the distribution and density of the very large reticular cells. For clarity the cell masses are not always drawn bilaterally. *Numbers on the scale* at the side of the drawing correspond to atlas figures of the same number and indicate the level of the photomicrographs in these figures. Distance between sections, indicated by *small horizontal bars,* equals 450 µm

Fig. 3. Topographical reconstruction showing the cell masses of the brain stem of the shark *Squalus acanthias* as projected upon a horizontal plane. To reduce crowding, the cell masses are not always drawn bilaterally. The *heavy continuous lines* represent the contours of external brain stem features. The taeniae are represented by *heavy dotted lines,* whereas *thin lines* represent the boundaries of cell masses within the brain stem. *Numbers on the scale* at the side of the drawing correspond to atlas figures of the same number and indicate the level of the photomicrographs in these figures. Distance between sections, indicated by *small horizontal bars,* equals 450 µm

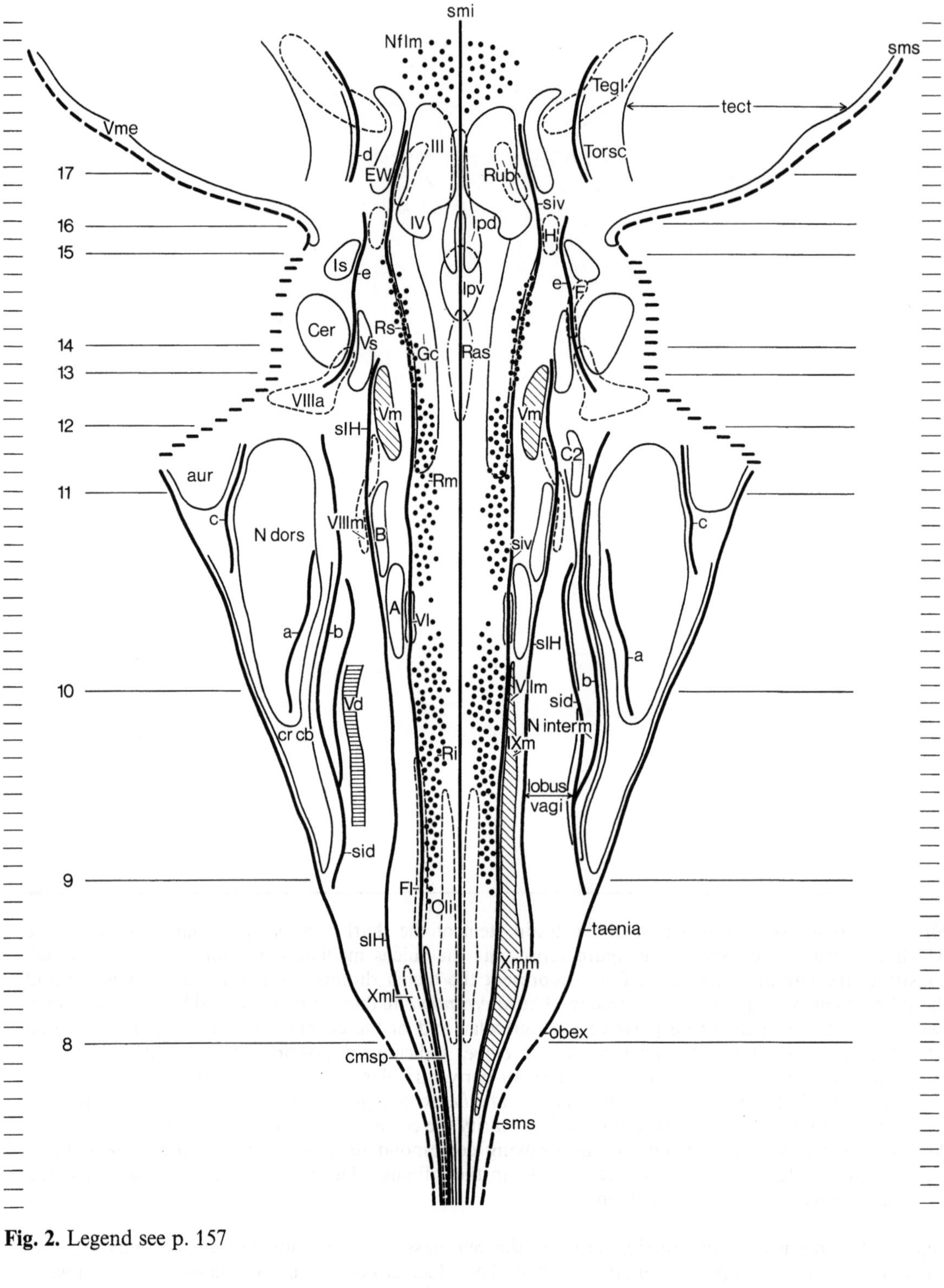

Fig. 2. Legend see p. 157

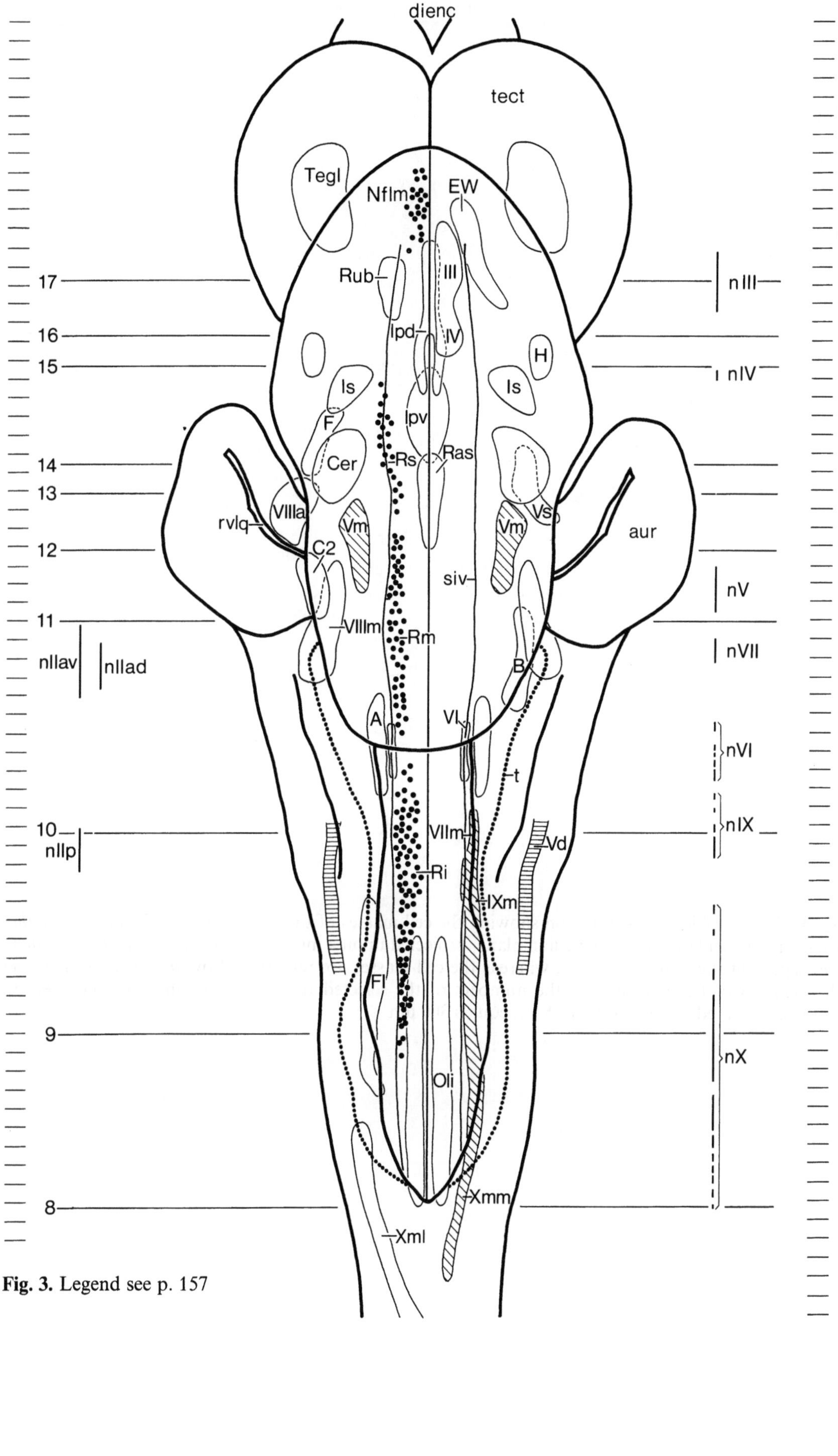

Fig. 3. Legend see p. 157

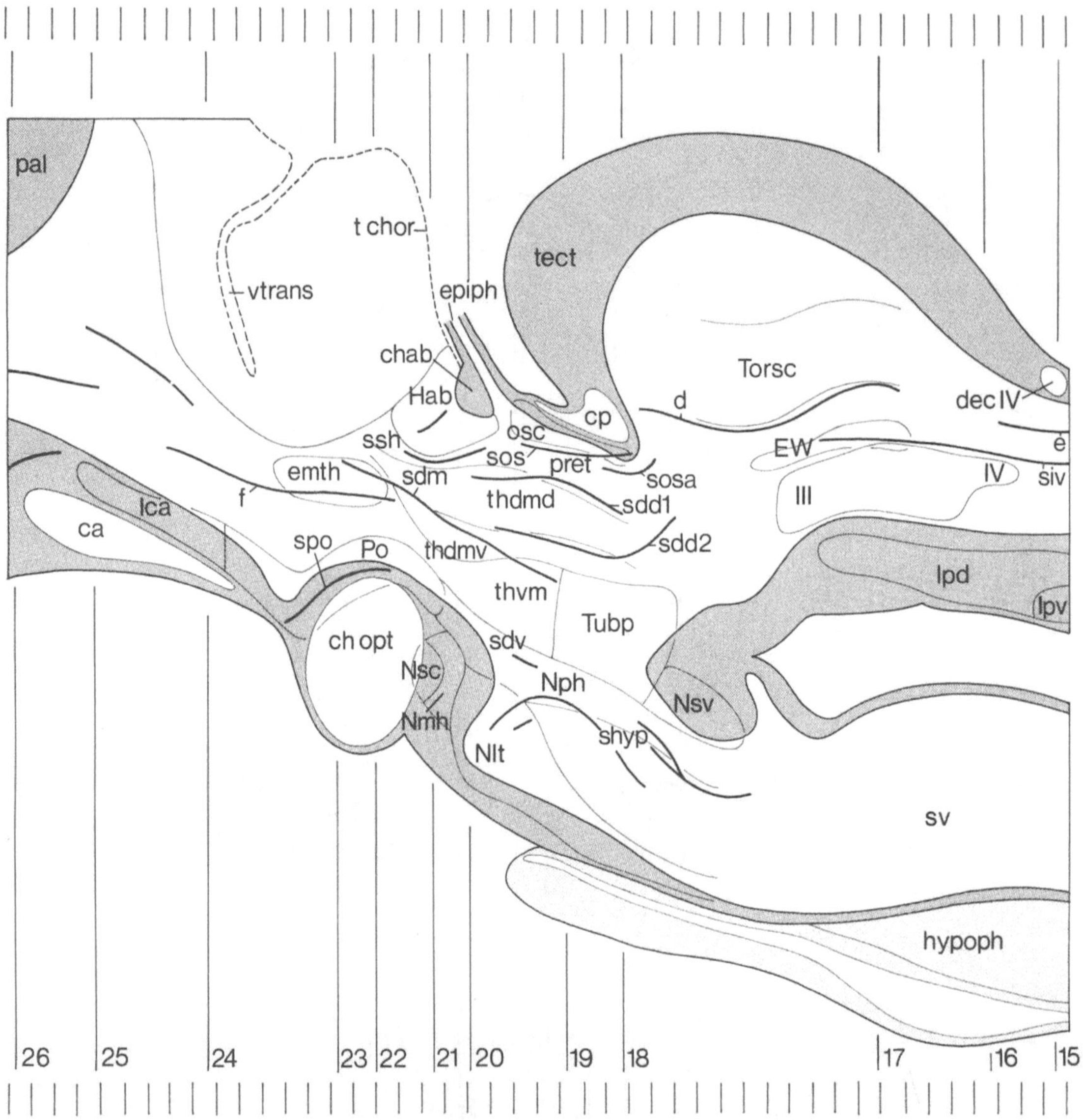

Fig. 4. Topographical reconstruction showing the cell masses of the diencephalon of *Squalus acanthias* lying periventricularly or in the median plane, projected upon a midsagittal plane. The *heavy continuous lines* represent the ventricular sulci, whereas the cell masses are represented by *thin, continuous curves.* The *numbers at the base* refer to the numbers of the selected transverse sections. Distance between sections, indicated by *small vertical bars,* equals 300 μm

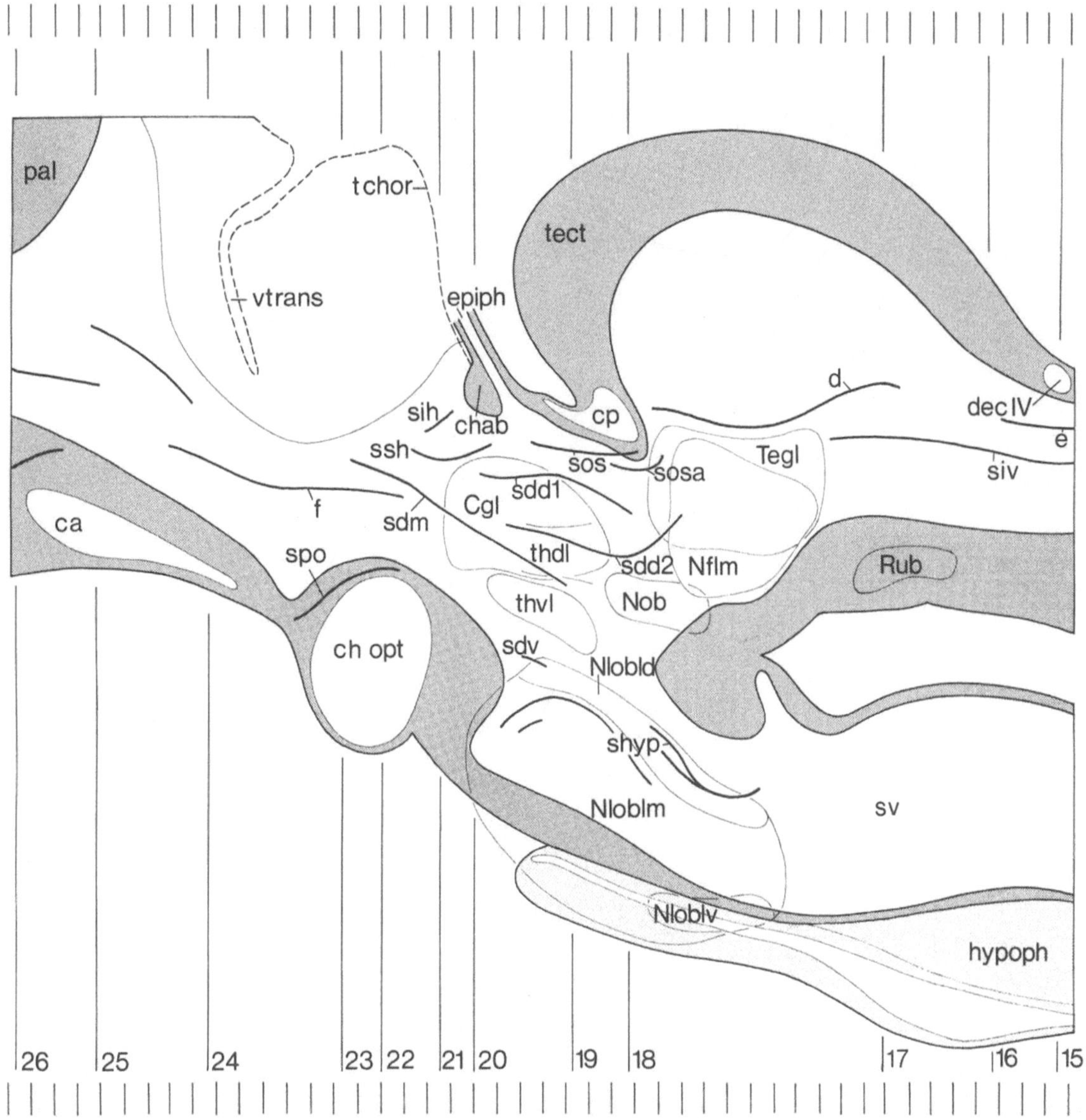

Fig. 5. Topographical reconstruction showing the migrated cell masses of the diencephalon of *Squalus acanthias* as projected upon a midsagittal plane. For further explanation see Fig. 4

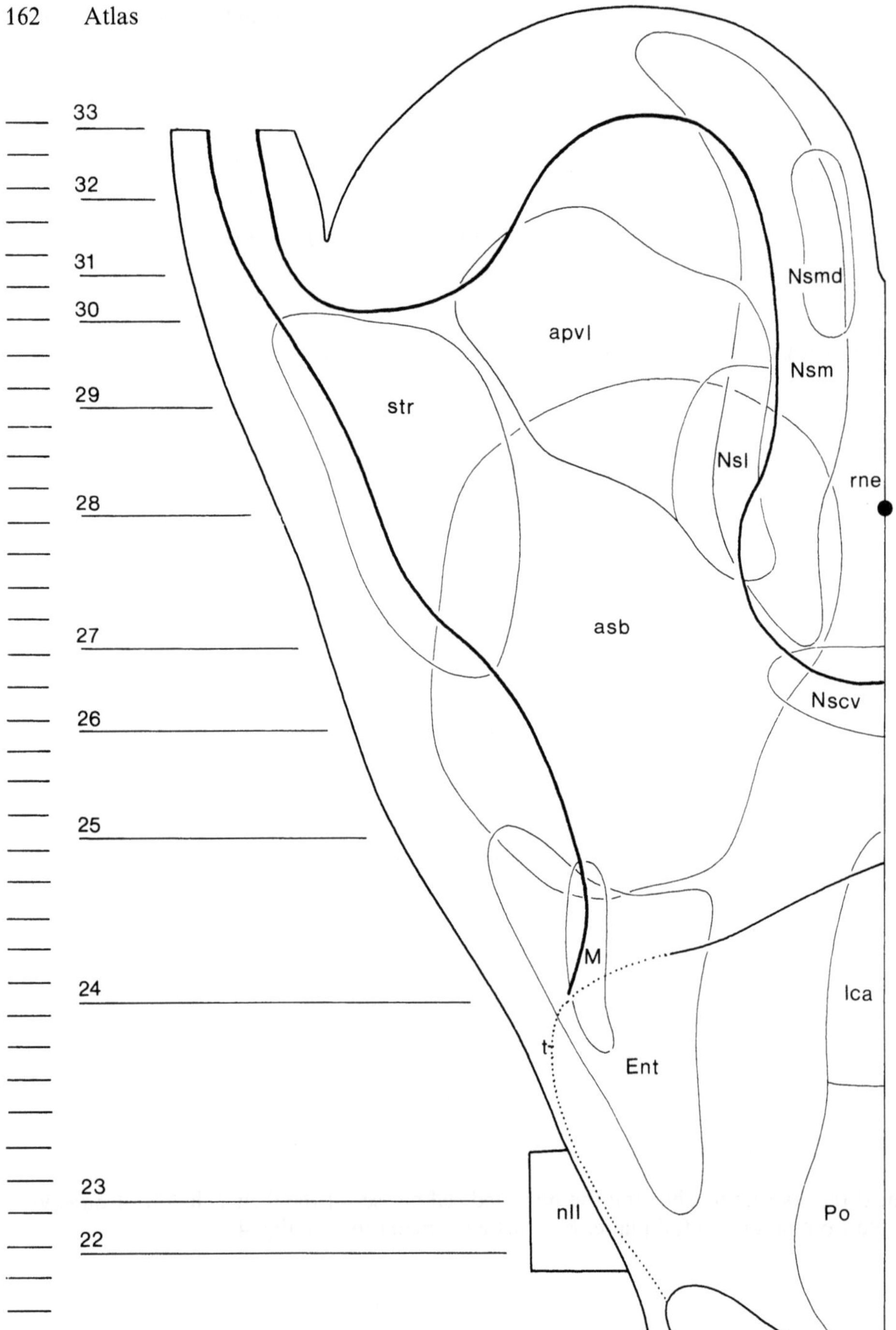

Fig. 6. Topographical reconstruction showing the cell masses of the subpallium and of the telencephalon impar of *Squalus acanthias* as projected upon a horizontal plane. *Heavy continuous lines* represent the outlines of the ventricle, *medium continuous curves* represent the contours of the external telencephalic features and the *dotted line* indicates the taeniae. The *remaining thin lines* represent the cell masses. *Numbers at the side* of the drawing refer to the numbers of the selected transverse sections. Distance between sections, indicated by *small horizontal bars*, equals 300 μm

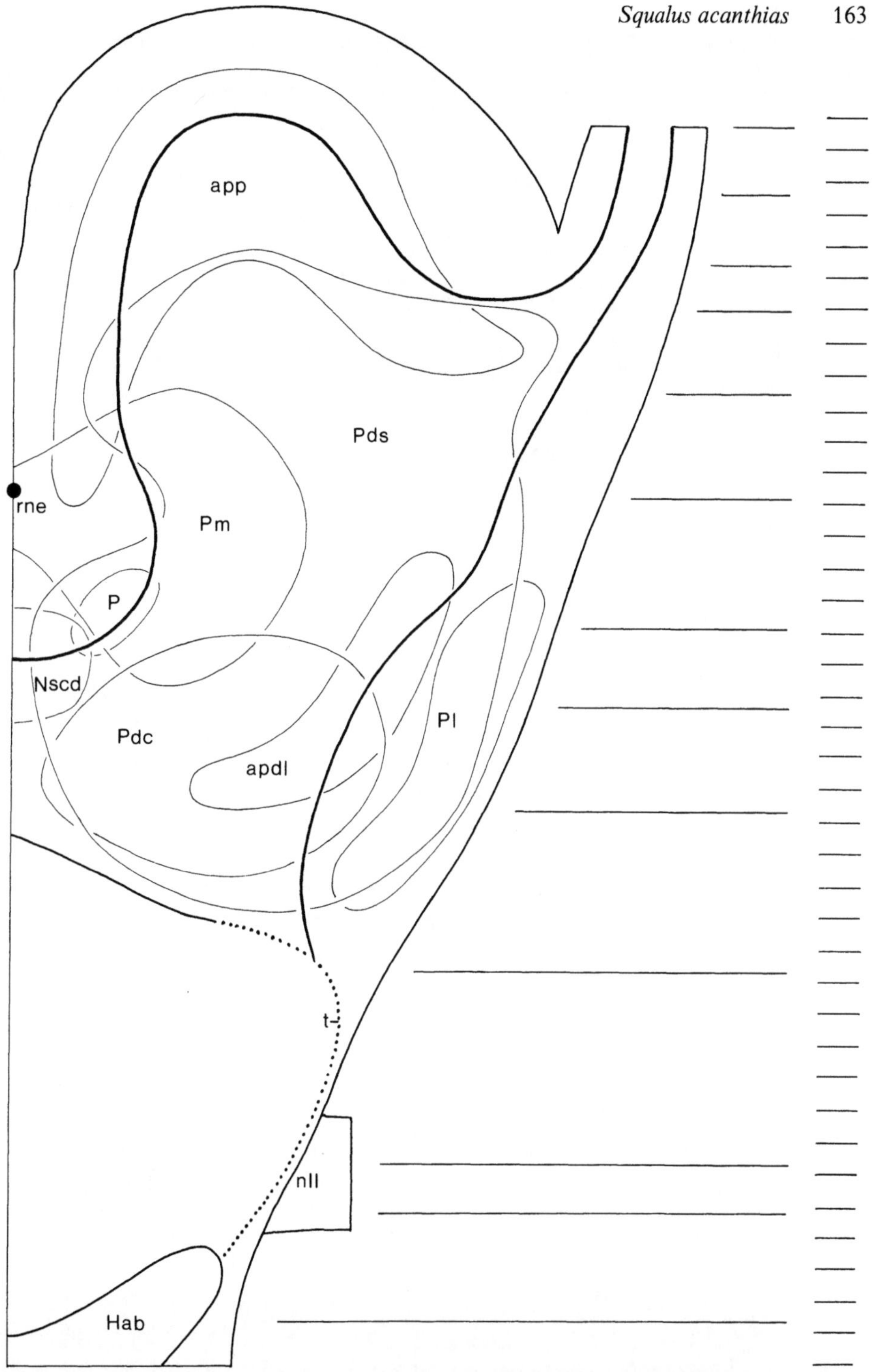

Fig. 7. Topographical reconstruction showing the pallial cell masses of the telencephalon of *Squalus acanthias* as projected upon a horizontal plane. For further explanation see Fig. 6

Figs. 8–33. Photomicrographs of transverse sections of the brain of the shark *Squalus acanthias*. The *left half* of each figure is a photomicrograph of a section stained according to Klüver and Barrera, whereas the *right half* is a photomicrograph taken from the adjacent Nissl stained section. *Bar scale* = 1 mm

8

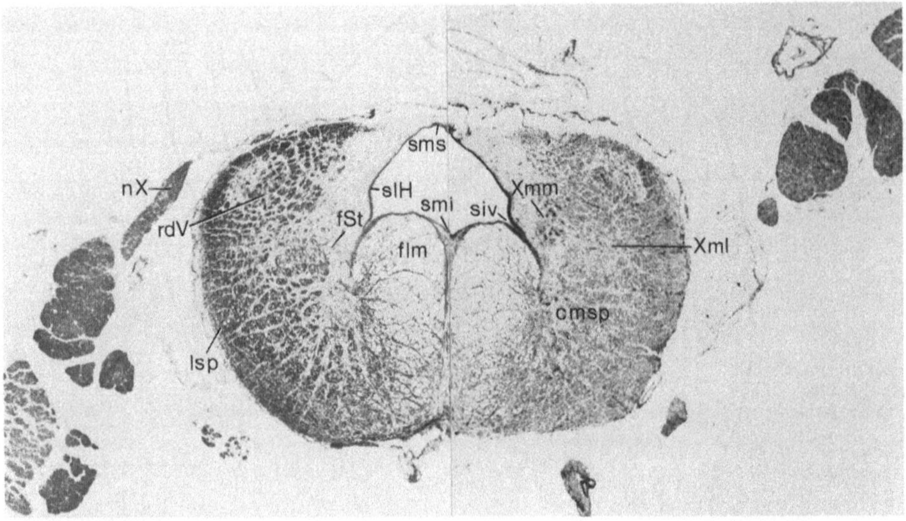
sms
slH
smi
siv
Xmm
nX
rdV
fSt
flm
Xml
cmsp
lsp

9

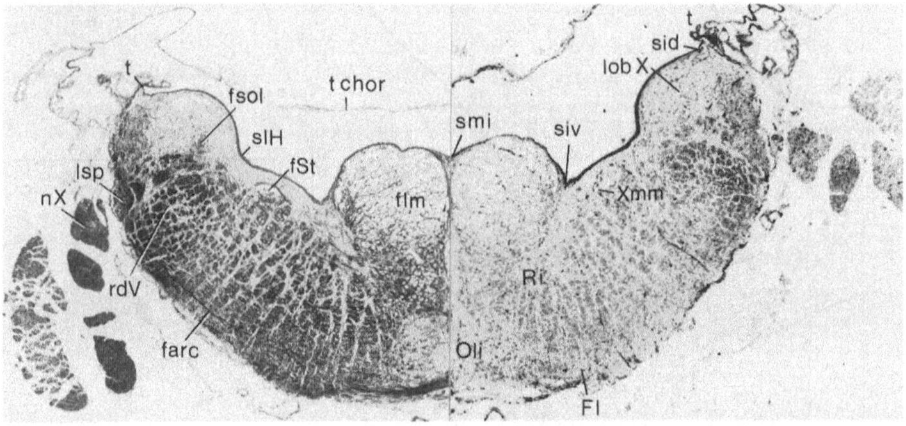
t
sid
lob X
t chor
fsol
slH
smi
siv
fSt
lsp
nX
flm
Xmm
Ri
rdV
farc
Oli
Fl

10

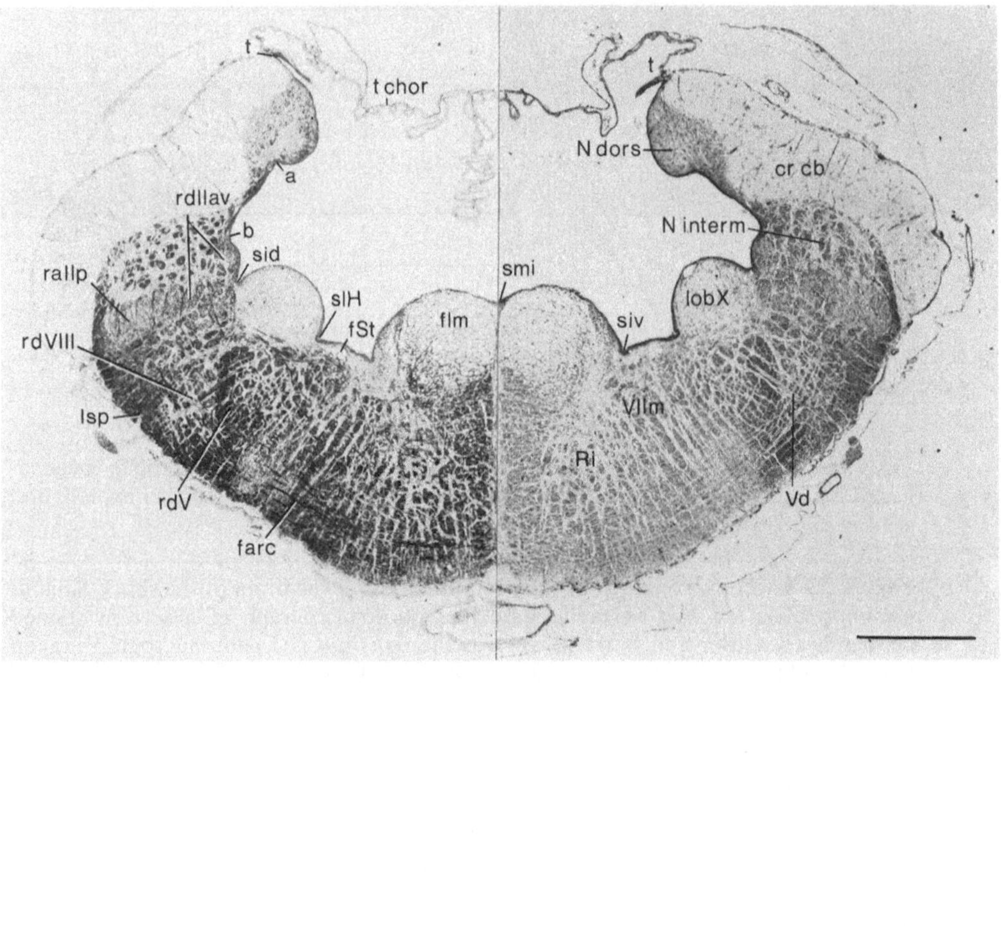
t
t chor
t
N dors
cr cb
a
rdllav
b
N interm
sid
rallp
smi
slH
lobX
fSt
flm
siv
rdVIII
VIIm
lsp
Ri
rdV
Vd
farc

11

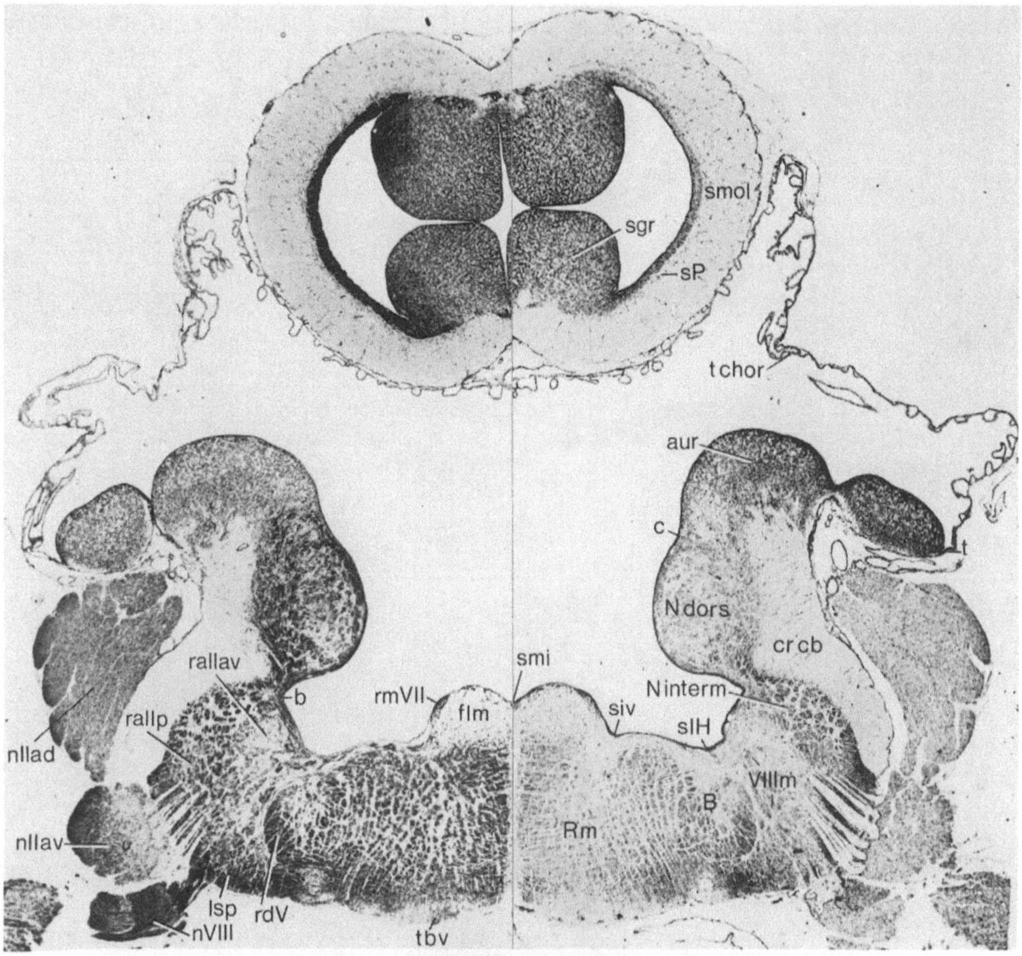
smol
sgr
sP
tchor
aur
c
t
Ndors
crcb
smi
rallav
b
rmVII
flm
Ninterm
siv
slH
rallp
nllad
VIIIm
B
Rm
nllav
lsp
rdV
nVIII
tbv

12

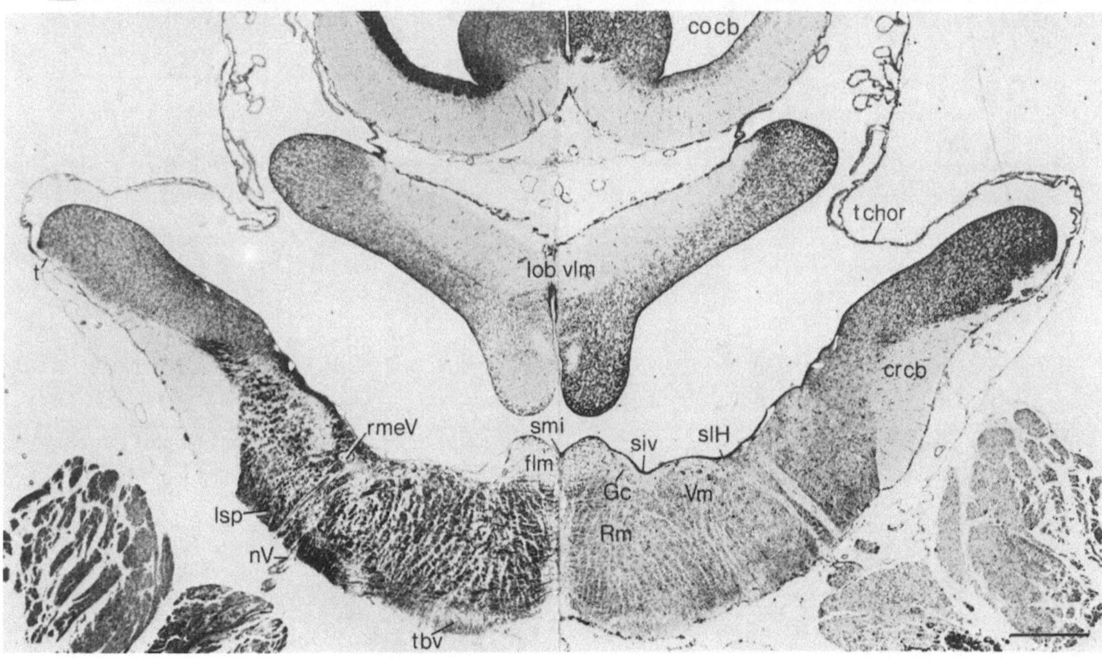
cocb
tchor
t
lob vlm
crcb
rmeV
smi
siv
slH
flm
Gc
Vm
Rm
lsp
nV
tbv

13

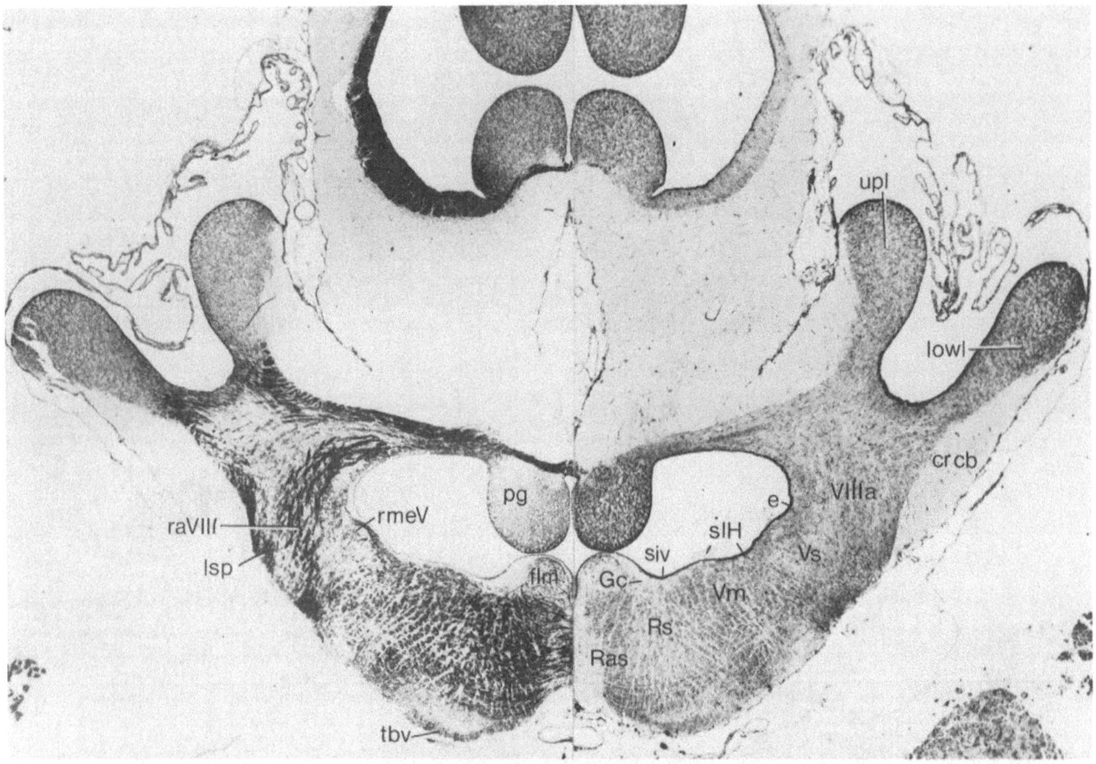
upl
lowl
crcb
VIIIa
e
Vs
sIH
siv
Gc
Vm
Rs
Ras
pg
rmeV
raVIII
lsp
flm
tbv

14

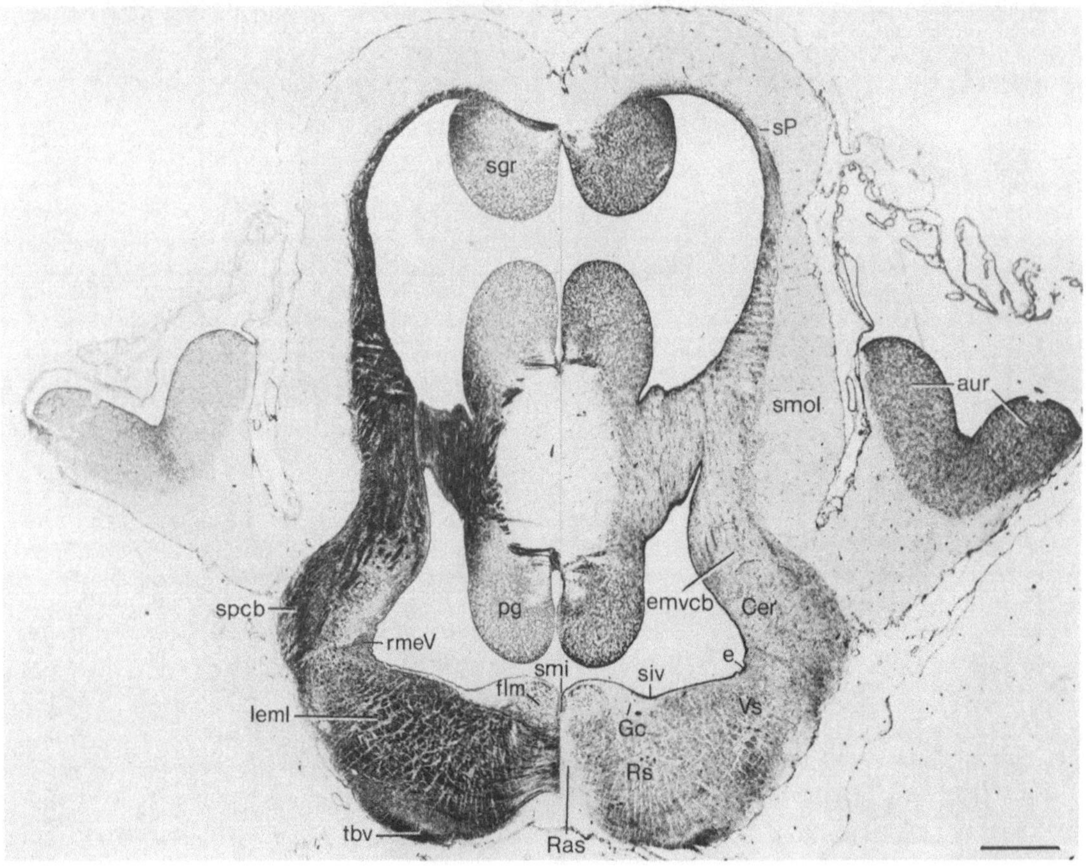
sP
sgr
aur
smol
emvcb
Cer
pg
spcb
rmeV
e
smi
siv
flm
Vs
leml
Gc
Rs
tbv
Ras

15

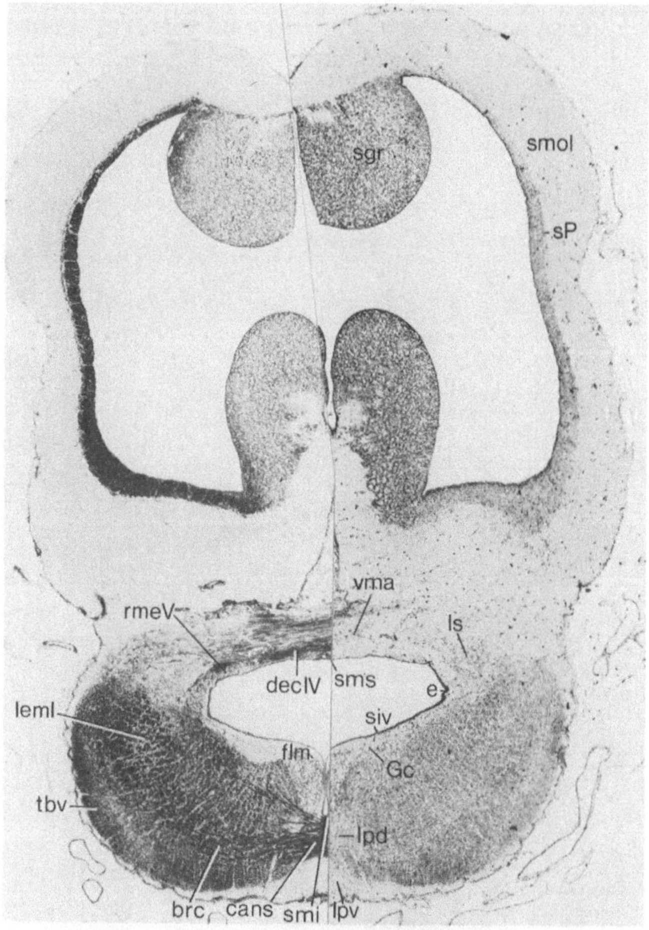
sgr
smol
sP
vma
rmeV
ls
declV
sms
e
leml
siv
flm
Gc
tbv
lpd
brc
cans
smi
lpv

16

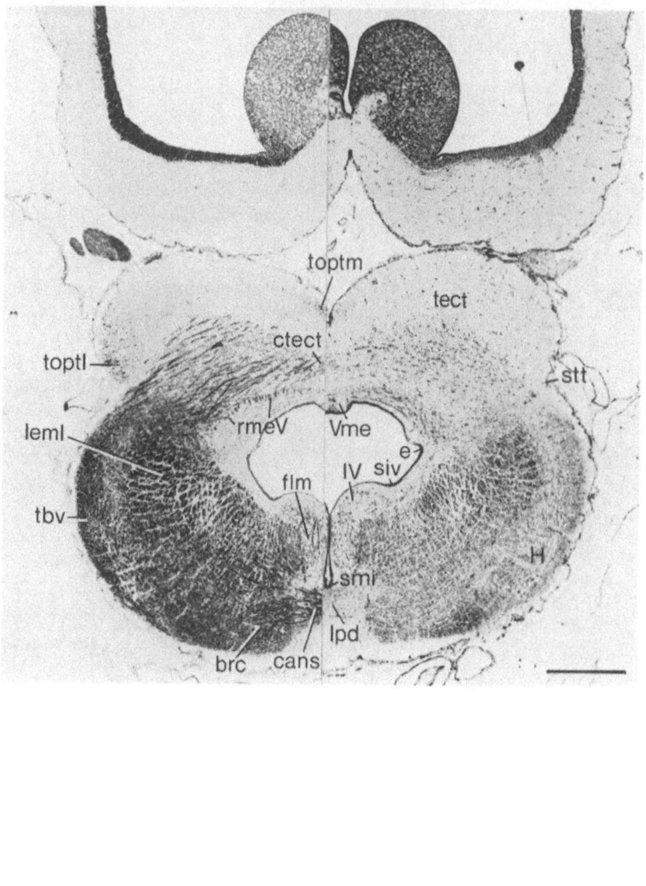
toptm
tect
ctect
toptl
stt
rmeV
Vme
leml
e
IV
siv
flm
tbv
H
smi
lpd
brc
cans

17

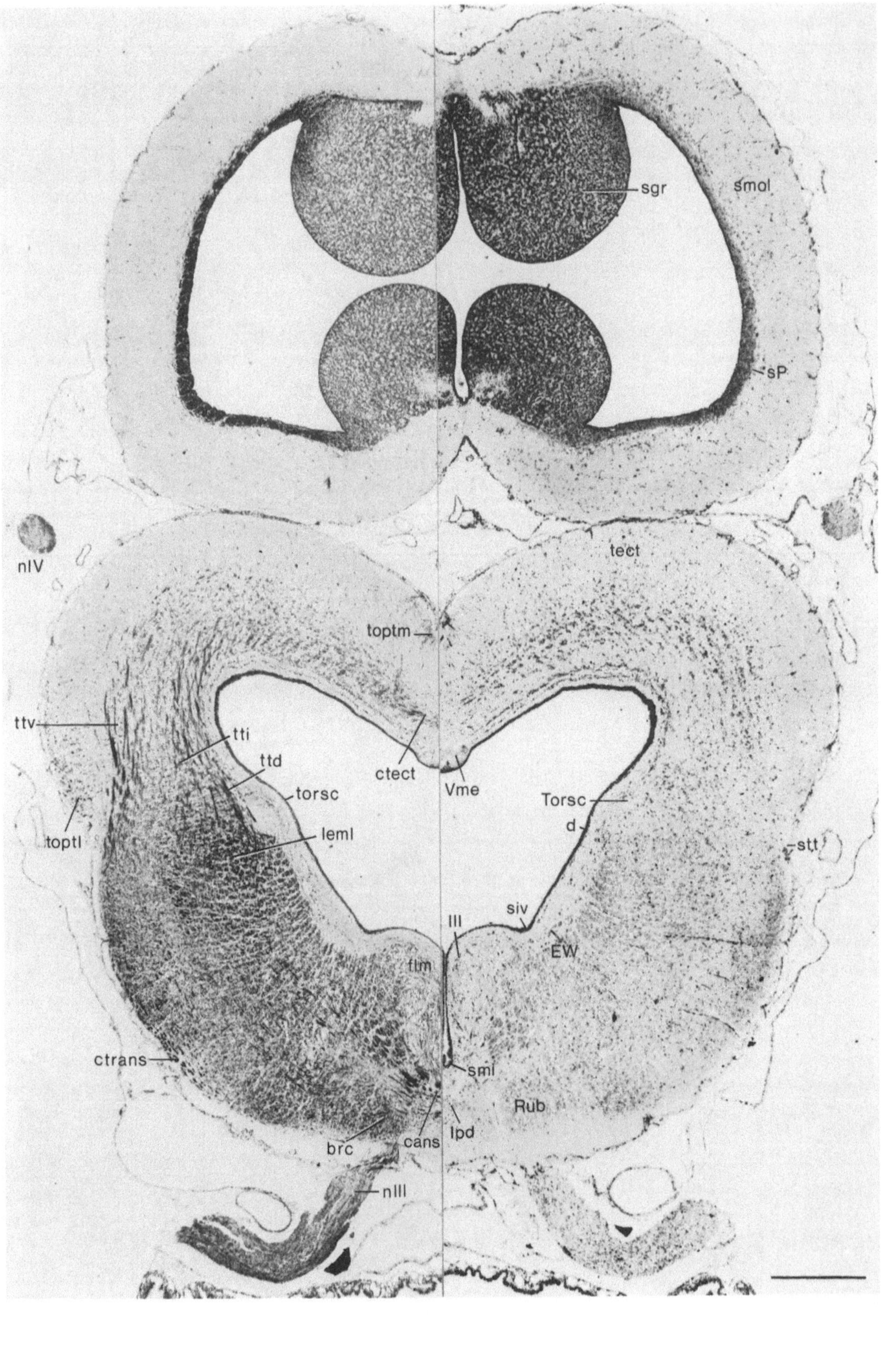
sgr
smol
sP
nIV
tect
toptm
ttv
tti
ttd
ctect
Vme
torsc
Torsc
toptl
leml
d
stt
III
siv
EW
flm
ctrans
sml
Rub
cans
lpd
brc
nIII

18

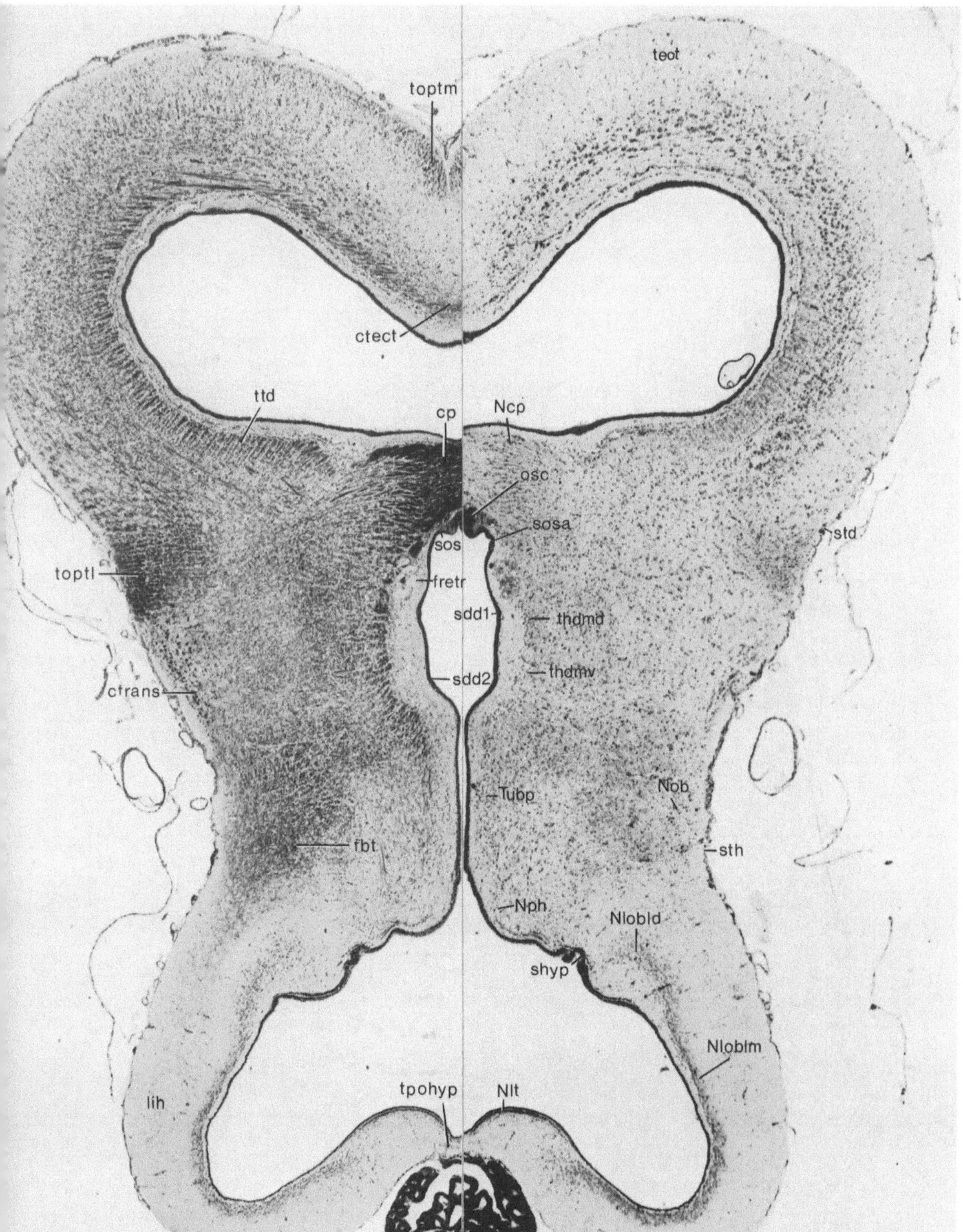
teot
toptm
ctect
ttd
cp
Ncp
osc
sosa
std
sos
toptl
fretr
sdd1
thdmd
sdd2
thdmv
ctrans
Tubp
Nob
fbt
sth
Nph
Nlobld
shyp
Nloblm
tpohyp
Nlt
lih
hypoph

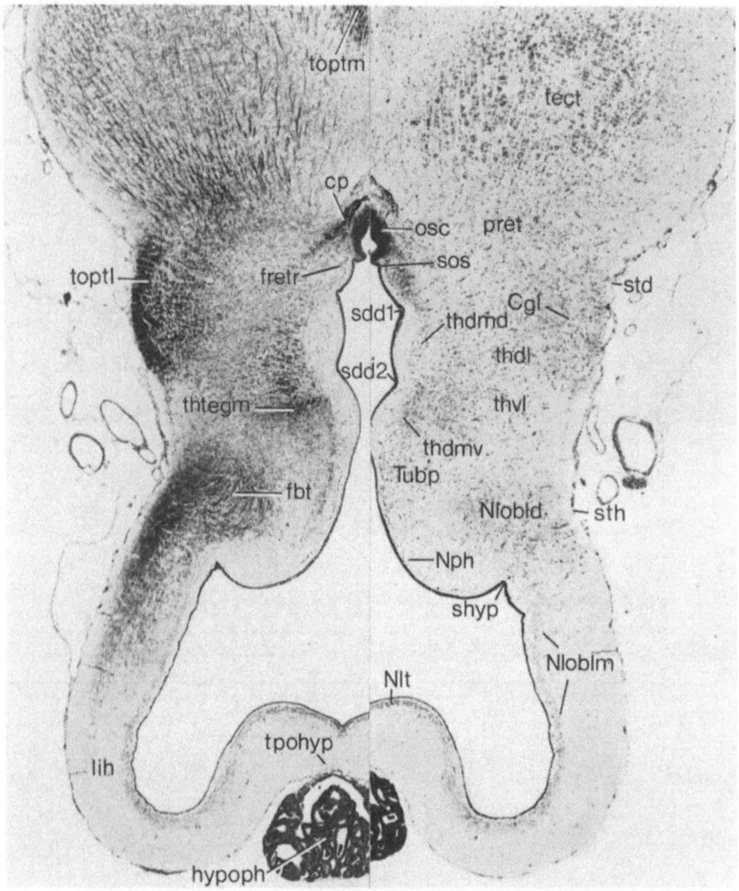
toptm
tect
cp
osc
pret
sos
toptl
fretr
std
Cgl
sdd1
thdmd
thdl
sdd2
thvl
thtegm
thdmv
Tubp
fbt
Nlobld
sth
Nph
shyp
Nloblm
Nlt
tpohyp
lih
hypoph

19

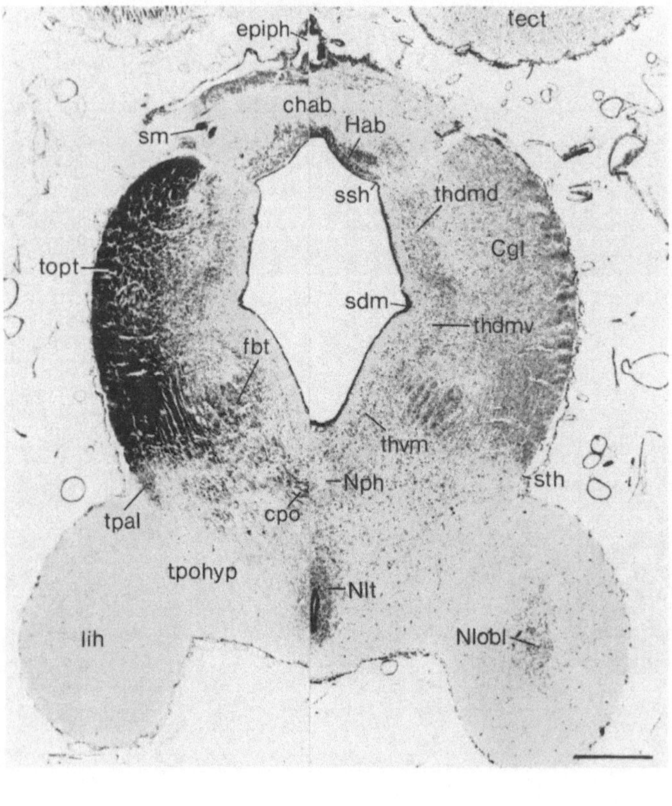
epiph
tect
chab
sm
Hab
ssh
thdmd
Cgl
topt
sdm
thdmv
fbt
thvm
Nph
sth
tpal
cpo
tpohyp
Nlt
lih
Nlobl

20

21

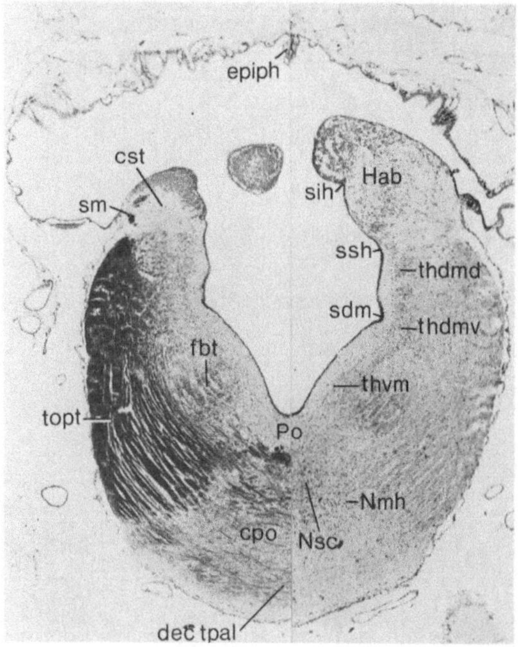
epiph
cst
sm
sih
Hab
ssh
thdmd
sdm
thdmv
fbt
thvm
topt
Po
Nmh
cpo
Nsc
dec tpal

22

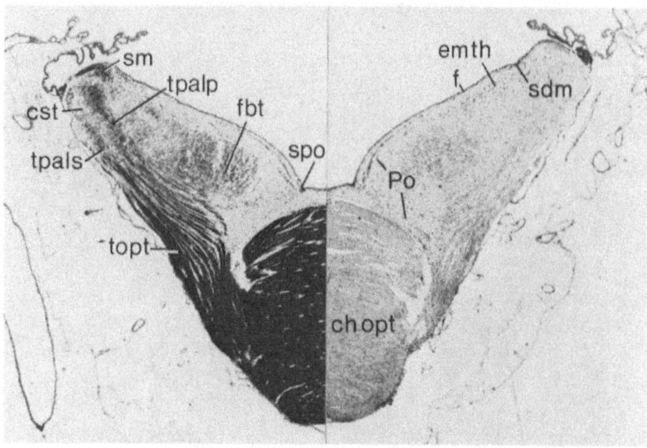
sm
emth
f
sdm
tpalp
cst
fbt
spo
tpals
Po
topt
ch opt

23

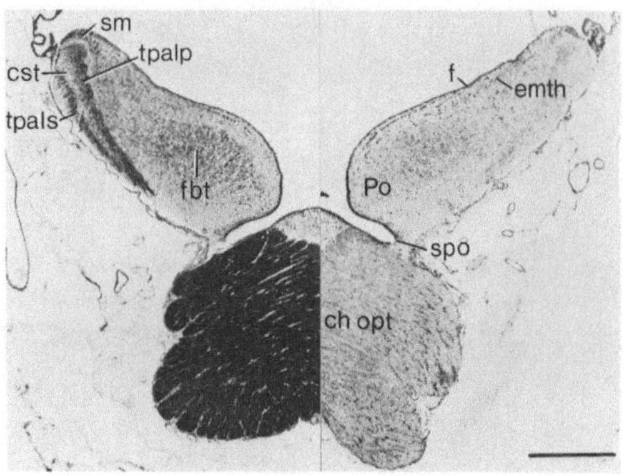
sm
tpalp
cst
f
emth
tpals
fbt
Po
spo
ch opt

24

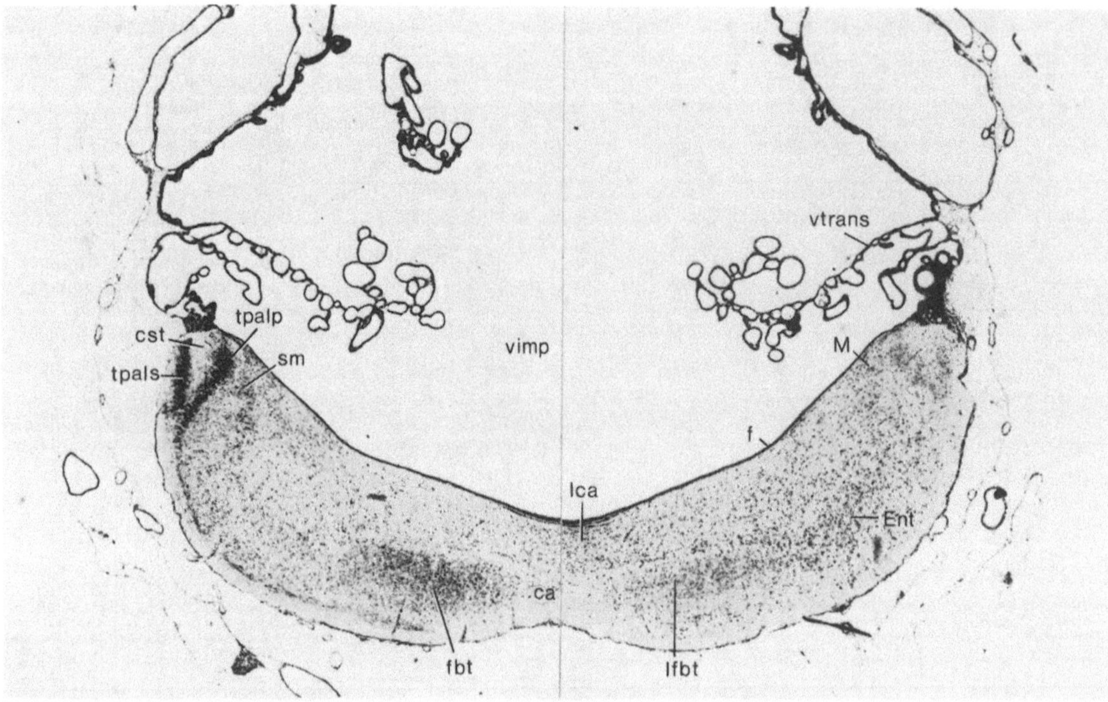
vtrans
cst
tpalp
tpals
sm
vimp
M
f
lca
Ent
ca
fbt
lfbt

25

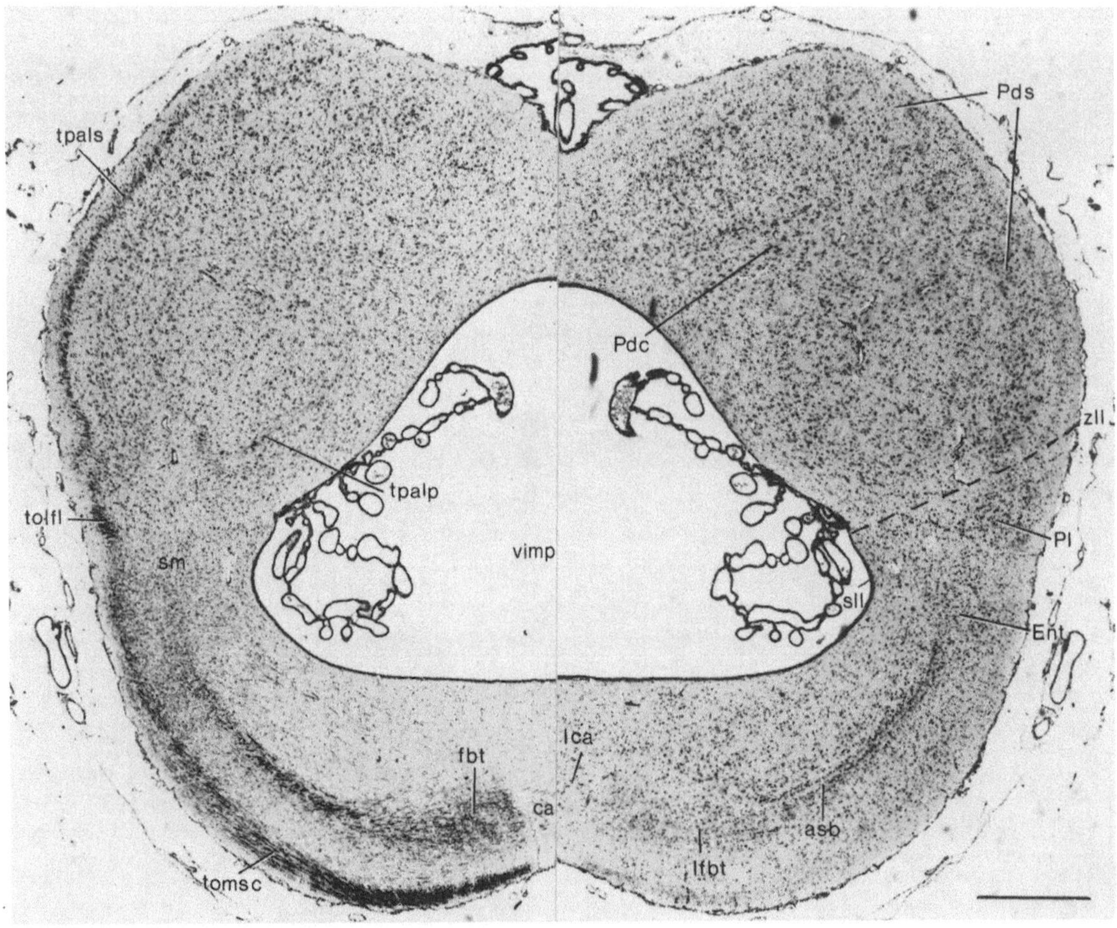
Pds
tpals
Pdc
zll
tpalp
tolfl
Pl
sm
vimp
sll
Ent
lca
fbt
ca
asb
lfbt
tomsc

26

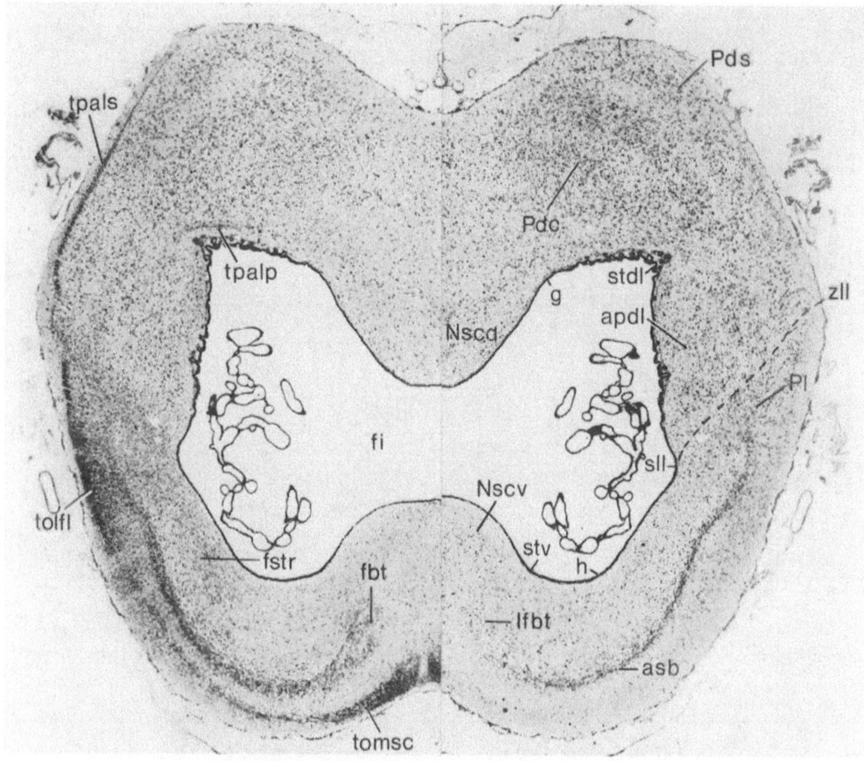
Pds
tpals
Pdc
tpalp
stdl
g
zll
apdl
Nscd
Pl
fi
sll
Nscv
tolfl
stv
h
fstr
fbt
lfbt
asb
tomsc

27

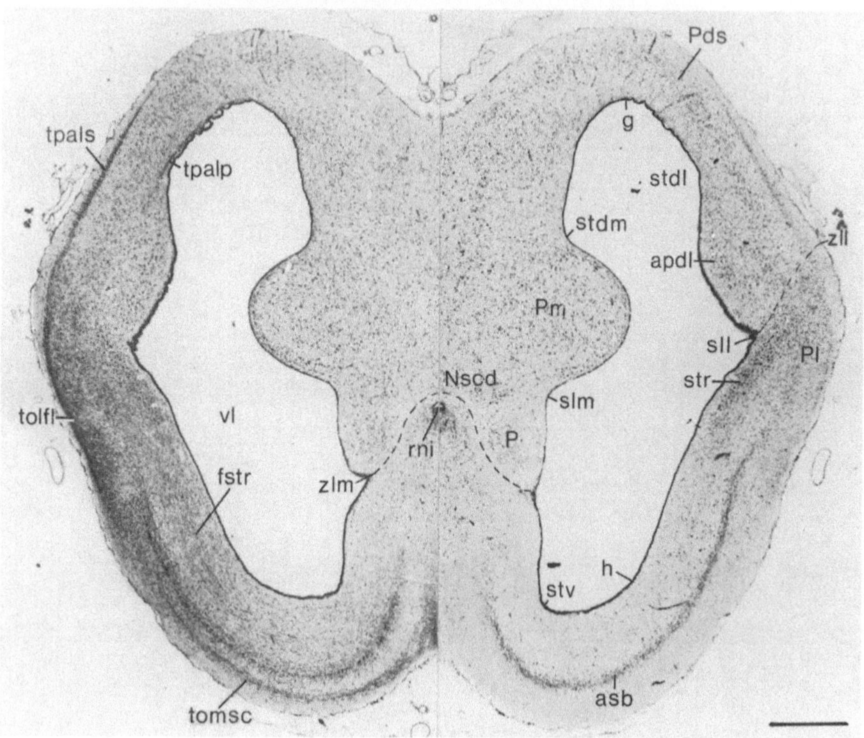
Pds
g
tpals
tpalp
stdl
stdm
zll
apdl
Pm
sll
Pl
Nscd
str
slm
tolfl
vl
P
rni
fstr
zlm
h
stv
asb
tomsc

28

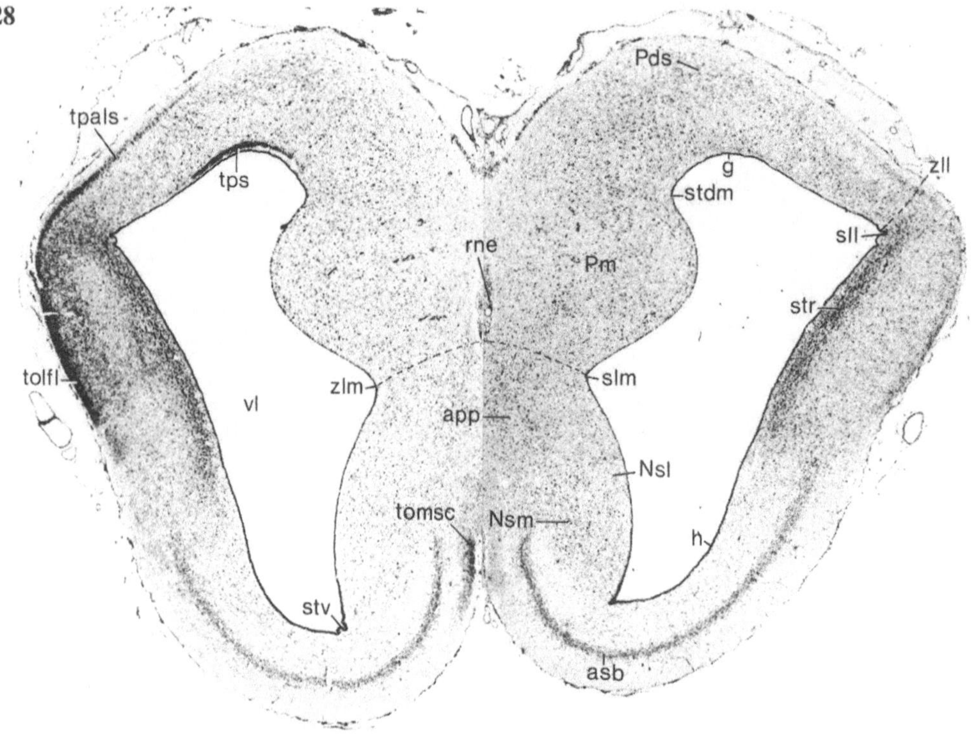
Pds
tpals
tps
g
zll
stdm
sll
rne
Pm
str
tolfl
zlm
slm
vl
app
Nsl
tomsc
Nsm
h
stv
asb

29

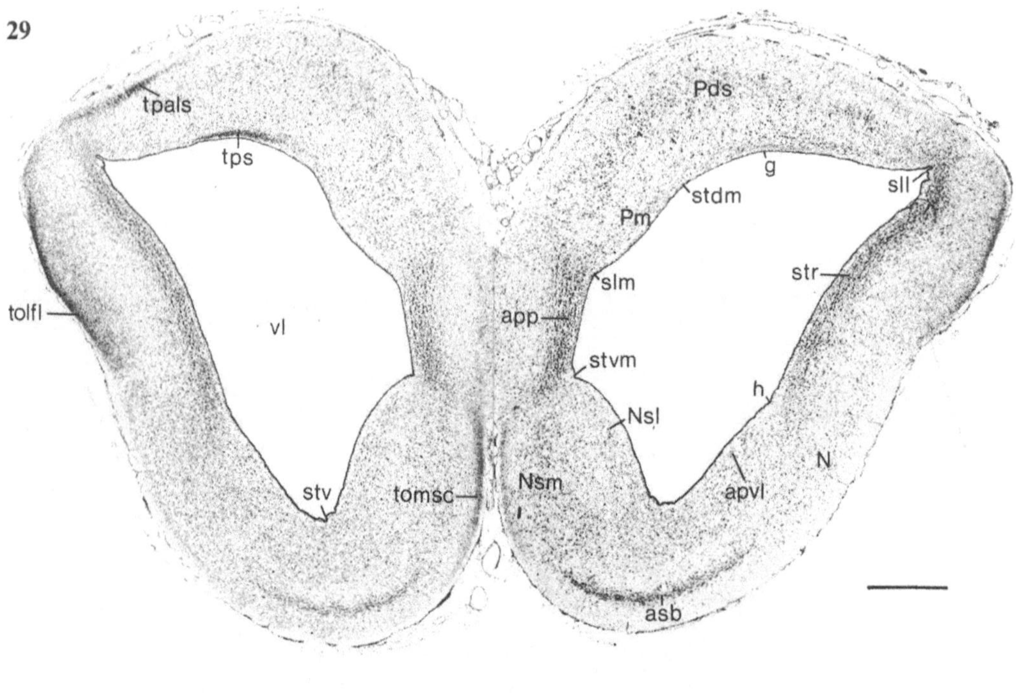
tpals
Pds
tps
g
sll
stdm
Pm
slm
str
tolfl
app
vl
stvm
h
Nsl
N
stv
tomsc
Nsm
apvl
asb

30

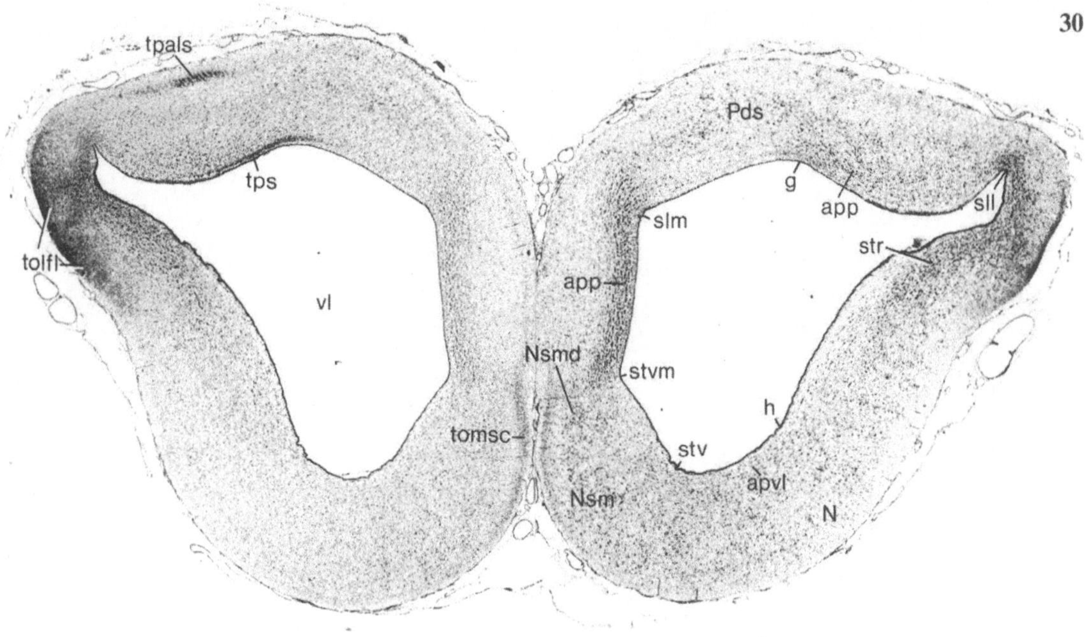
tpals
Pds
tps
g
app
sll
slm
str
tolfl
app
vl
Nsmd
stvm
h
tomsc
stv
apvl
Nsm
N

31

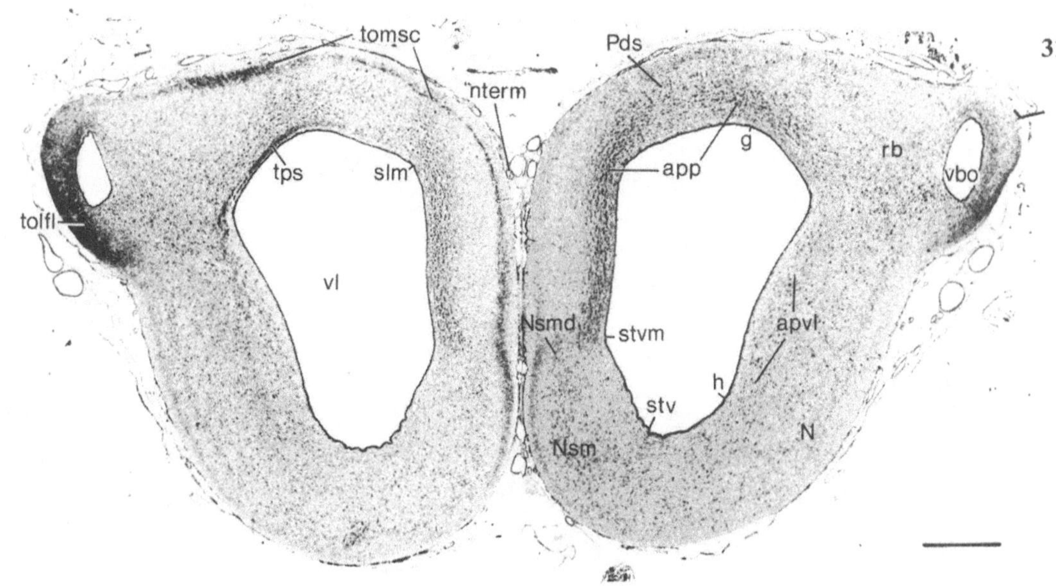
tomsc
Pds
nterm
g
rb
tps
slm
app
vbo
tolfl
vl
Nsmd
stvm
apvl
h
stv
N
Nsm

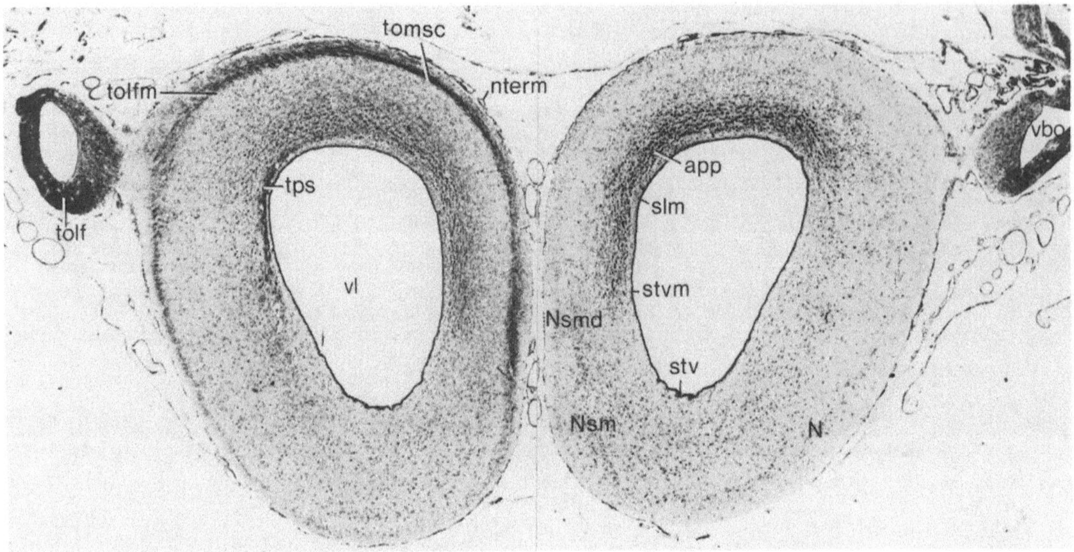
tomsc
tolfm
nterm
vbo
app
tps
slm
tolf
vl
stvm
Nsmd
stv
Nsm
N

32

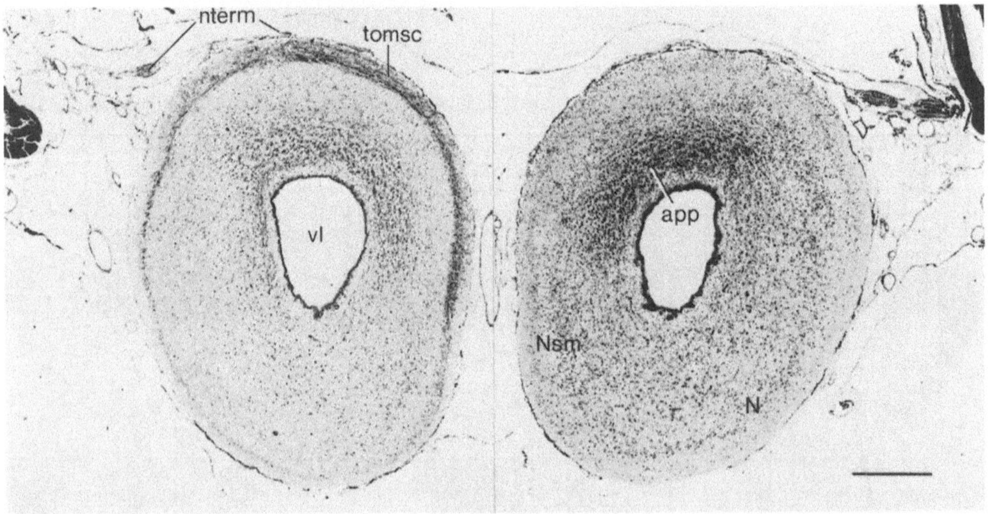
nterm
tomsc
app
vl
Nsm
N

33

Scyliorhinus canicula

(Fig. 34–65)

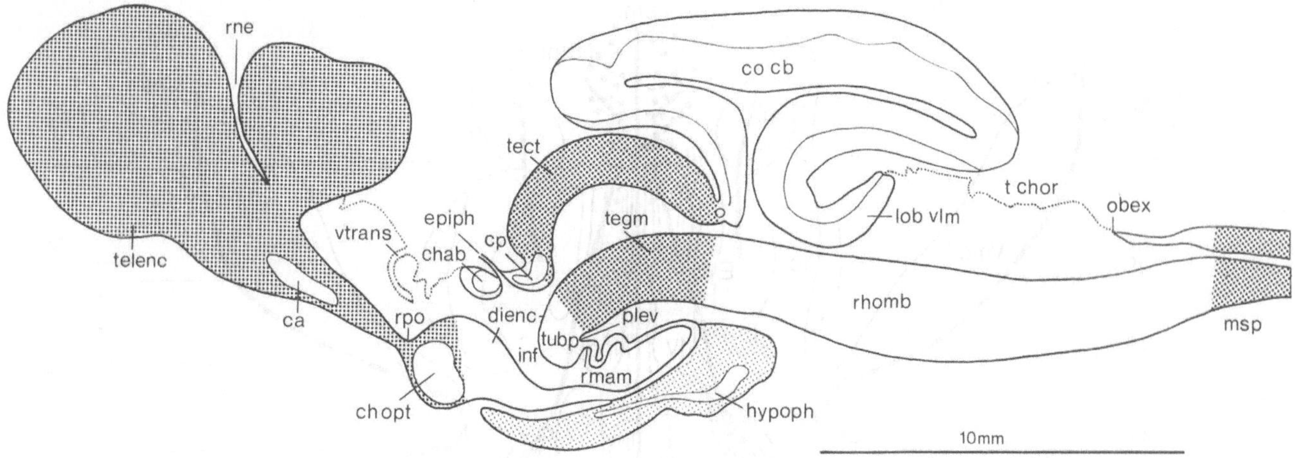

Fig. 34. Midsagittal section of the brain of *Scyliorhinus canicula*

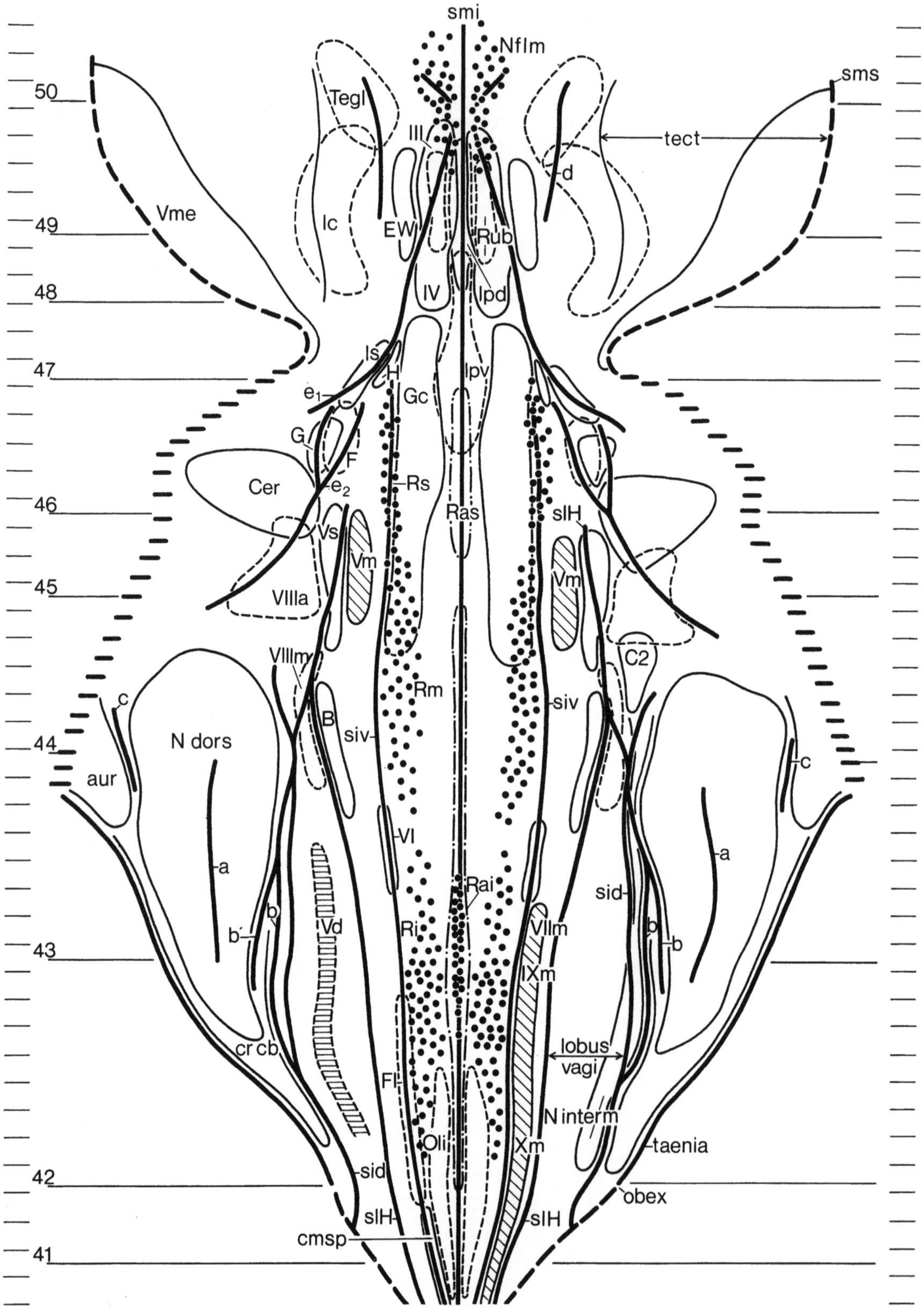

Fig. 35. Topological reconstruction of the brain stem of the shark *Scyliorhinus canicula*. Distance between sections, indicated by *small horizontal bars*, equals 300 μm. For further explanation see Fig. 2

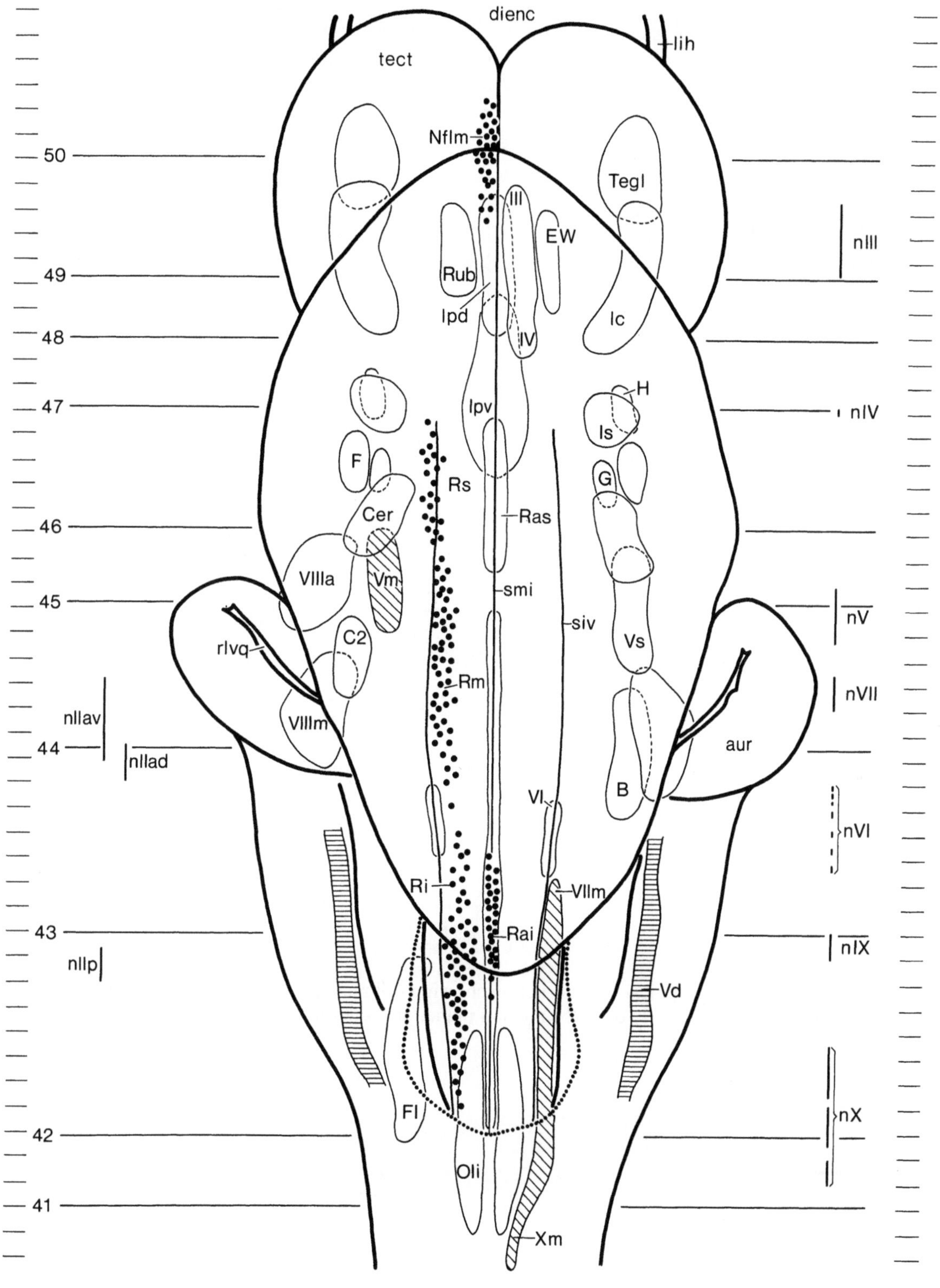

Fig. 36. Topographical reconstruction showing the cell masses of the brain stem of *Scyliorhinus canicula* as projected upon a horizontal plane. Distance between sections, indicated by *small horizontal bars*, equals 300 μm. For further explanation see Fig. 3

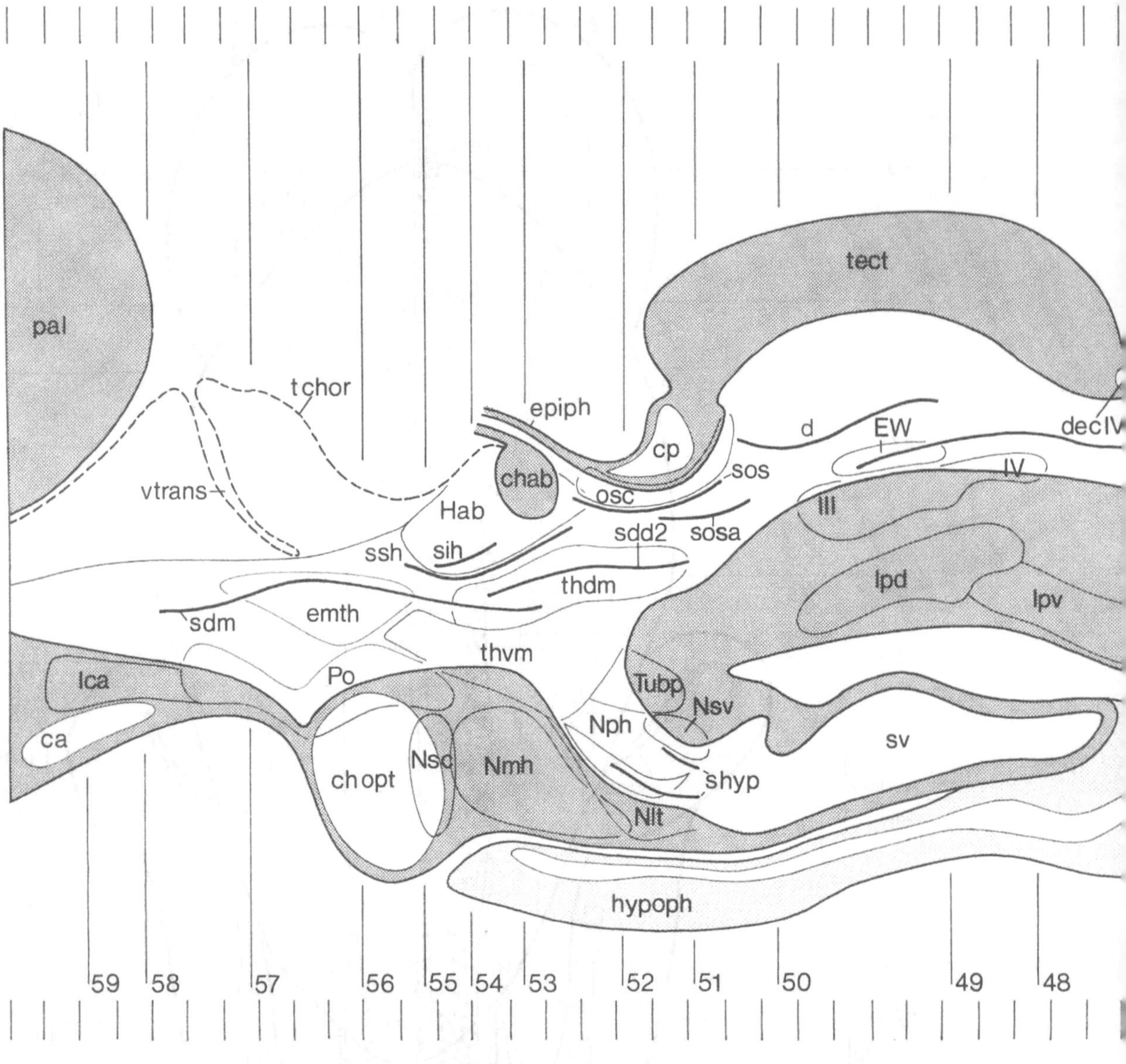

Fig. 37. Topographical reconstruction showing cell masses of the diencephalon of *Scyliorhinus canicula* lying periventricularly or in the median plane, projected upon a midsagittal plane. For further explanation see Fig. 4

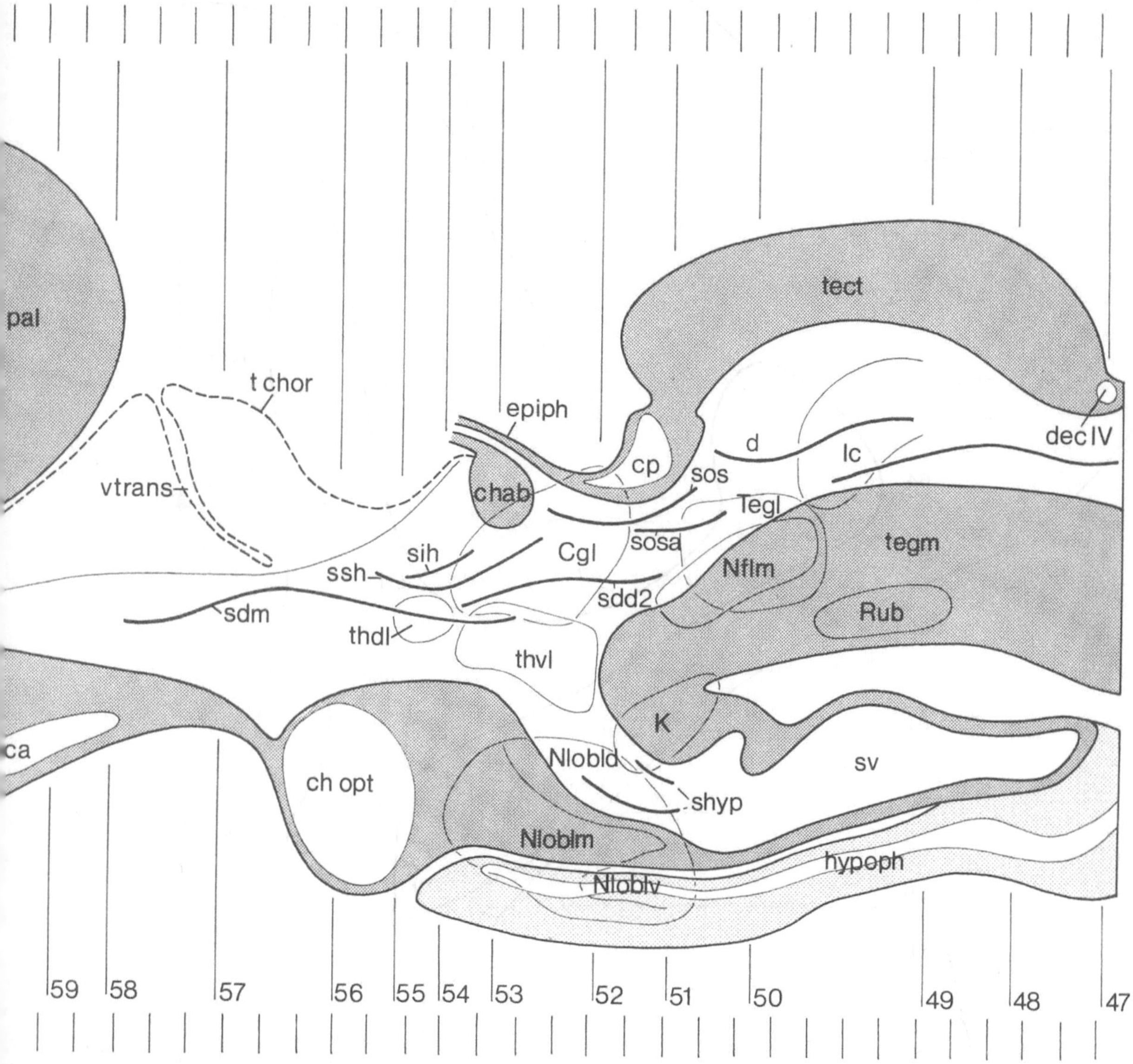

Fig. 38. Topographical reconstruction showing the migrated cell masses of the diencephalon in *Scyliorhinus canicula* as projected upon a midsagittal plane. For further explanation see Fig. 4

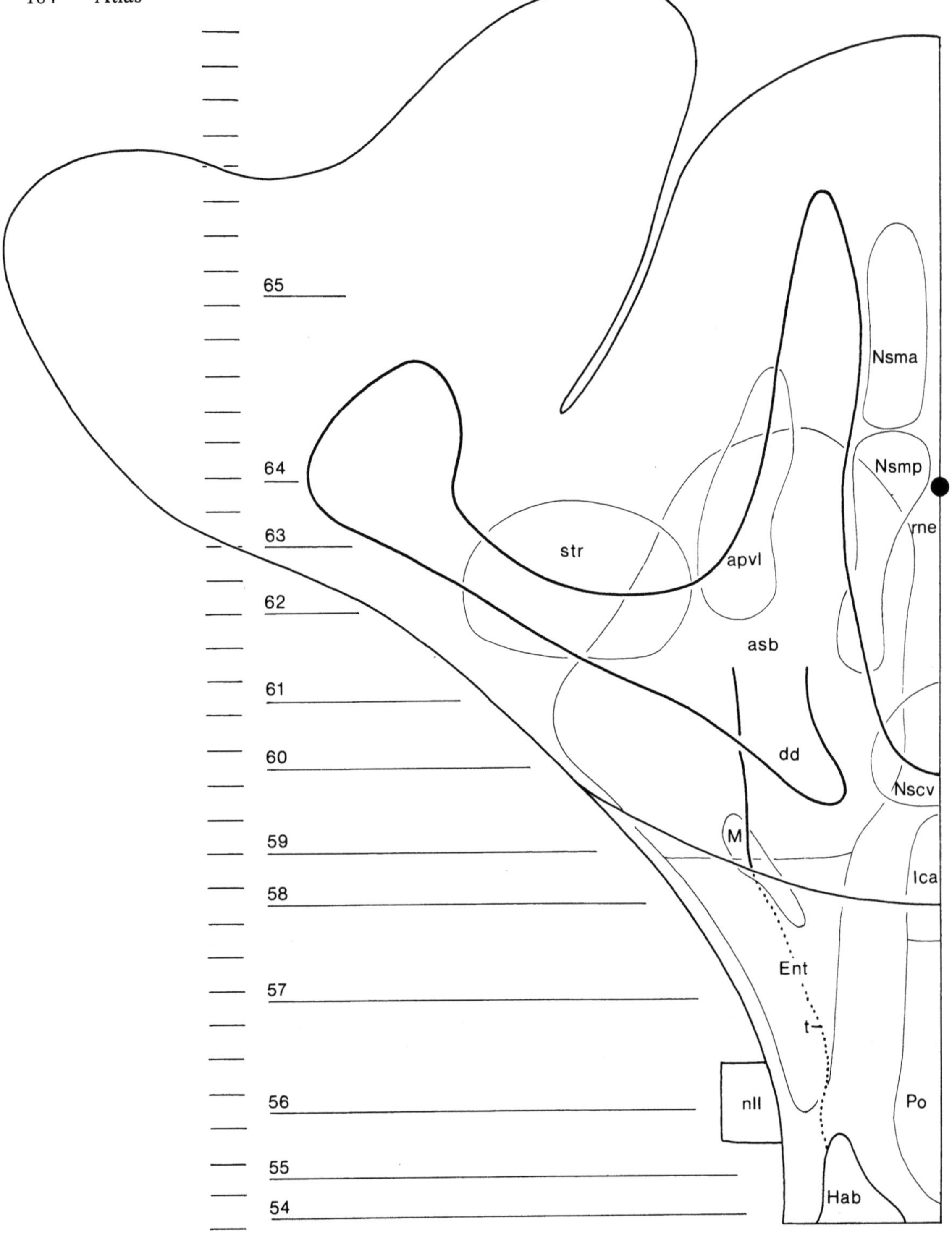

Fig. 39. Topographical reconstruction showing the cell masses of the subpallium and of the telencephalon impar of *Scyliorhinus canicula* as projected upon a horizontal plane. For further explanation see Fig. 6

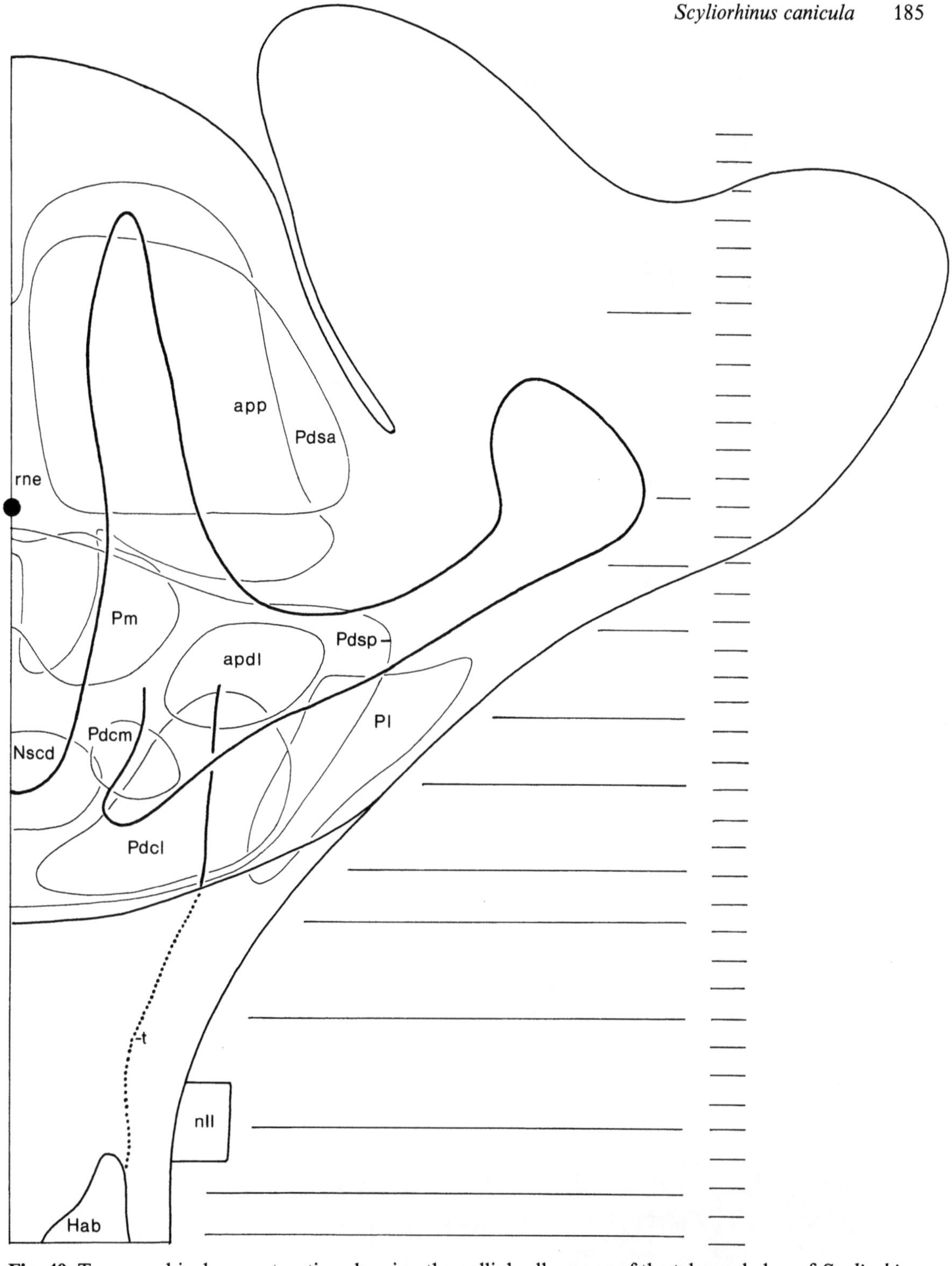

Fig. 40. Topographical reconstruction showing the pallial cell masses of the telencephalon of *Scyliorhinus canicula* as projected upon a horizontal plane. For further explanation see Fig. 6

Figs. 41–65. Photomicrographs of transverse sections of the brain of *Scyliorhinus canicula*. For further explanation see Figs. 8–33

41

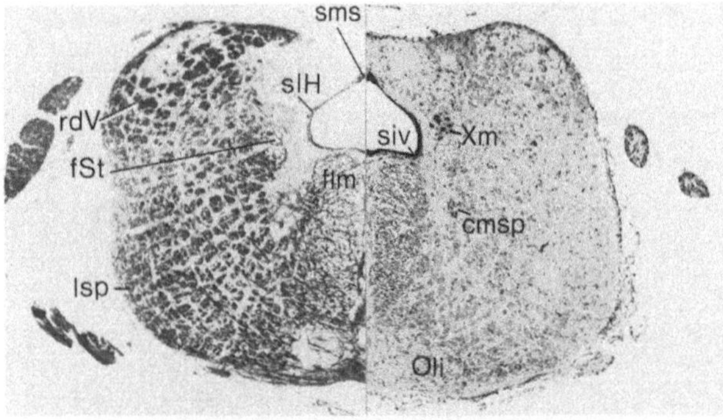
sms
slH
rdV
siv
Xm
fSt
flm
cmsp
lsp
Oli

42

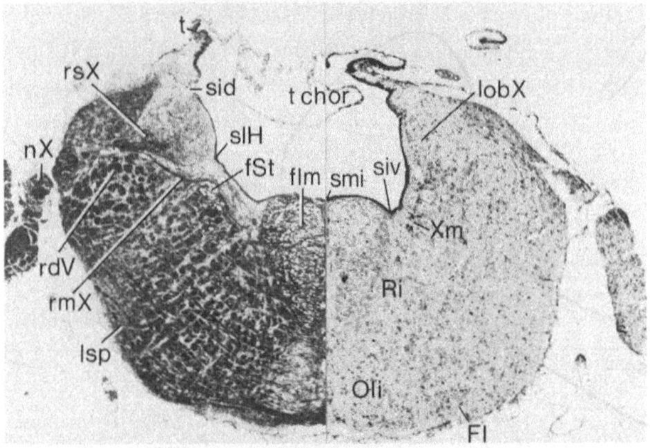
t
rsX
sid
t chor
lobX
slH
nX
fSt
flm
smi
siv
Xm
rdV
Ri
rmX
lsp
Oli
Fl

43

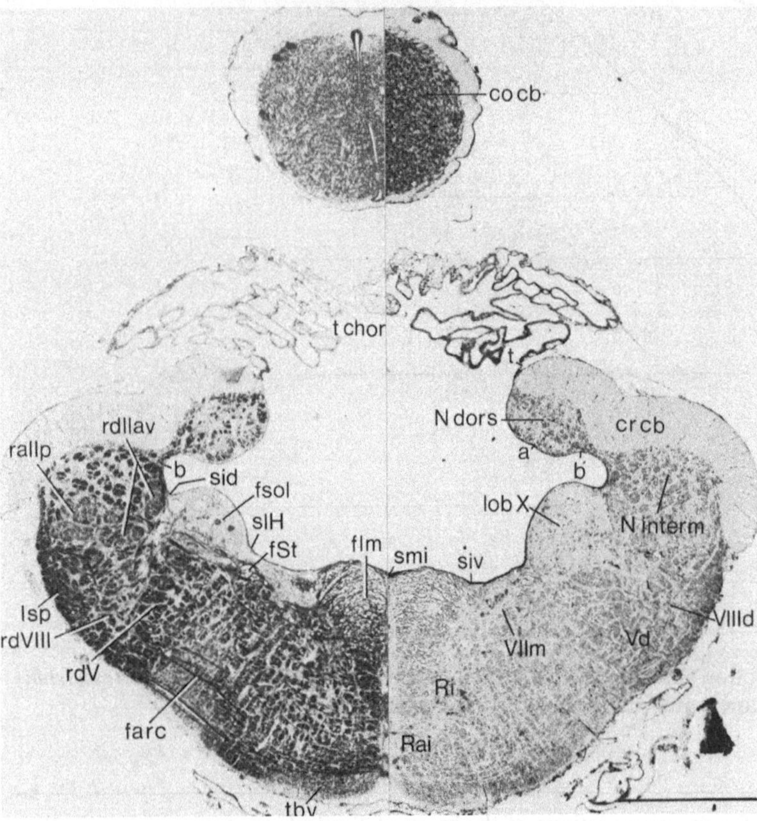
co cb
t chor
t
N dors
cr cb
rdllav
rallp
a
b
b
sid
fsol
lob X
slH
N interm
fSt
flm
smi
siv
lsp
VIIId
rdVIII
VIIm
Vd
rdV
Ri
farc
Rai
tbv

44

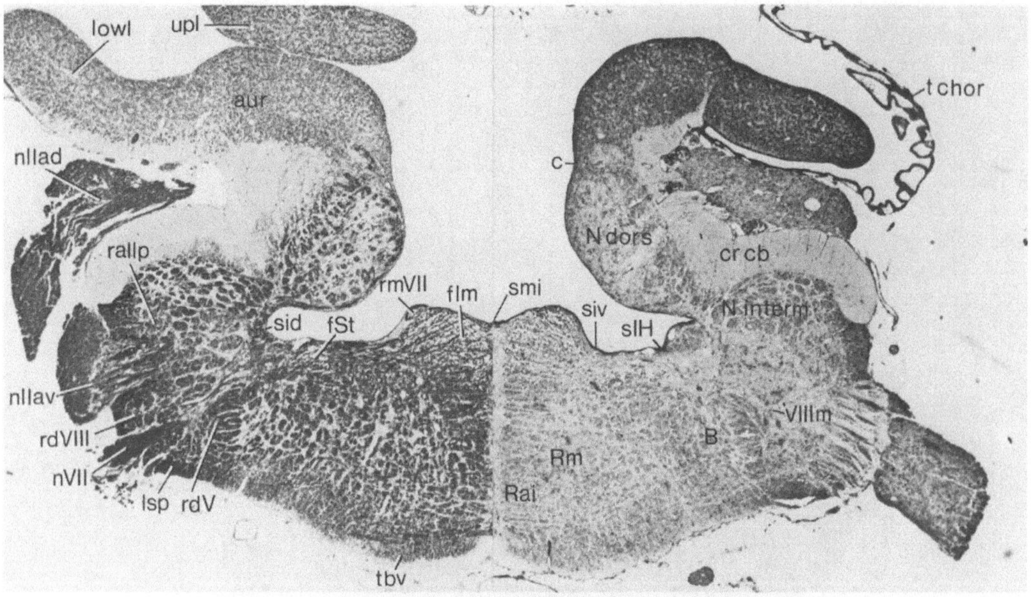
lowl
upl
aur
nIIad
rallp
rmVII
flm
smi
sid
fSt
siv
slH
nIIav
rdVIII
nVII
lsp
rdV
Rm
Rai
B
VIIIm
tbv
tchor
c
N dors
cr cb
N interm

45

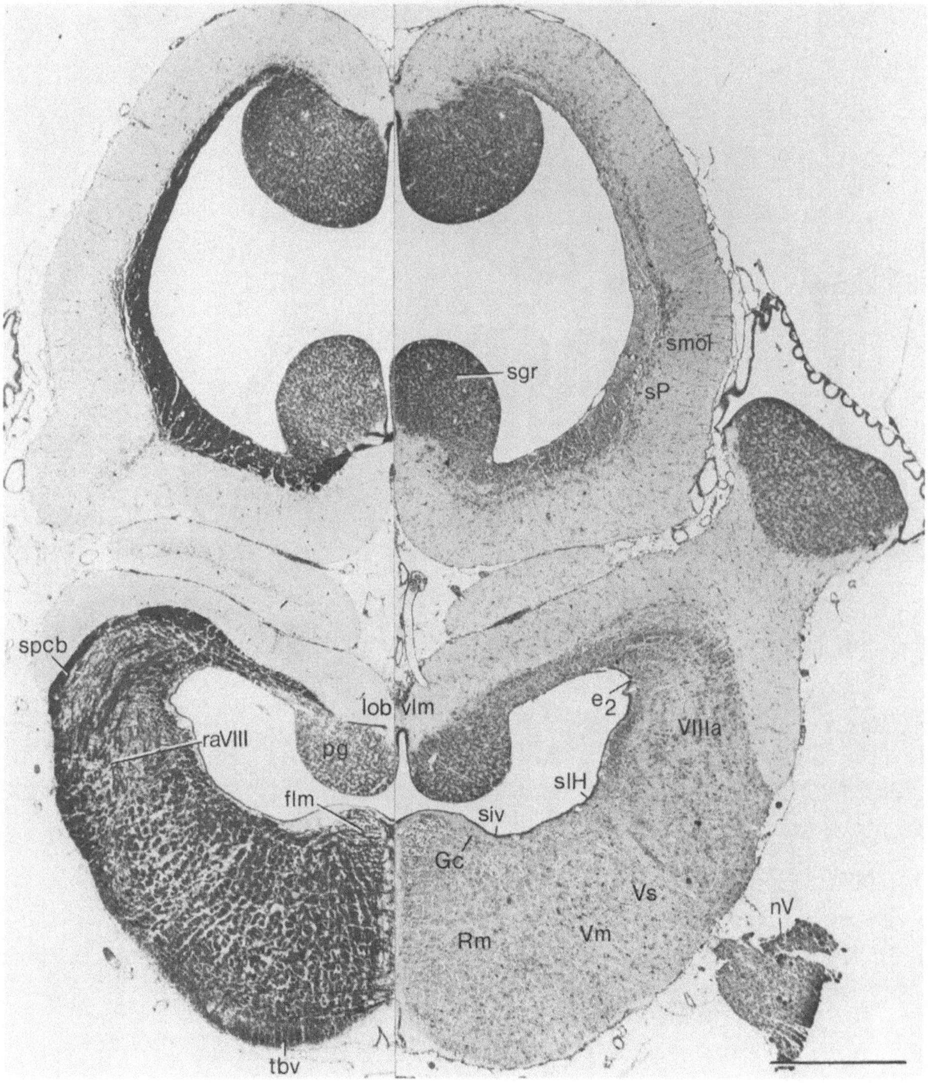
smol
sgr
sP
spcb
lob
vlm
e2
VIIIa
raVIII
pg
slH
flm
siv
Gc
Vs
nV
Rm
Vm
tbv

46

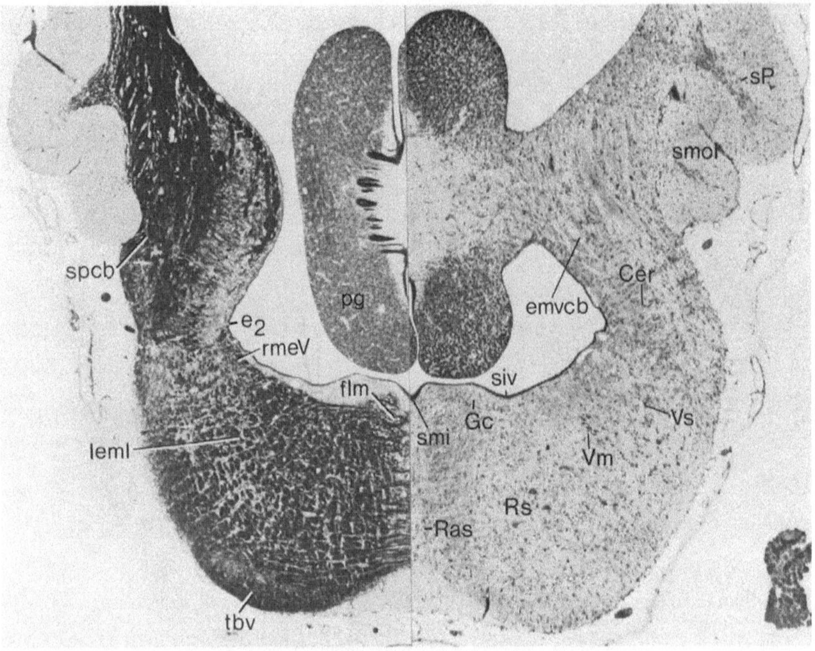
sP
smol
spcb
Cer
pg
emvcb
e2
rmeV
flm
siv
Gc
Vs
smi
leml
Vm
Rs
Ras
tbv

47

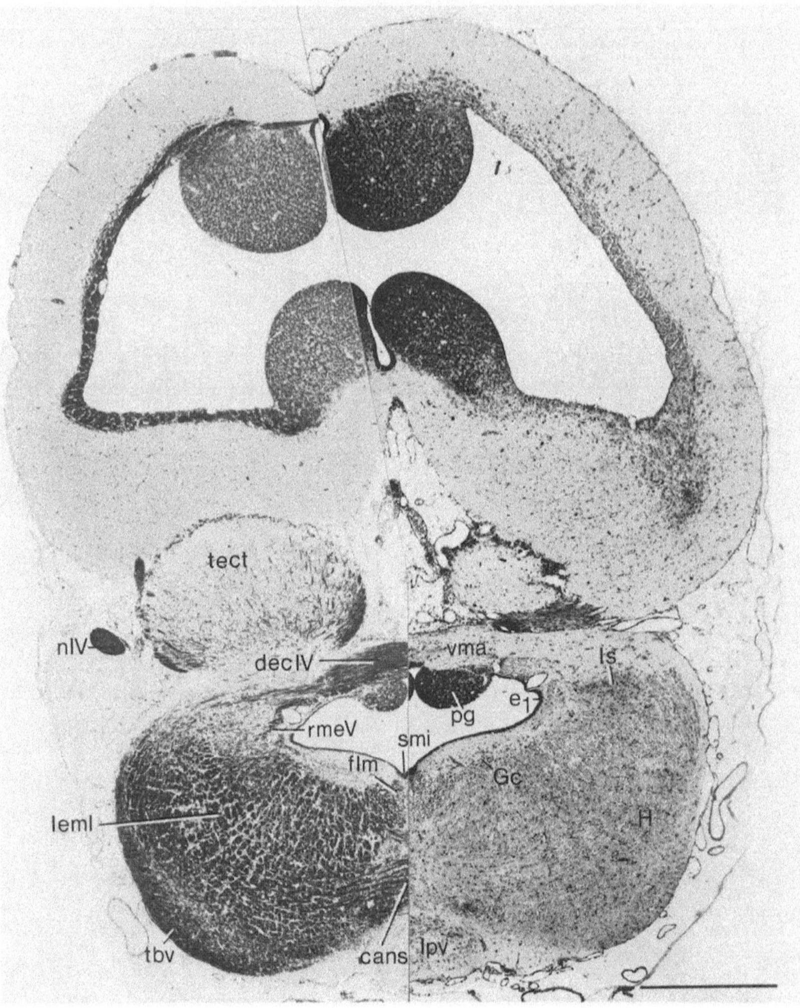
tect
nIV
decIV
vma
ls
pg
e1
rmeV
smi
flm
Gc
leml
H
tbv
cans
lpv

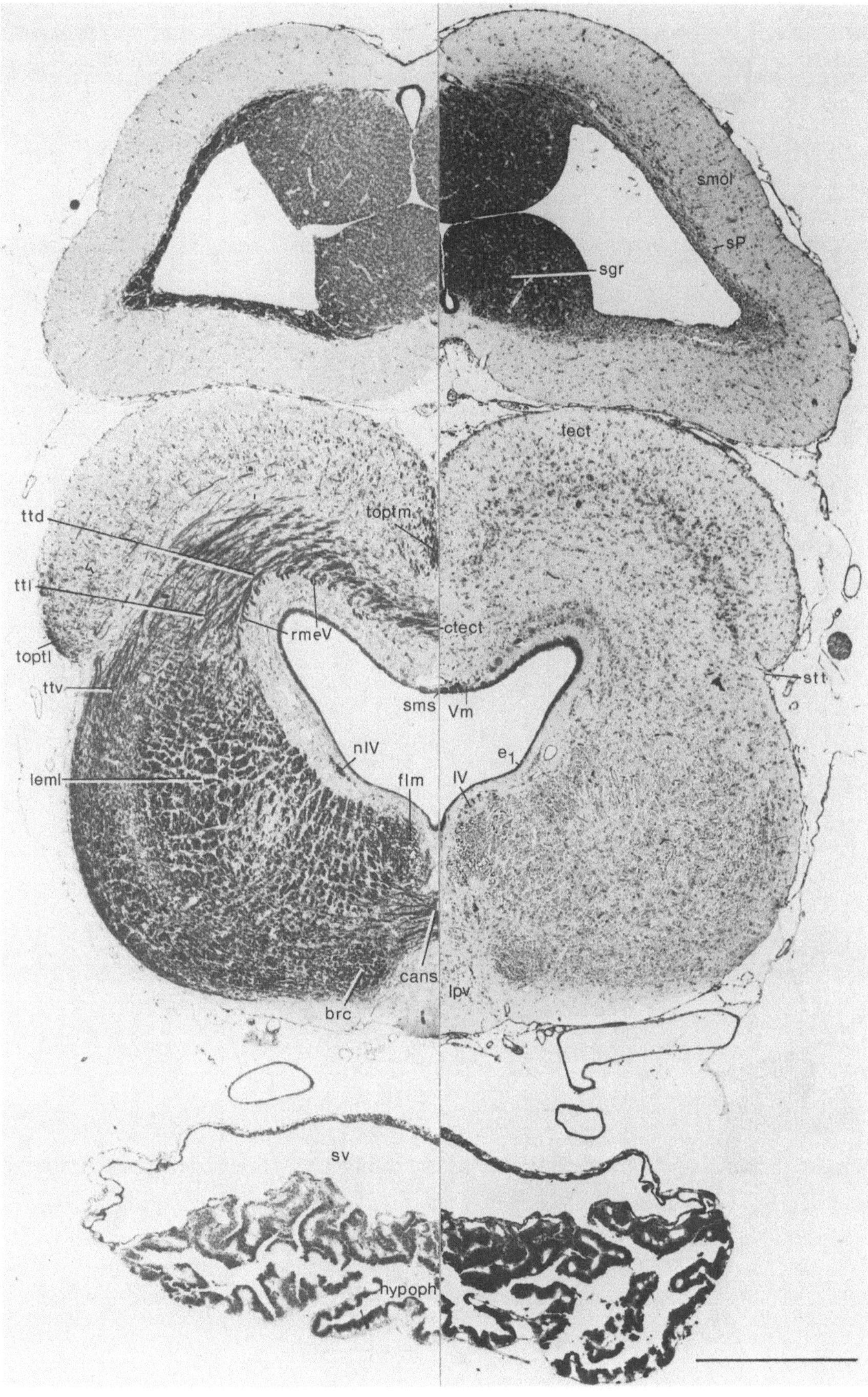
smol
sP
sgr
tect
ttd
toptm
ttl
rmeV
ctect
toptl
ttv
stt
sms
Vm
nIV
e1
leml
flm
IV
cans
lpv
brc
sv
hypoph

49

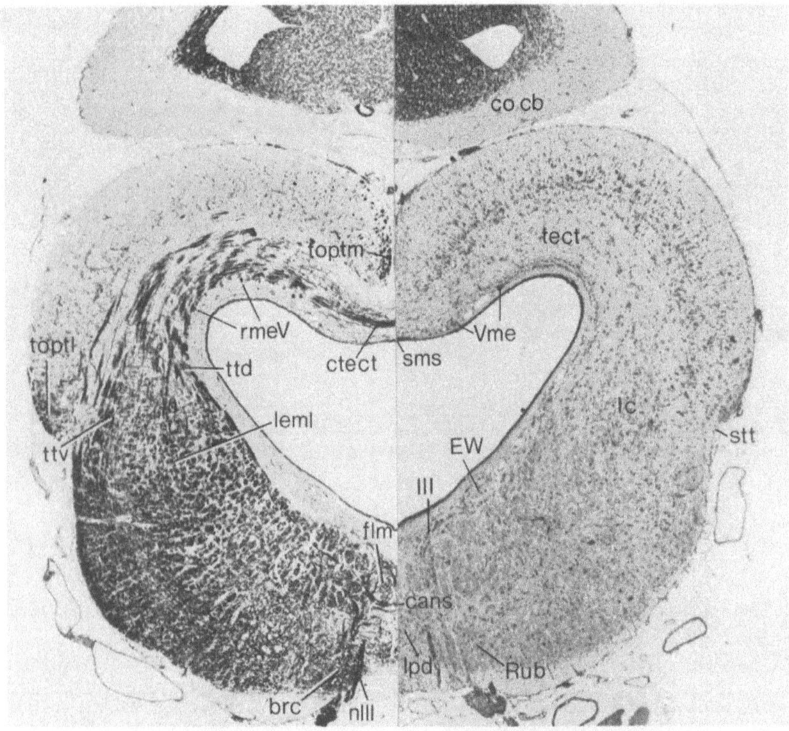
co cb
toptm
tect
rmeV
Vme
toptl
ctect
sms
ttd
lc
leml
stt
ttv
EW
III
flm
cans
lpd
Rub
brc
nIII

50

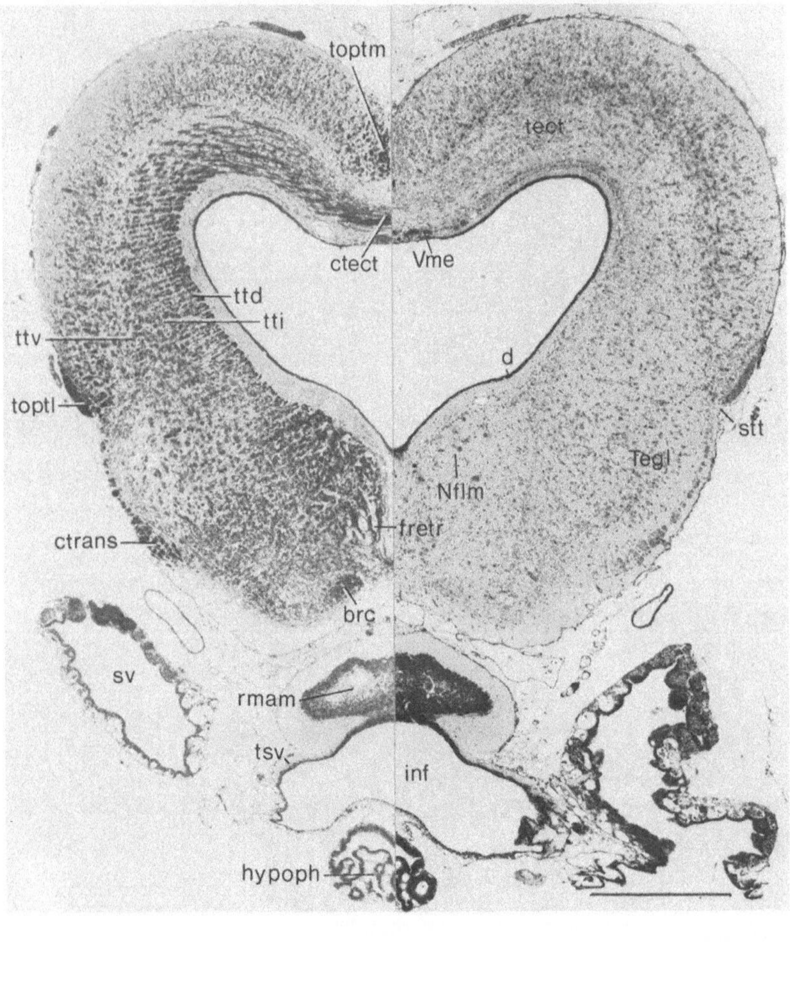
toptm
tect
ctect
Vme
ttd
tti
ttv
d
toptl
stt
Tegl
Nflm
fretr
ctrans
brc
sv
rmam
tsv
inf
hypoph

51

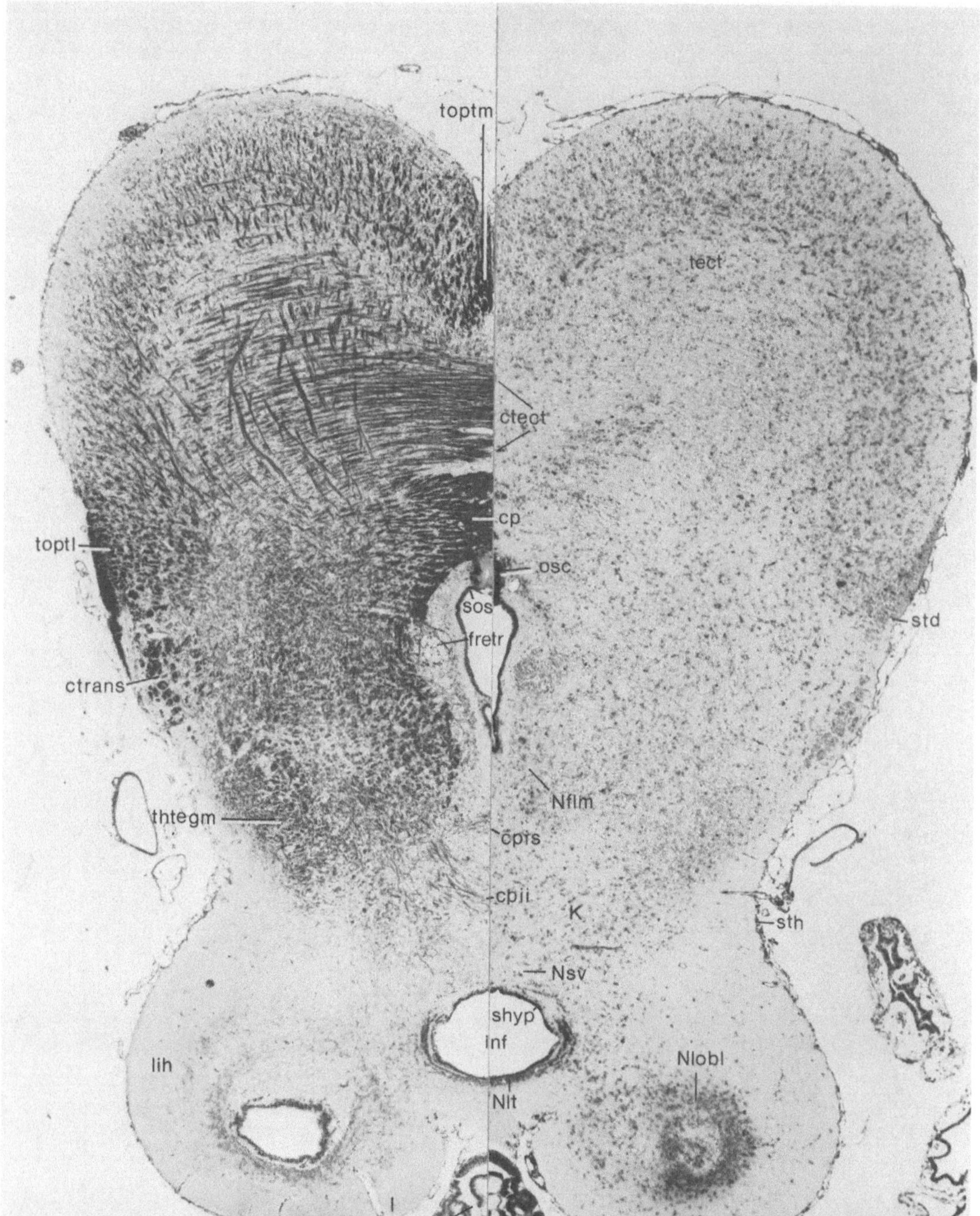
toptm
tect
ctect
cp
toptl
osc
sos
fretr
std
ctrans
Nflm
thtegm
cprs
cpii
K
sth
Nsv
shyp
inf
lih
Nlobl
Nlt
l
hypoph

52

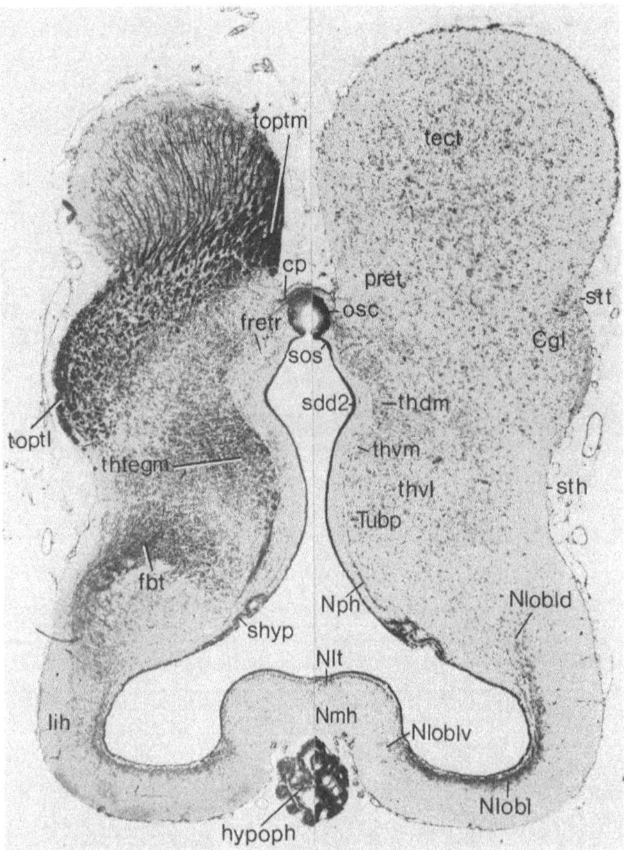
toptm
tect
cp
pret
stt
osc
fretr
Cgl
sos
sdd2
thdm
toptl
thvm
thtegm
thvl
sth
Tubp
fbt
Nph
Nlobld
shyp
Nlt
Nmh
lih
Nloblv
Nlobl
hypoph

53

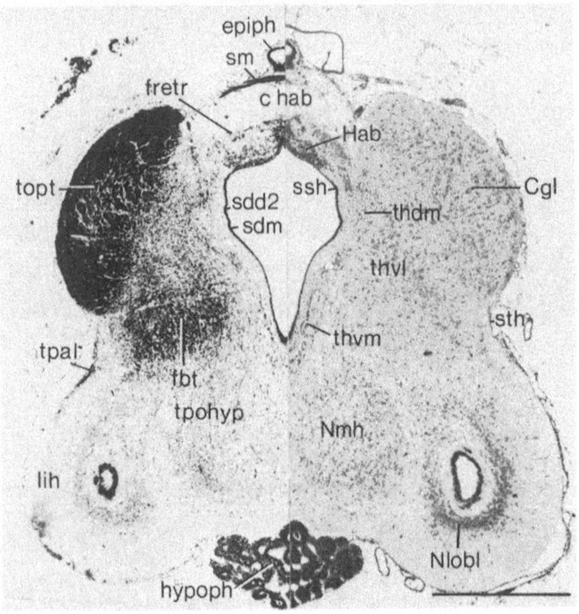
epiph
sm
fretr
c hab
Hab
topt
ssh
Cgl
sdd2
sdm
thdm
thvl
sth
thvm
tpal
fbt
tpohyp
Nmh
lih
Nlobl
hypoph

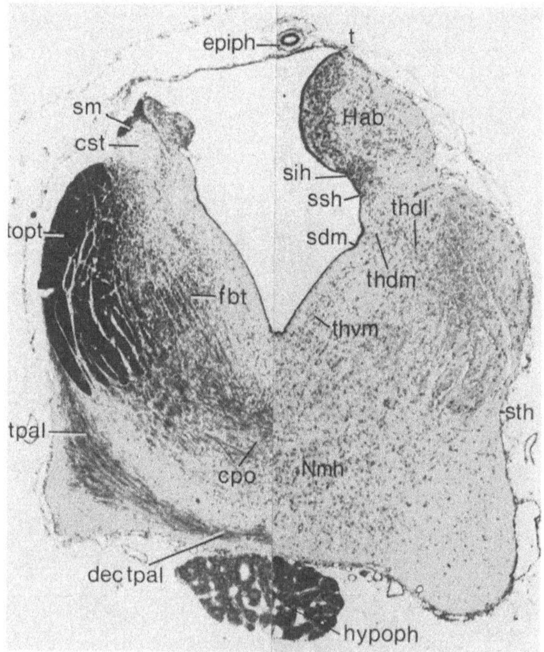
epiph
t
sm
cst
Hab
sih
ssh
thdl
sdm
topt
thdm
fbt
thvm
sth
tpal
cpo
Nmh
dec tpal
hypoph

54

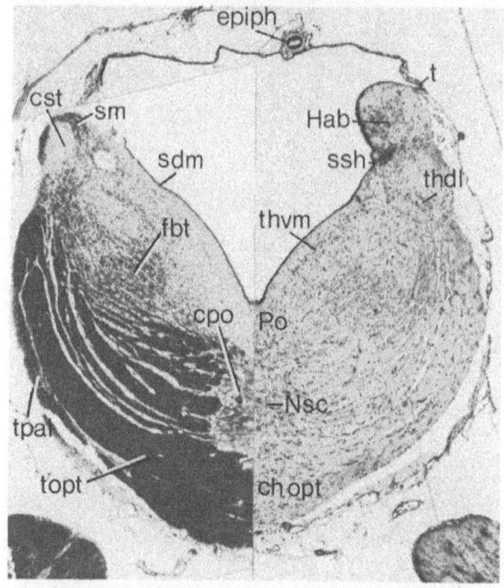
epiph
t
cst
sm
Hab
sdm
ssh
thdl
fbt
thvm
cpo
Po
Nsc
tpal
topt
ch opt

55

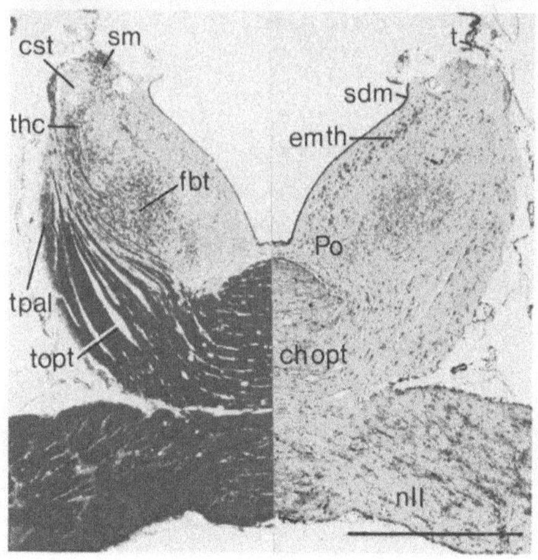
cst
sm
t
sdm
thc
emth
fbt
Po
tpal
topt
ch opt
nII

56

57

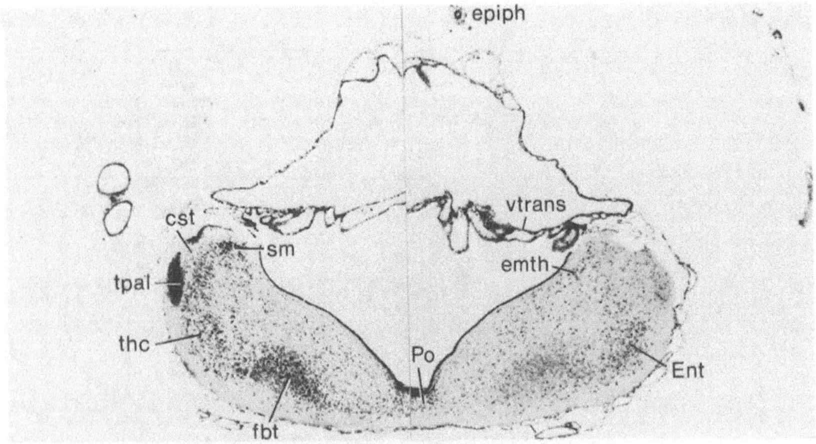
epiph
vtrans
cst
sm
emth
tpal
thc
Po
Ent
fbt

58

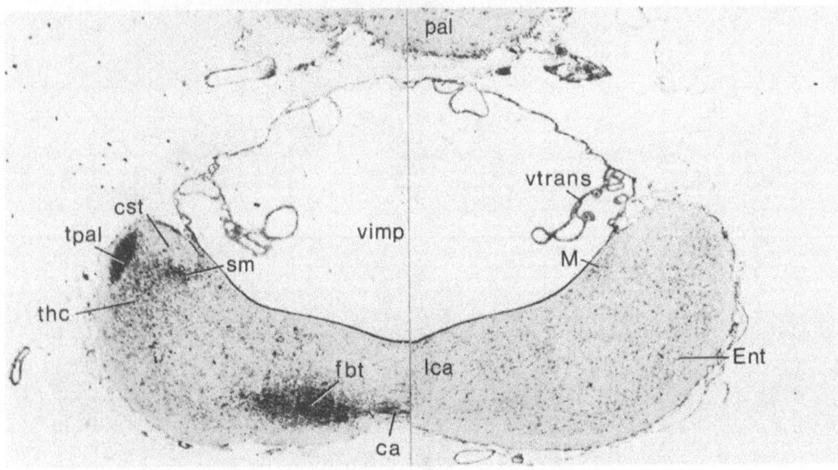
pal
vtrans
cst
tpal
vimp
sm
M
thc
Ent
fbt
lca
ca

59

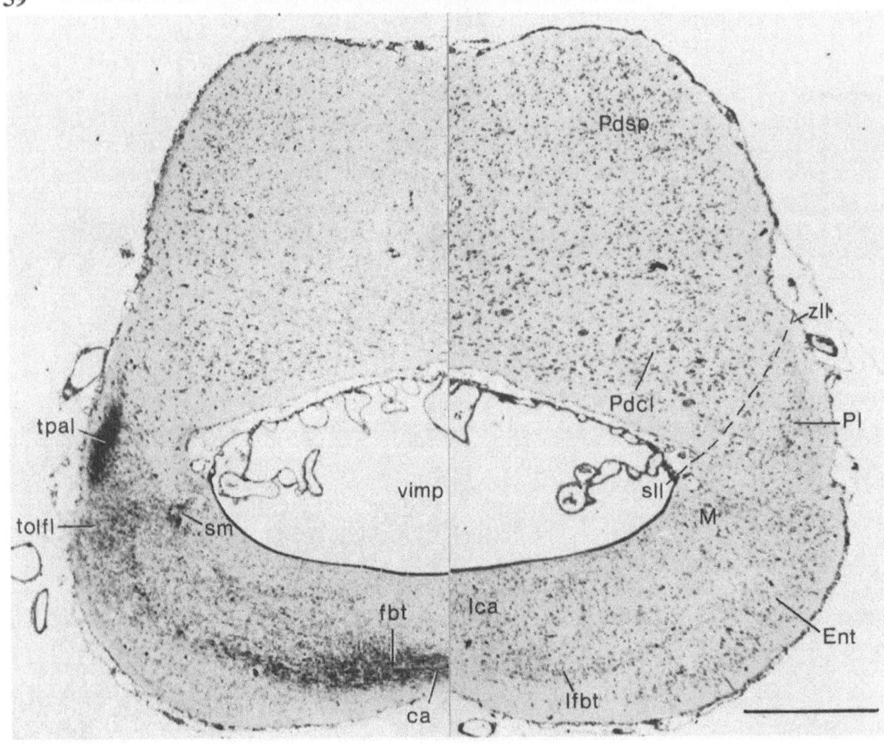
Pdsp
zll
Pdcl
tpal
Pl
vimp
sll
tolfl
sm
M
fbt
lca
Ent
ca
lfbt

60

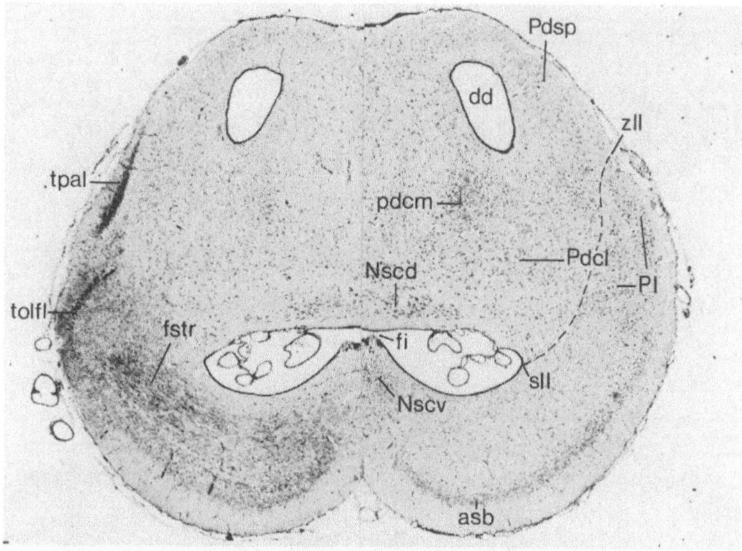
Pdsp
dd
zll
tpal
pdcm
Pdcl
Nscd
Pl
tolfl
fstr
fi
sll
Nscv
asb

61

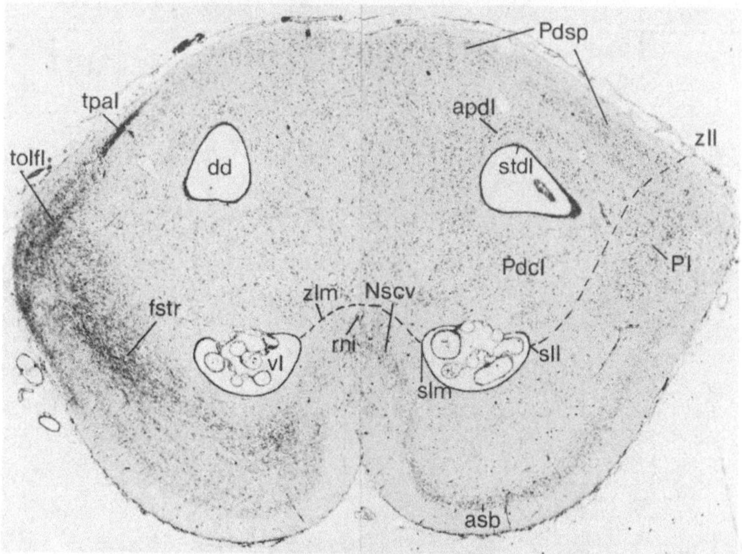
Pdsp
tpal
apdl
zll
tolfl
dd
stdl
Pl
Pdcl
zlm
Nscv
fstr
rni
vl
sll
slm
asb

62

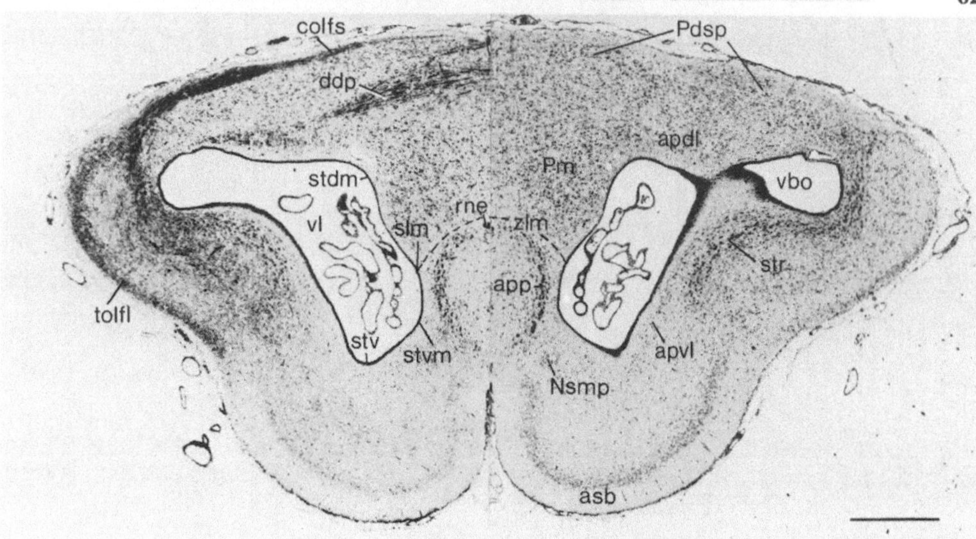
colfs
Pdsp
ddp
apdl
Pm
vbo
stdm
rne
vl
slm
zlm
str
app
tolfl
stv
stvm
apvl
Nsmp
asb

63

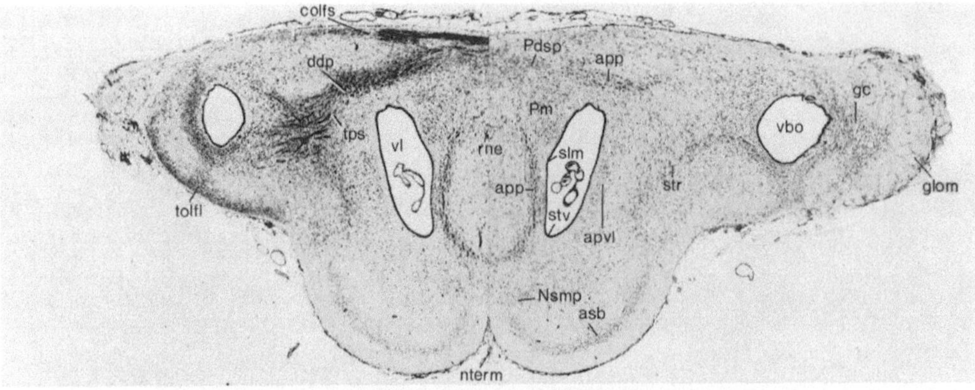
colfs
Pdsp
app
ddp
gc
Pm
tps
vl
rne
vbo
slm
app
str
glom
stv
tolfl
apvl
Nsmp
asb
nterm

64

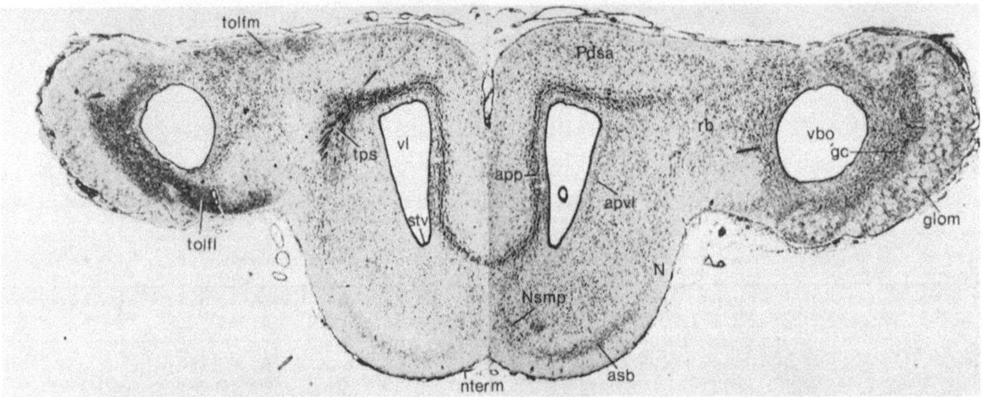
tolfm
Pdsa
rb
vbo
gc
tps
vl
app
apvl
stv
glom
tolfl
N
Nsmp
asb
nterm

65

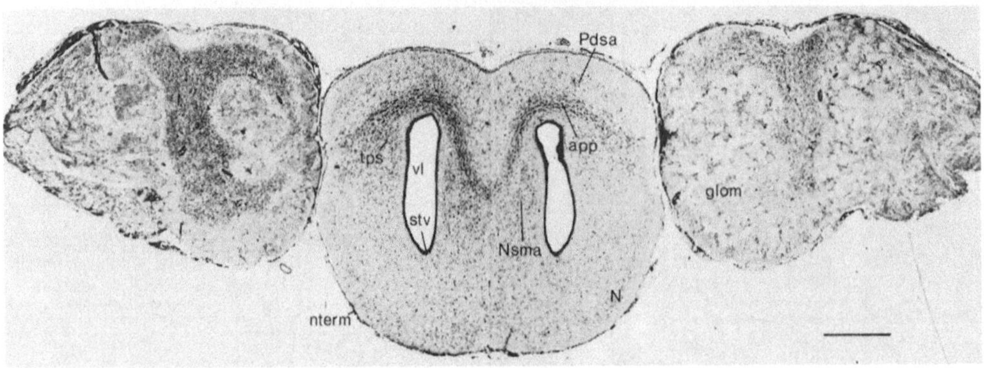
Pdsa
app
tps
vl
glom
stv
Nsma
N
nterm

Raja clavata

(Fig. 66–99)

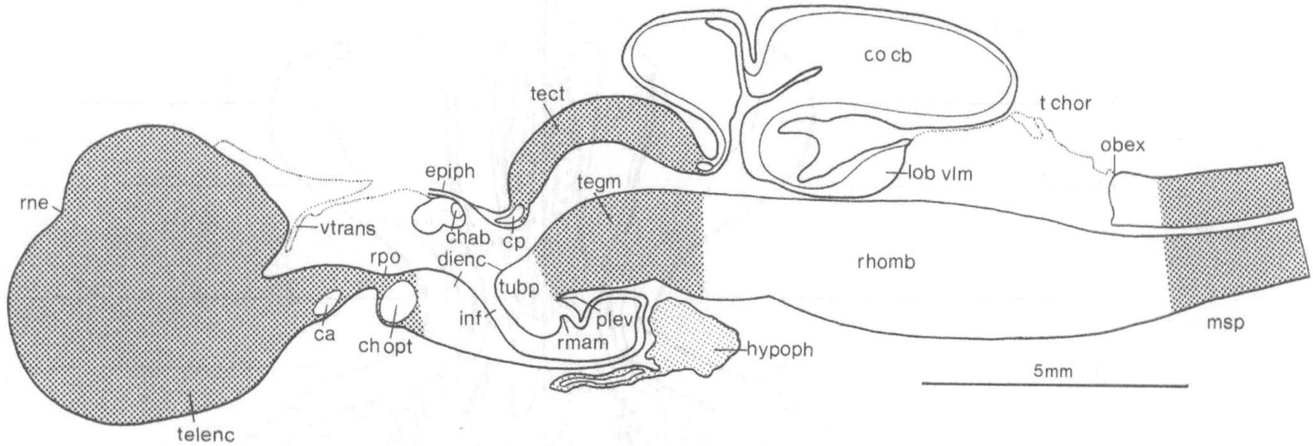

Fig. 66. Midsagittal section of the brain of *Raja clavata*

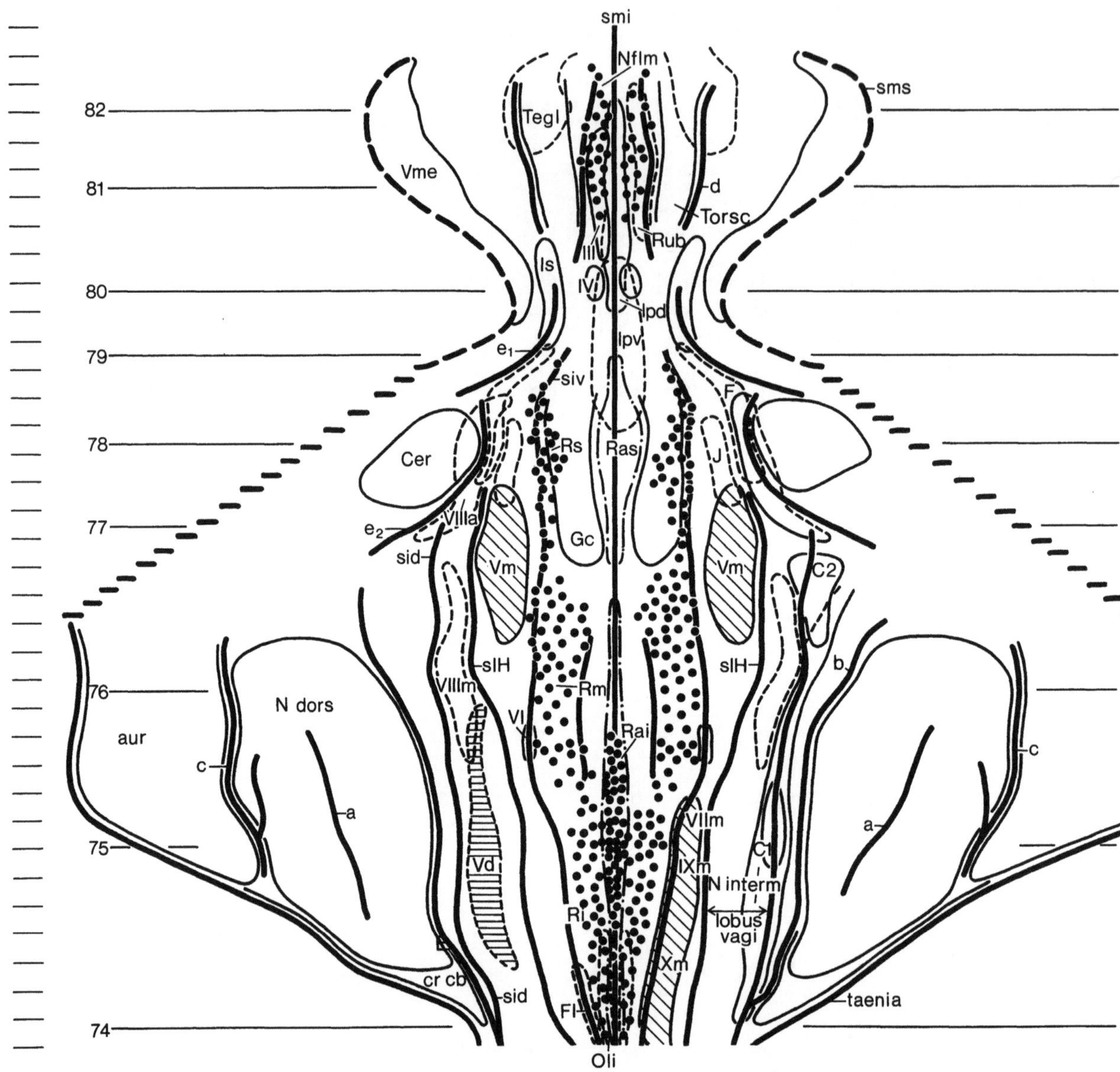

Fig. 67. Topological reconstruction of the brain stem of the ray *Raja clavata*. Distance between sections, indicated by *small horizontal bars,* equals 300 μm. For further explanation see Fig. 2

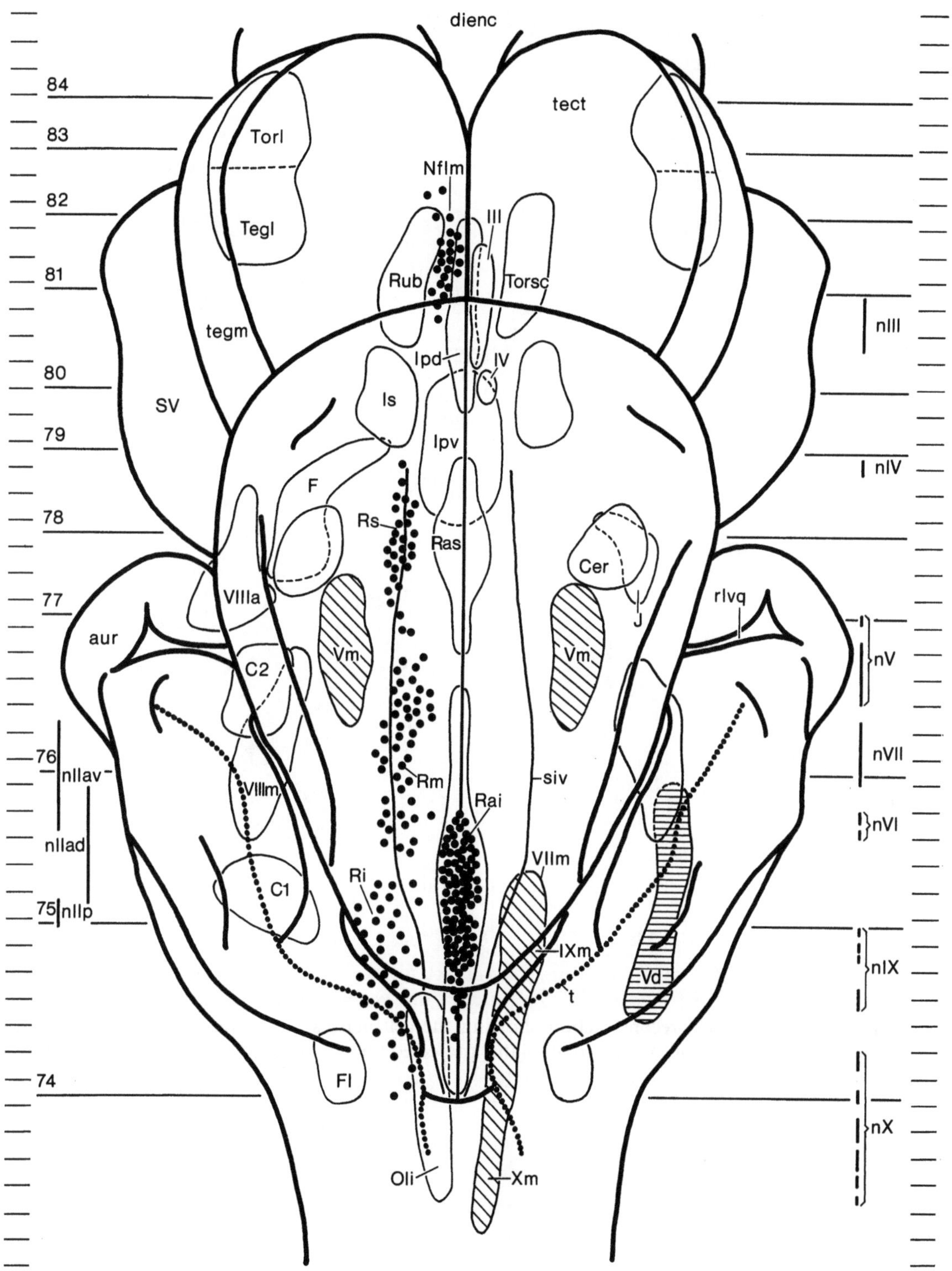

Fig. 68. Topographical reconstruction showing the cell masses of the brain stem of *Raja clavata* as projected upon a horizontal plane. Distance between sections, indicated by *small horizontal bars*, equals 300 µm. For further explanation see Fig. 3

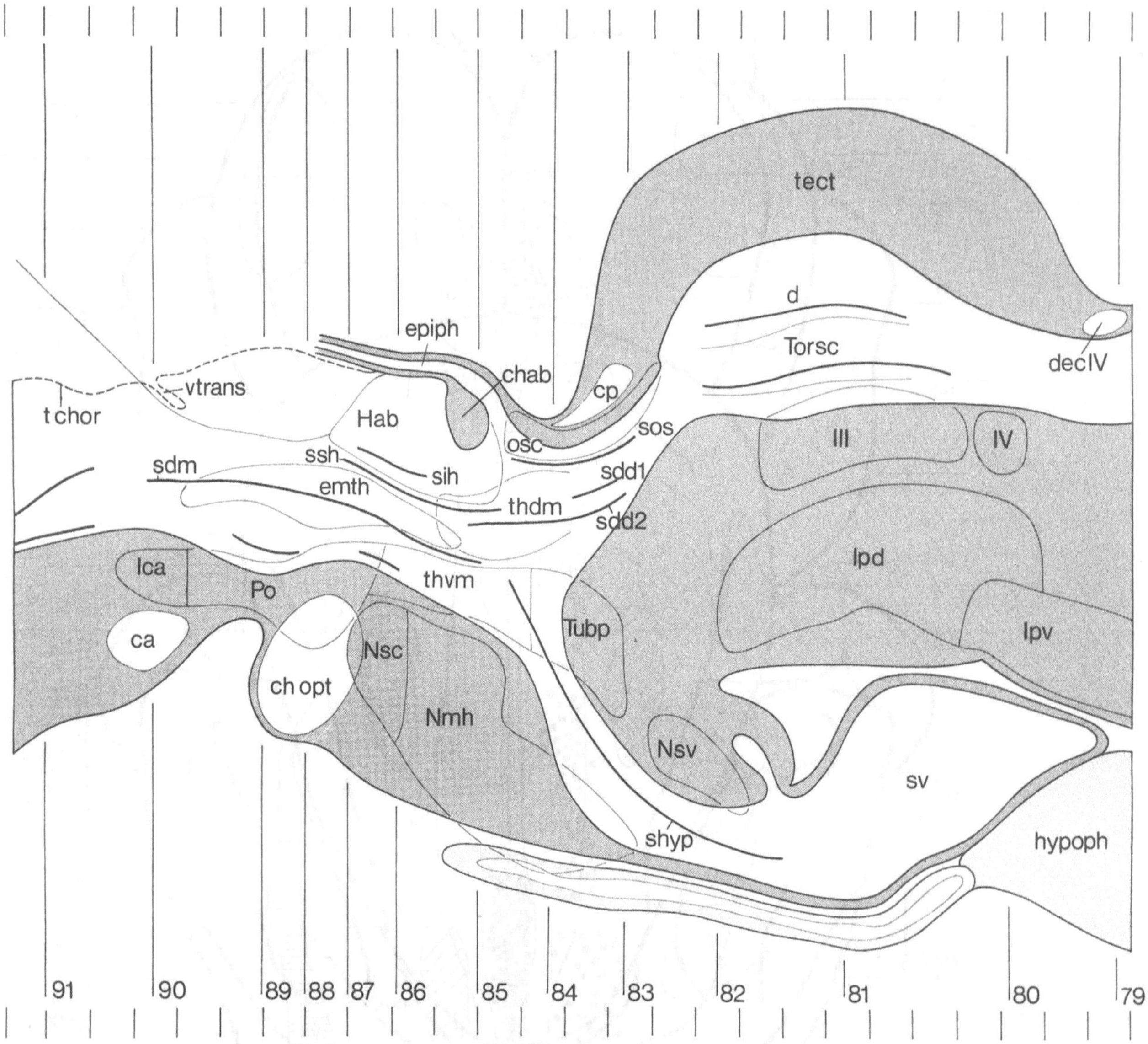

Fig. 69. Topographical reconstruction showing the cell masses of the diencephalon of *Raja clavata* lying periventricularly or in the median plane, projected upon a midsagittal plane. For further explanation see Fig. 4

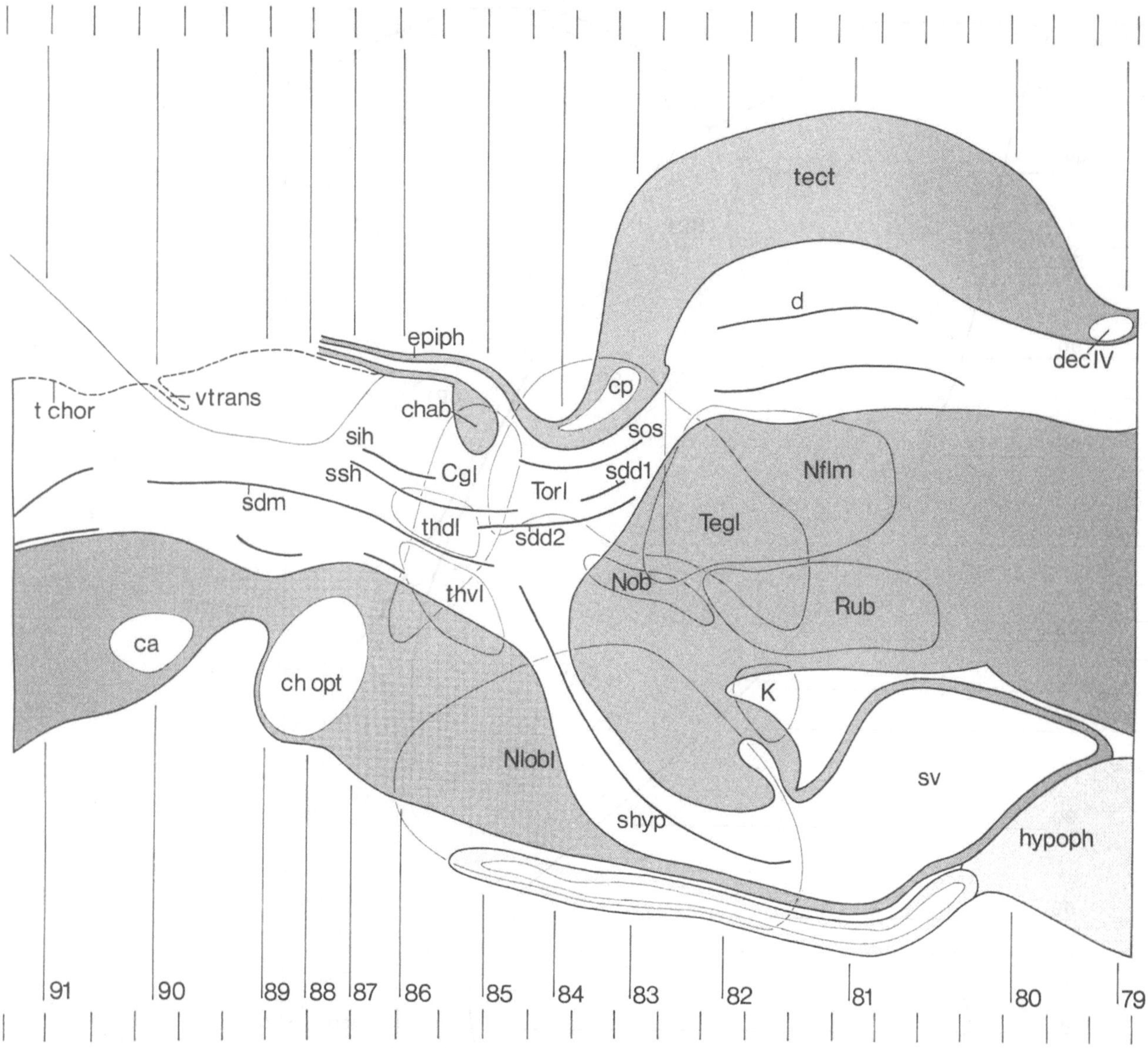

Fig. 70. Topographical reconstruction showing the migrated cell masses of the diencephalon of *Raja clavata* as projected upon a midsagittal plane. For explanation see Fig. 4

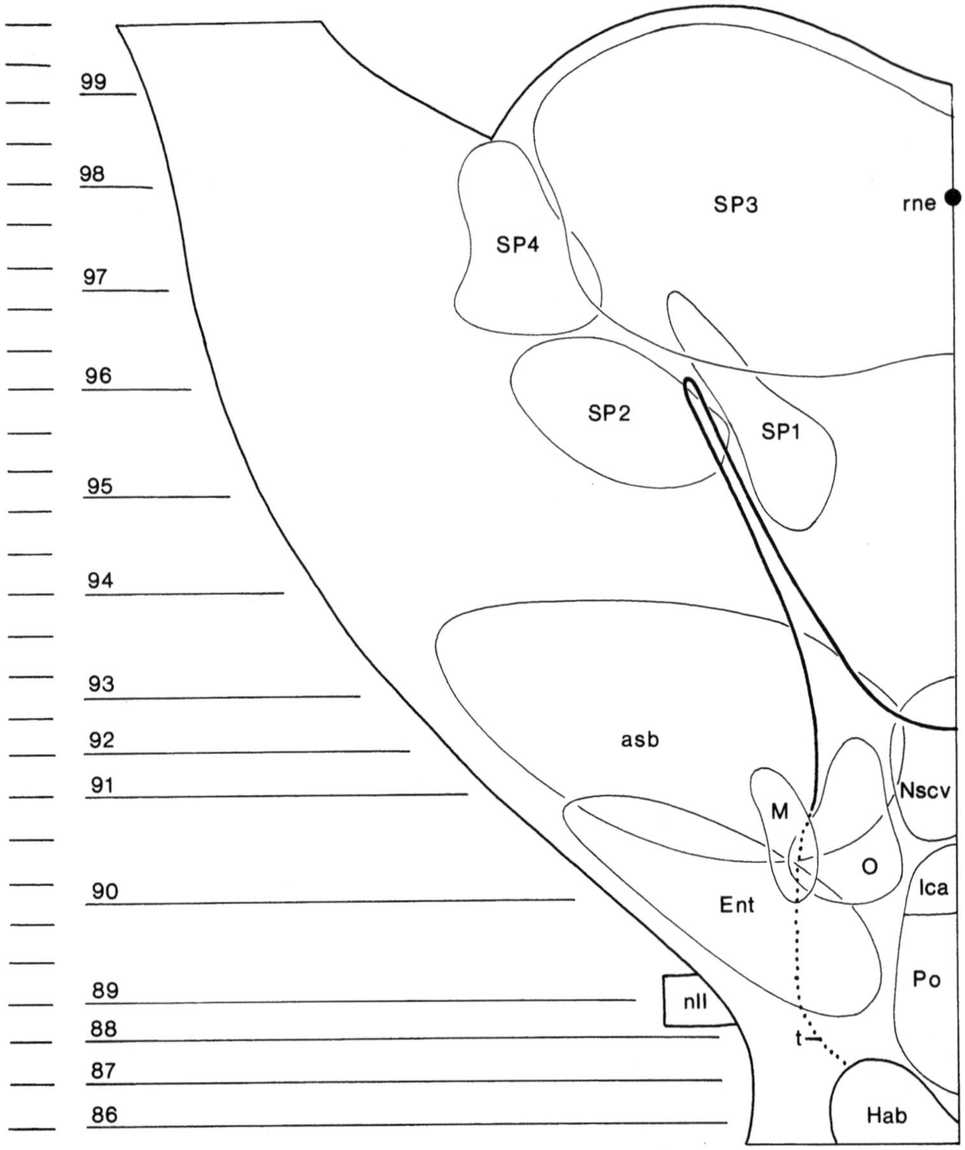

Fig. 71. Topographical reconstruction showing the cell masses of the subpallium and of the telencephalon impar of *Raja clavata* as projected upon a horizontal plane. For further explanation see Fig. 6

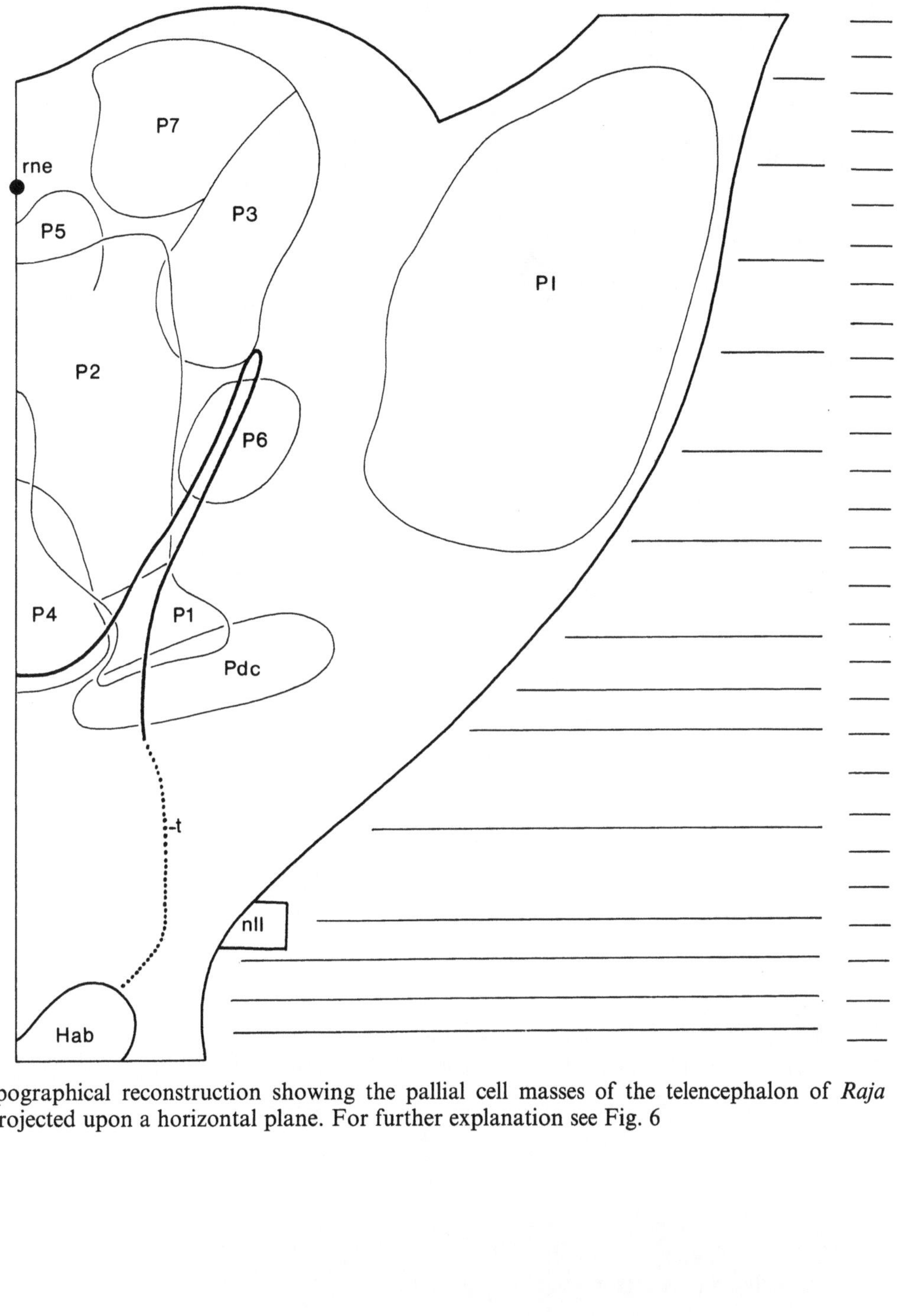

Fig. 72. Topographical reconstruction showing the pallial cell masses of the telencephalon of *Raja clavata* as projected upon a horizontal plane. For further explanation see Fig. 6

Figs. 73–99. Photomicrographs of transverse sections of the brain of *Raja clavata*. The *left half* of each figure is a photomicrograph of a section stained according to Klüver and Barrera, whereas the *right half* is a photomicrograph taken from the adjacent Nissl stained section. In Figs. **94–96** the *left half* is a photomicrograph of a Bodian stained section. *Bar scale* = 1 mm

73

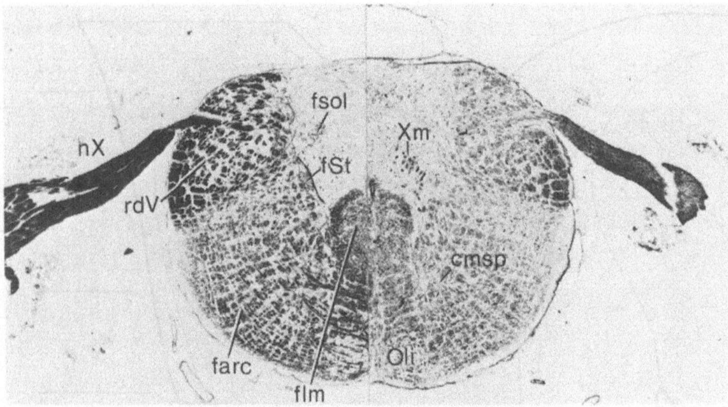
fsol
nX
Xm
fSt
rdV
cmsp
farc
Oli
flm

74

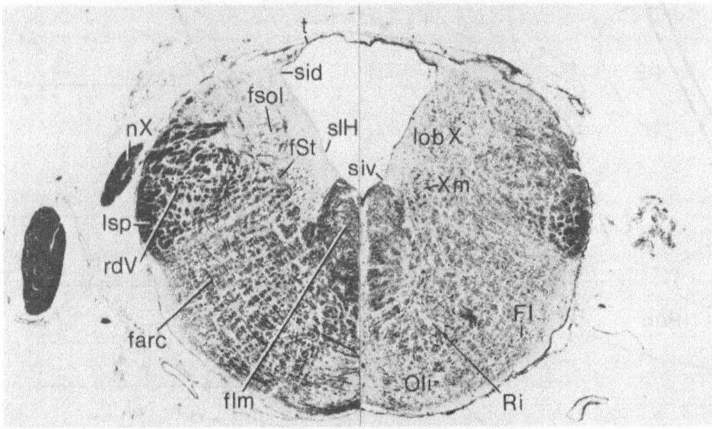
t
sid
fsol
nX
slH
lob X
fSt
siv
Xm
lsp
rdV
Fl
farc
flm
Oli
Ri

75

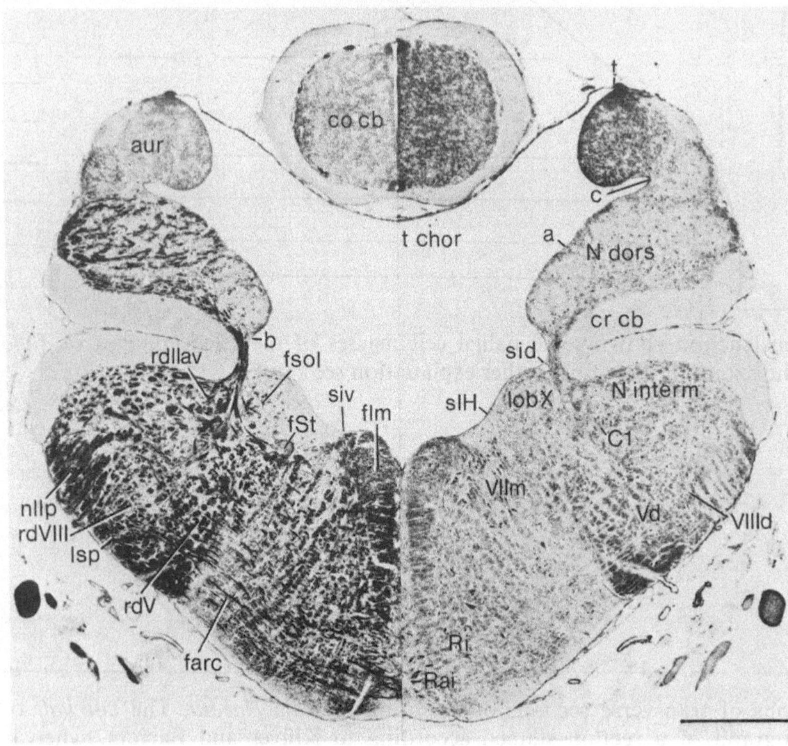
t
co cb
aur
c
t chor
a
N dors
cr cb
b
rdllav
fsol
sid
N interm
siv
flm
slH
lobX
fSt
C1
VIIm
nIIp
Vd
VIIId
rdVIII
lsp
rdV
farc
Ri
Rai

76

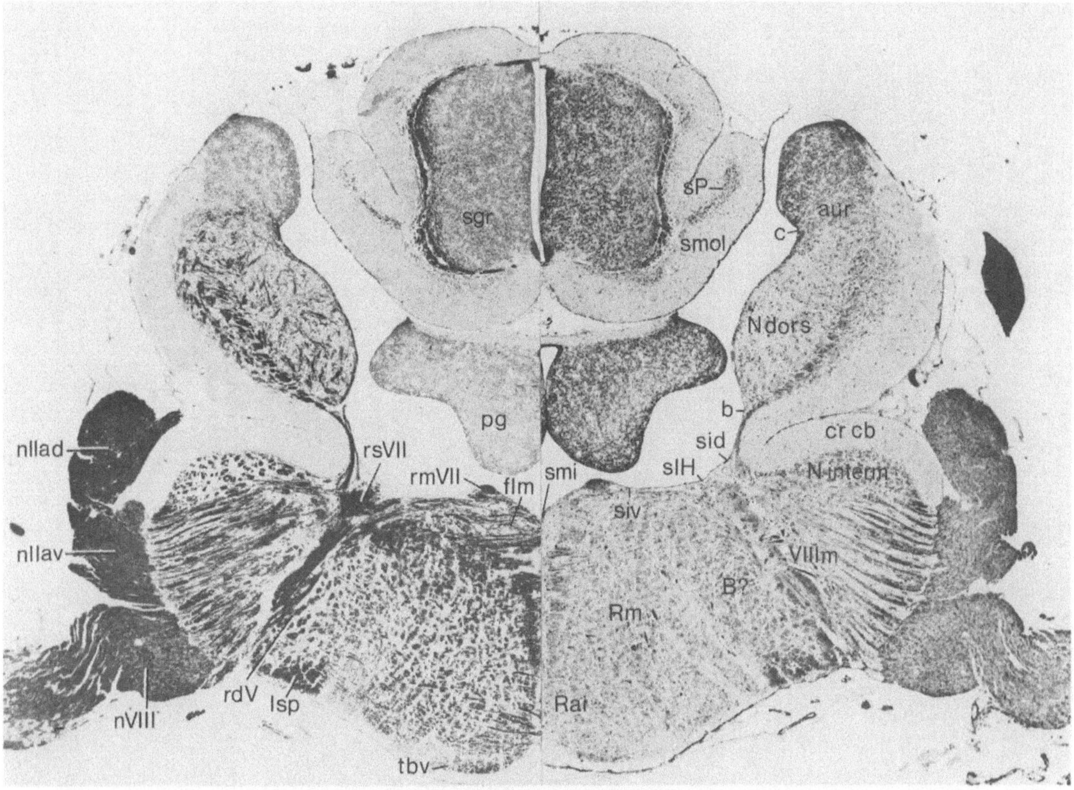
sP
sgr
aur
smol
c
Ndors
pg
b
sid
cr cb
nllad
rsVII
slH
N intern
rmVII
flm
smi
siv
nllav
VIIIm
B?
Rm
nVIII
rdV
lsp
Ral
tbv

77

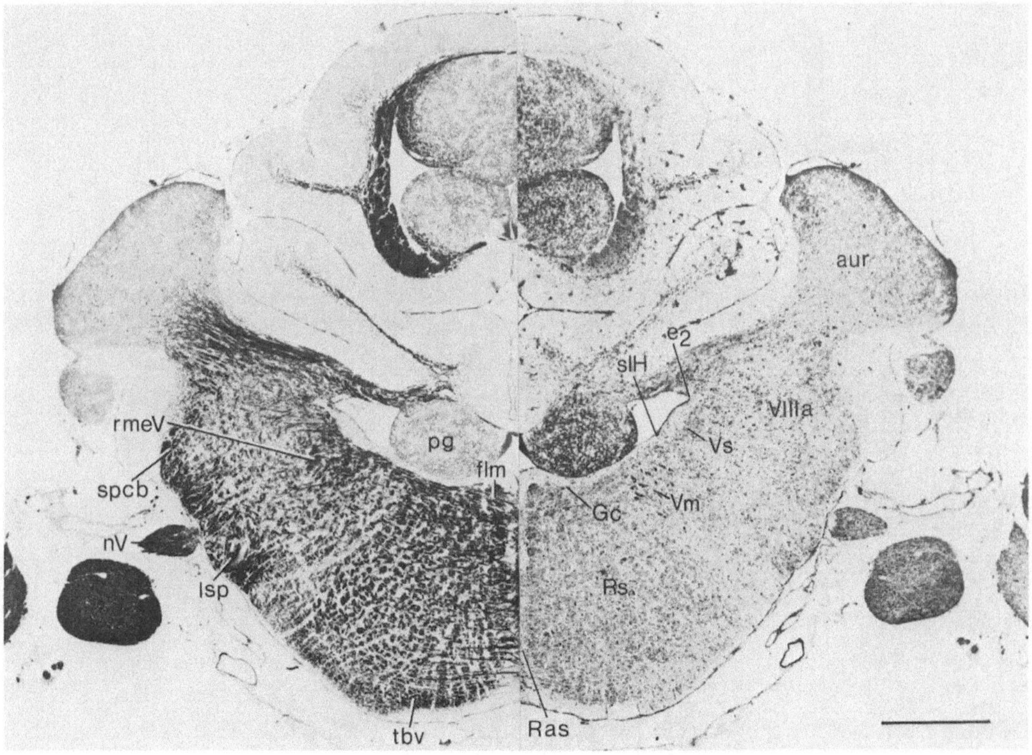
aur
e2
slH
VIIIa
rmeV
pg
Vs
flm
spcb
Gc
Vm
nV
lsp
Rs
tbv
Ras

78

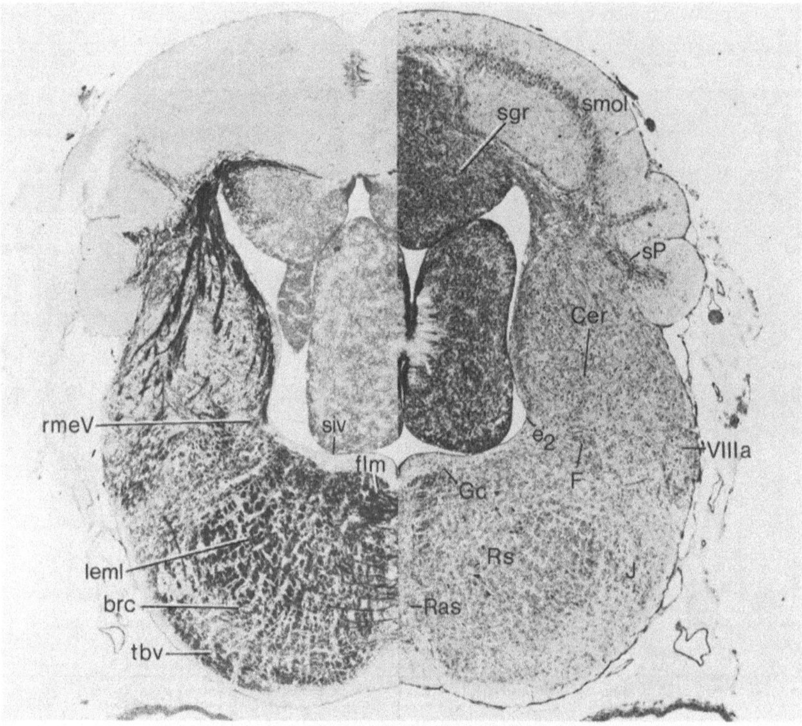
sgr
smol
sP
Cer
rmeV
siv
e2
VIIIa
flm
Gc
F
leml
Rs
J
brc
Ras
tbv

79

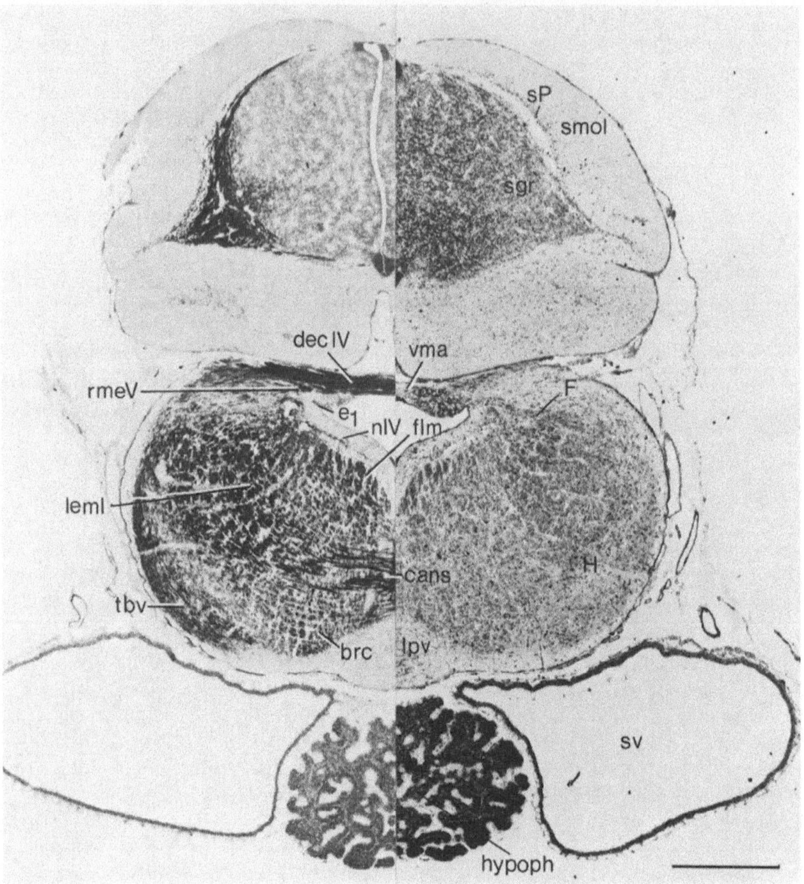
sP
smol
sgr
dec IV
vma
rmeV
F
e1
nIV
flm
leml
H
cans
tbv
brc
lpv
sv
hypoph

80

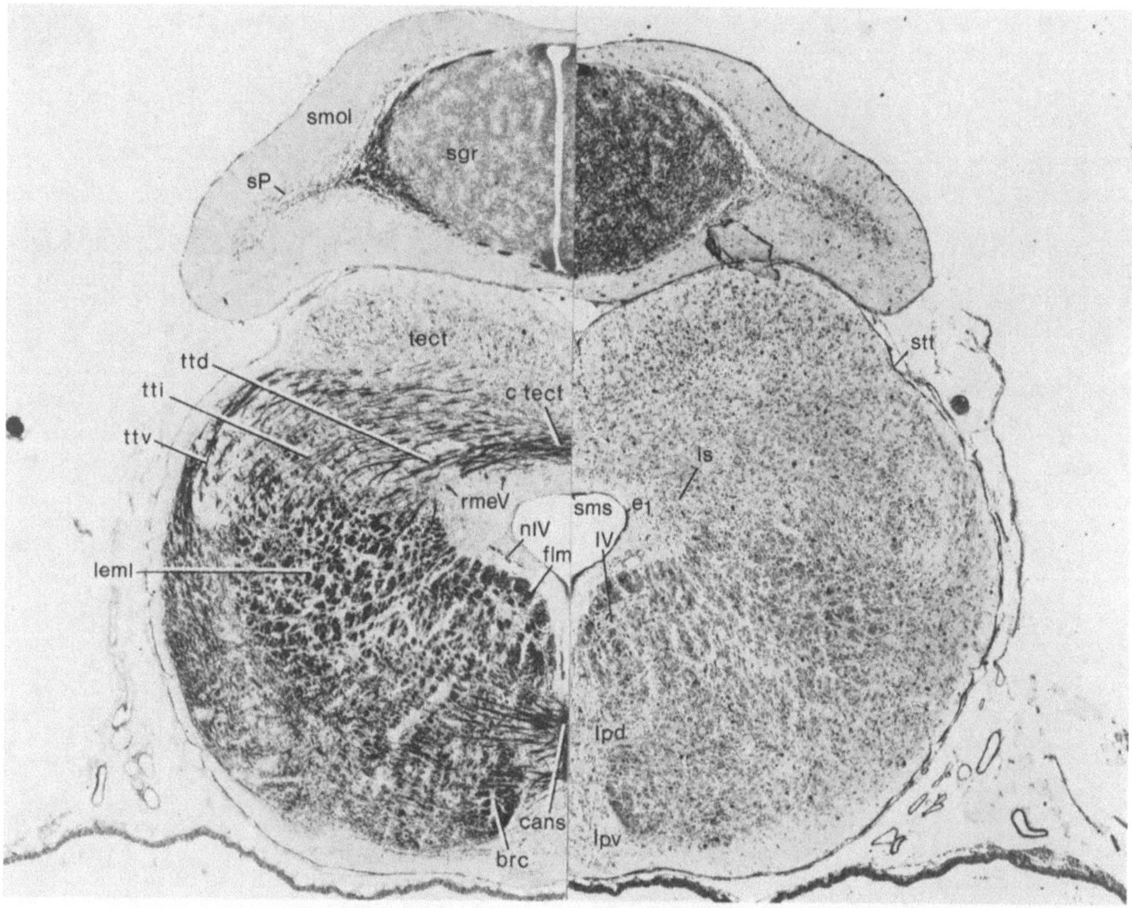
smol
sgr
sP
tect
stt
ttd
tti
c tect
ttv
ls
rmeV
sms
e1
nIV
IV
flm
leml
lpd
cans
lpv
brc

81

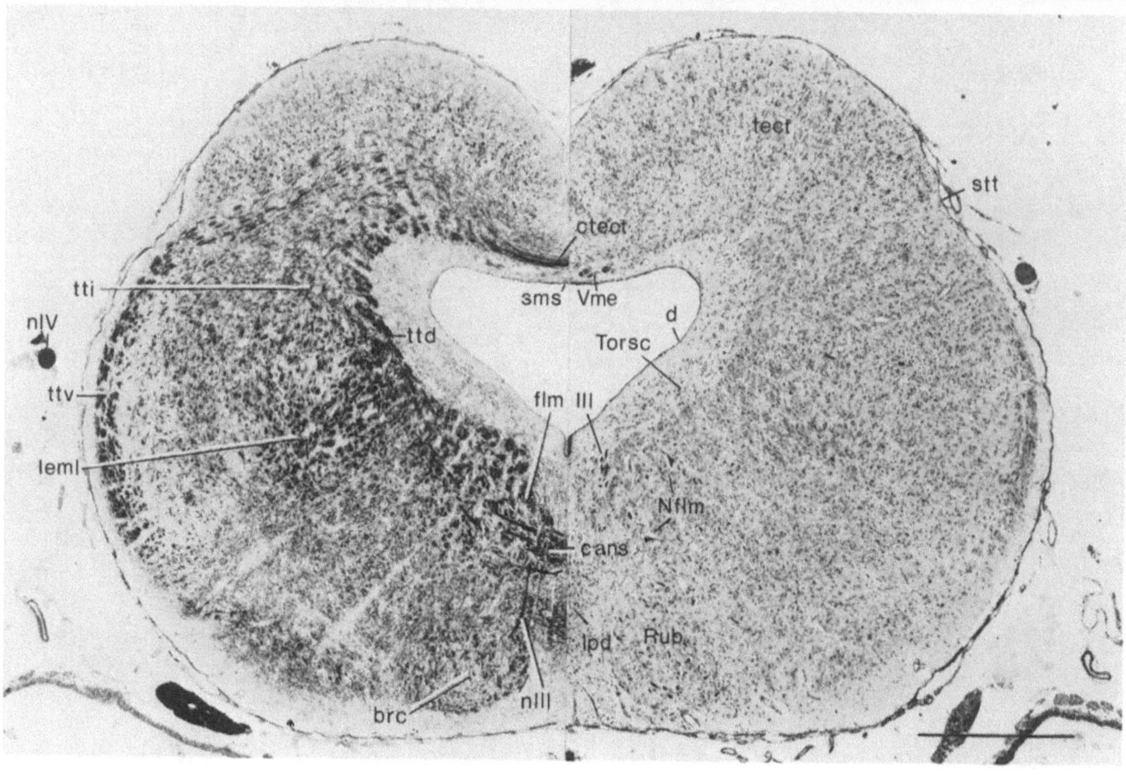
tect
stt
ctect
tti
sms
Vme
nIV
d
ttd
Torsc
ttv
flm
III
leml
Nflm
cans
lpd
Rub
brc
nIII

82

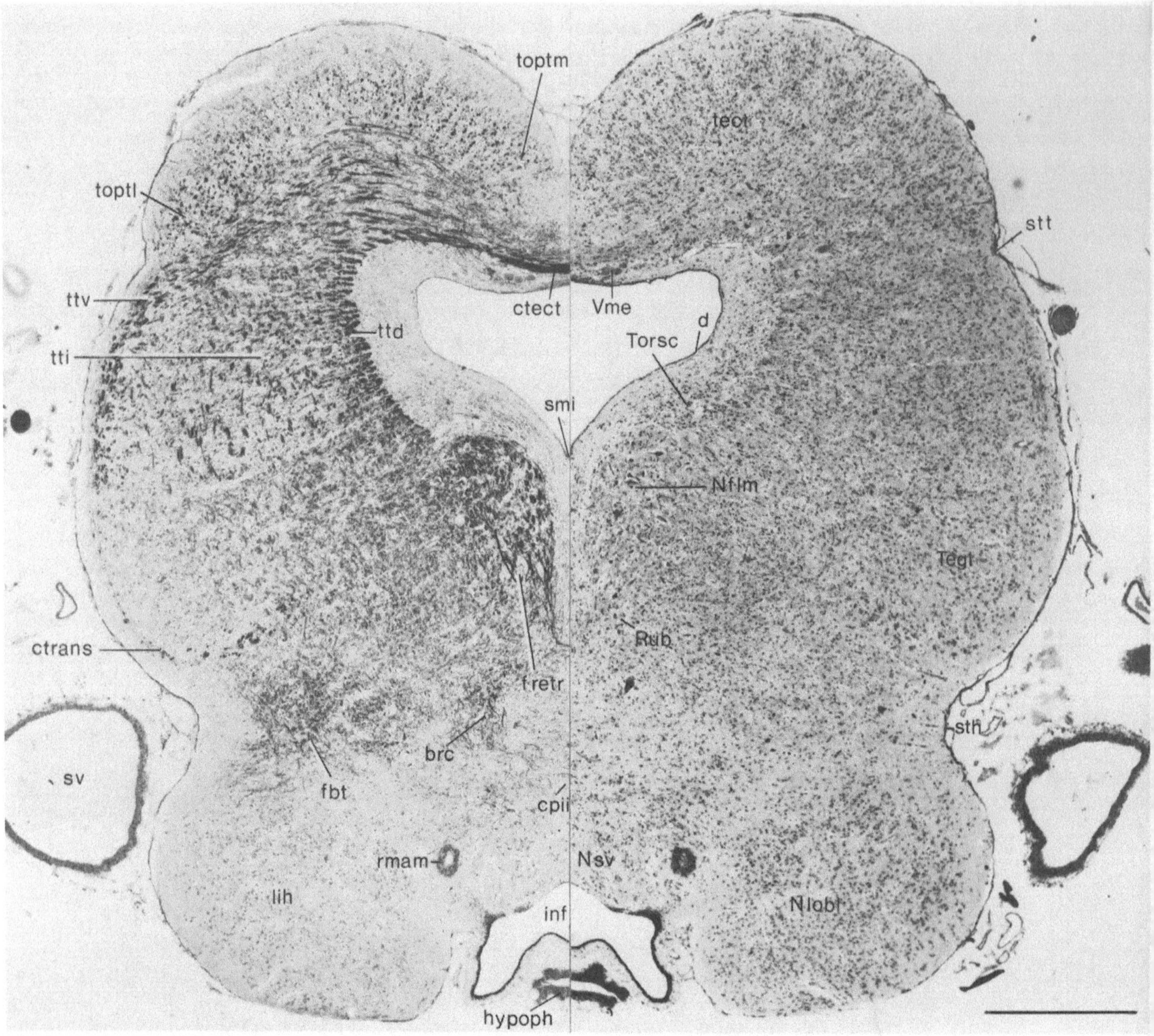
toptm
teol
toptl
stt
ttv
ctect
Vme
d
ttd
Torsc
tti
smi
Nflm
Tegl
ctrans
Rub
fretr
sth
brc
sv
fbt
cpii
rmam
Nsv
lih
Nlobl
inf
hypoph

83

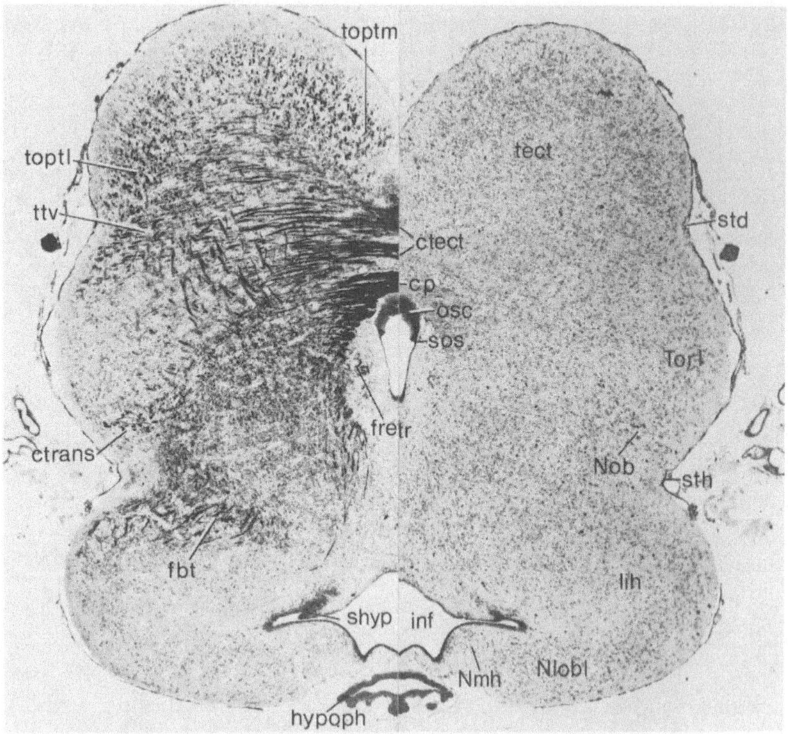
toptm
toptl
ttv
ctect
cp
osc
sos
fretr
ctrans
fbt
shyp
inf
Nmh
hypoph
tect
std
Torl
Nob
sth
lih
Nlobl

84

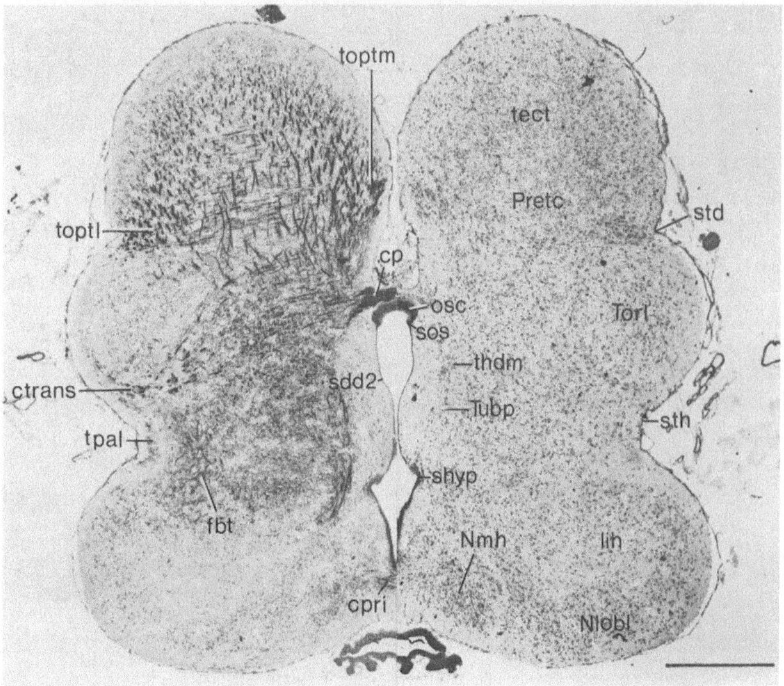
toptm
tect
Pretc
std
toptl
cp
osc
sos
Torl
thdm
sdd2
ctrans
Tubp
sth
tpal
shyp
fbt
Nmh
lih
cpri
Nlobl

85

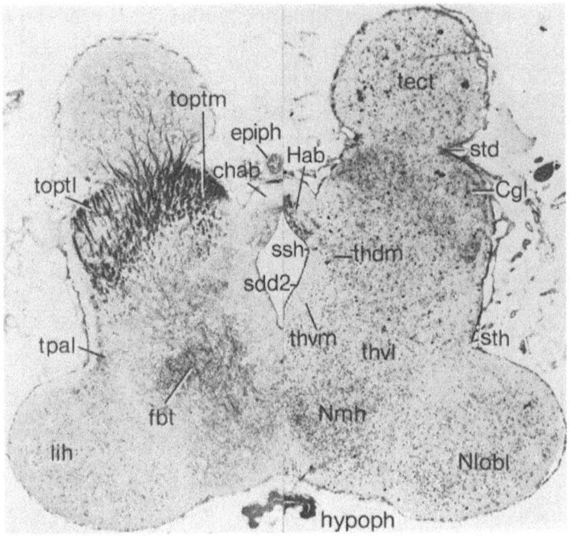
toptm
epiph
Hab
tect
std
chab
toptl
Cgl
ssh
thdm
sdd2
thvm
thvl
sth
tpal
fbt
Nmh
lih
Nlobl
hypoph

86

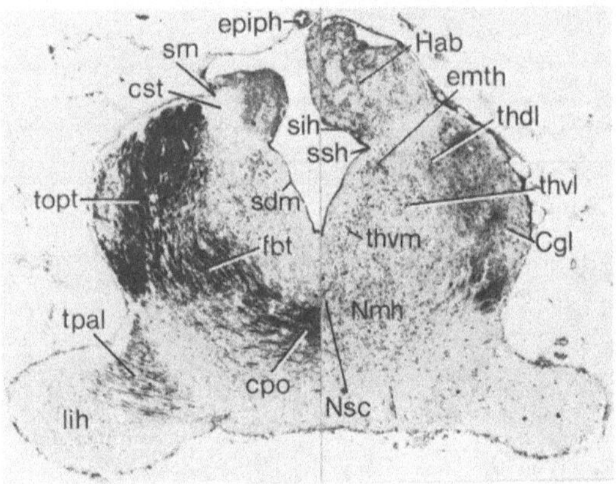
epiph
Hab
sm
emth
cst
sih
thdl
ssh
thvl
topt
sdm
thvm
Cgl
fbt
Nmh
tpal
cpo
Nsc
lih

87

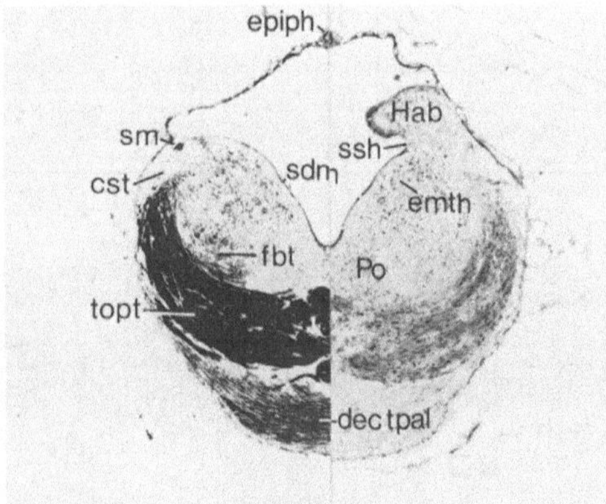
epiph
Hab
sm
ssh
cst
sdm
emth
fbt
Po
topt
dec tpal

88

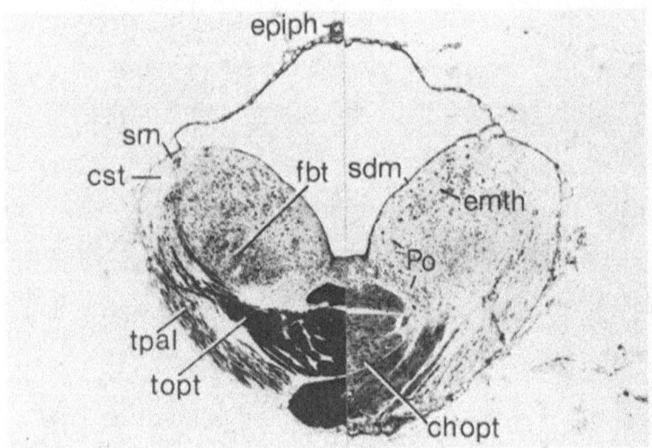
epiph
sm
cst
fbt
sdm
emth
Po
tpal
topt
chopt

89

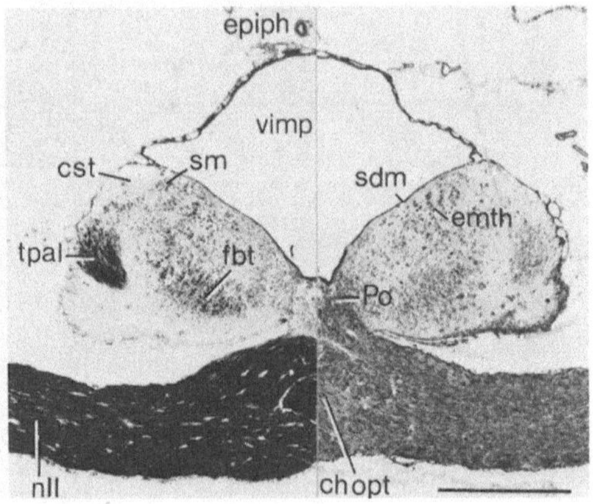
epiph
vimp
cst
sm
sdm
emth
tpal
fbt
Po
nII
chopt

90

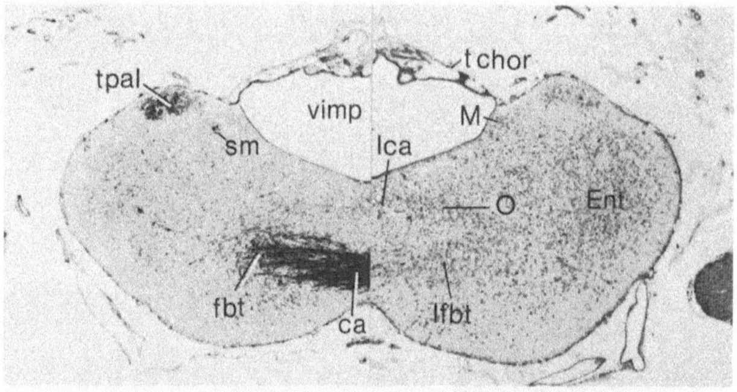
tpal
t chor
vimp
M
sm
lca
O
Ent
fbt
ca
lfbt

91

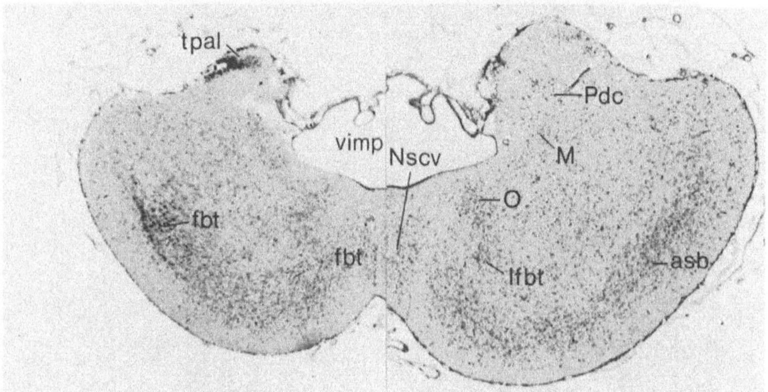
tpal
Pdc
vimp
Nscv
M
O
fbt
fbt
lfbt
asb

92

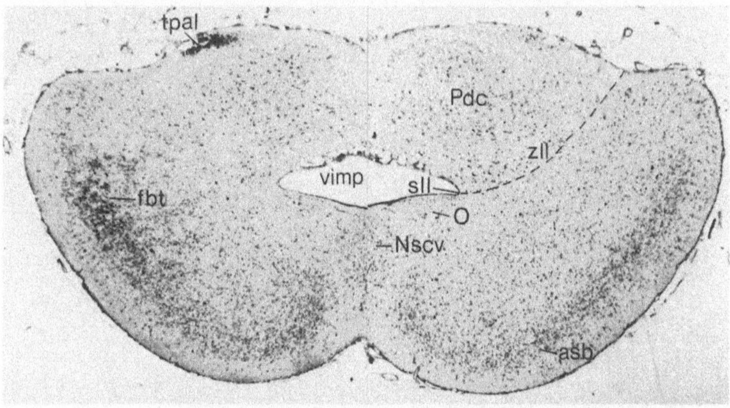
tpal
Pdc
zll
vimp
sll
fbt
O
Nscv
asb

93

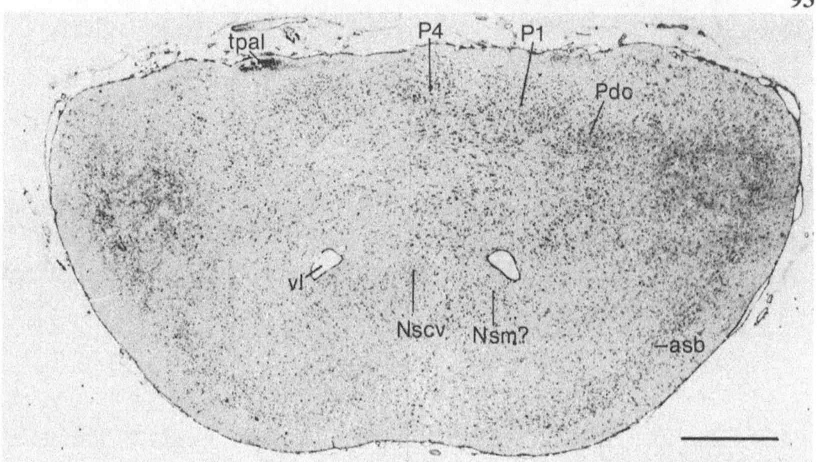
tpal
P4
P1
Pdc
vl
Nscv
Nsm?
asb

94

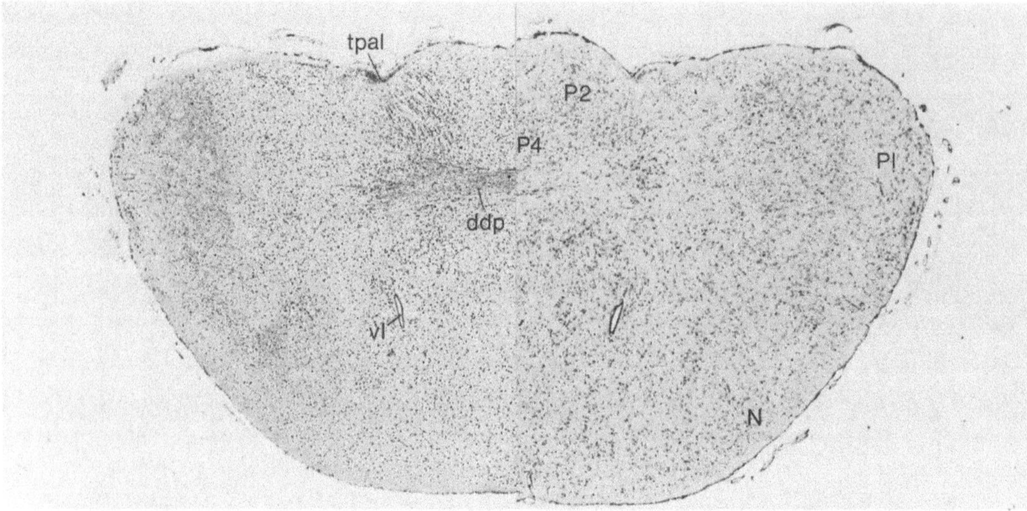
tpal
P2
P4
Pl
ddp
vl
N

95

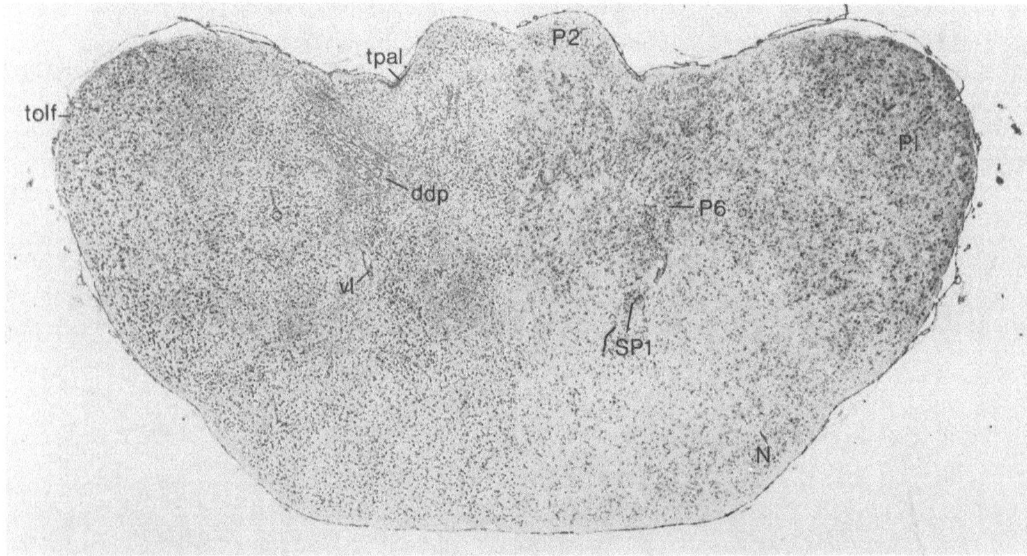
P2
tpal
tolf
Pl
ddp
P6
vl
SP1
N

96

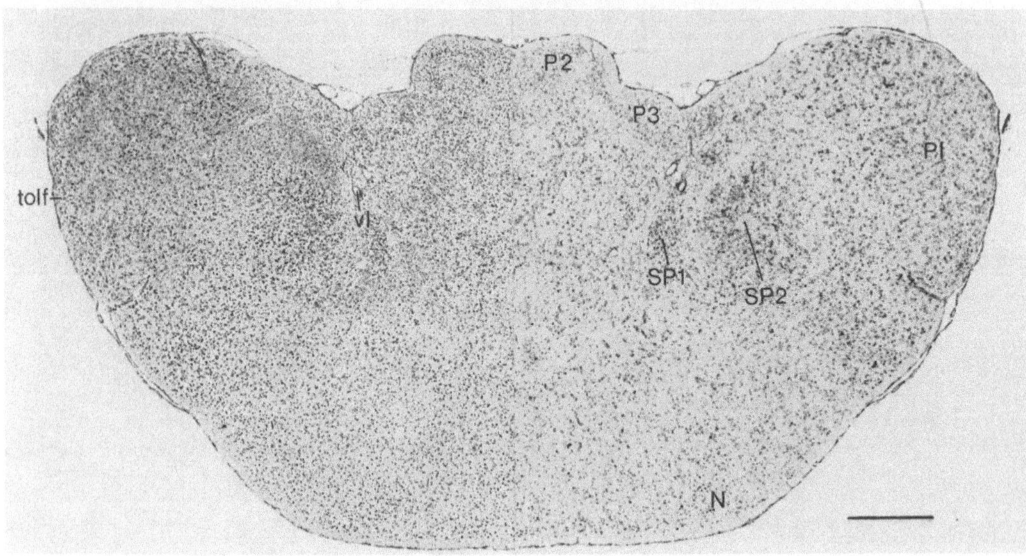
P2
P3
Pl
tolf
vl
SP1
SP2
N

97

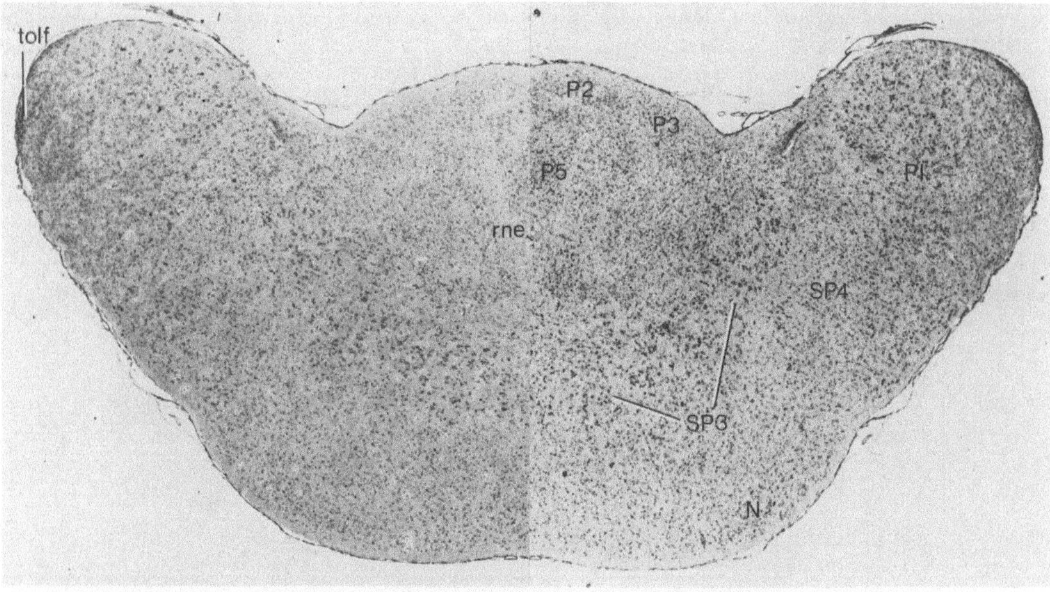

98

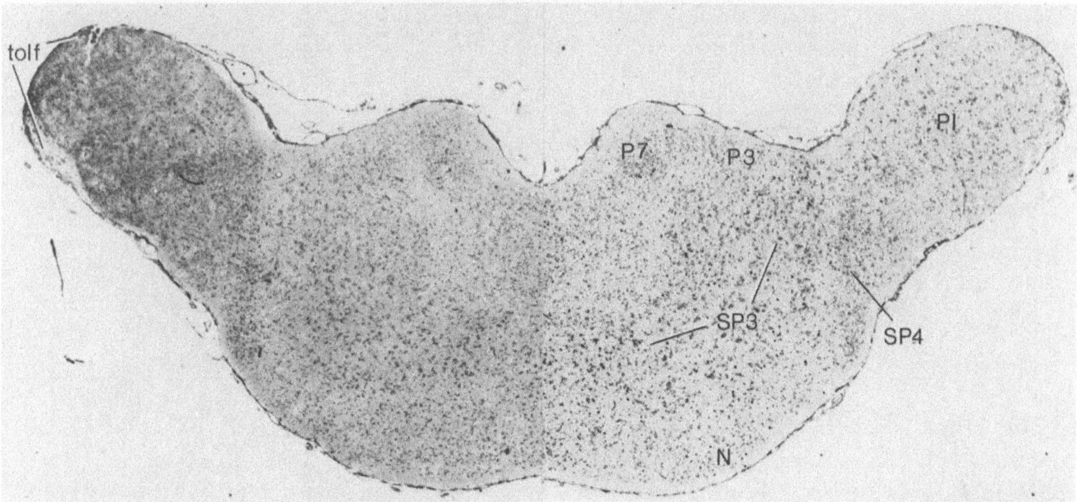

99

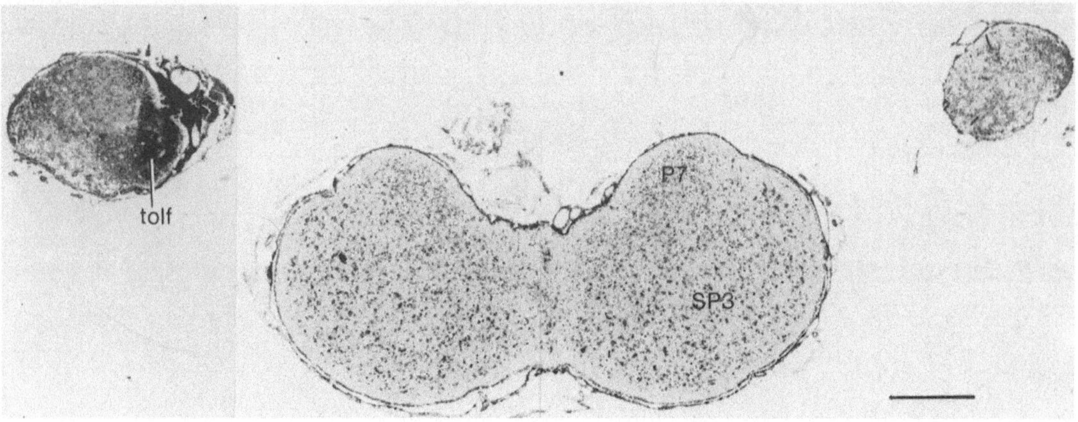

Hydrolagus collei

(Fig. 100–139)

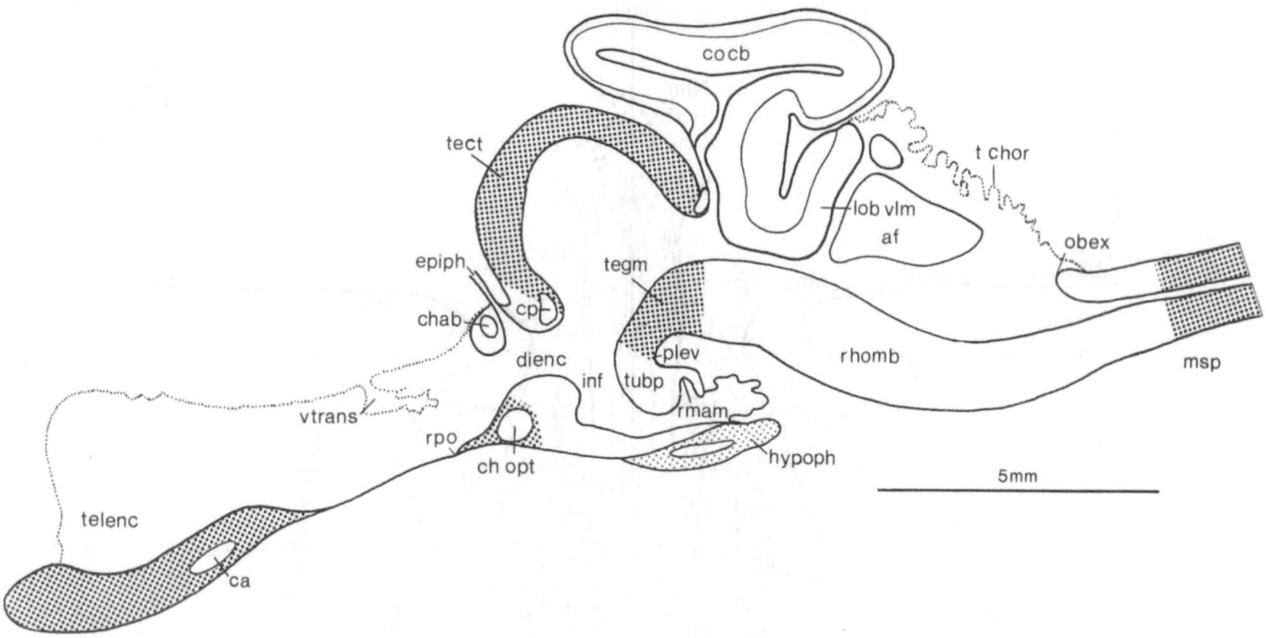

Fig. 100. Midsagittal section of the brain of *Hydrolagus collei*

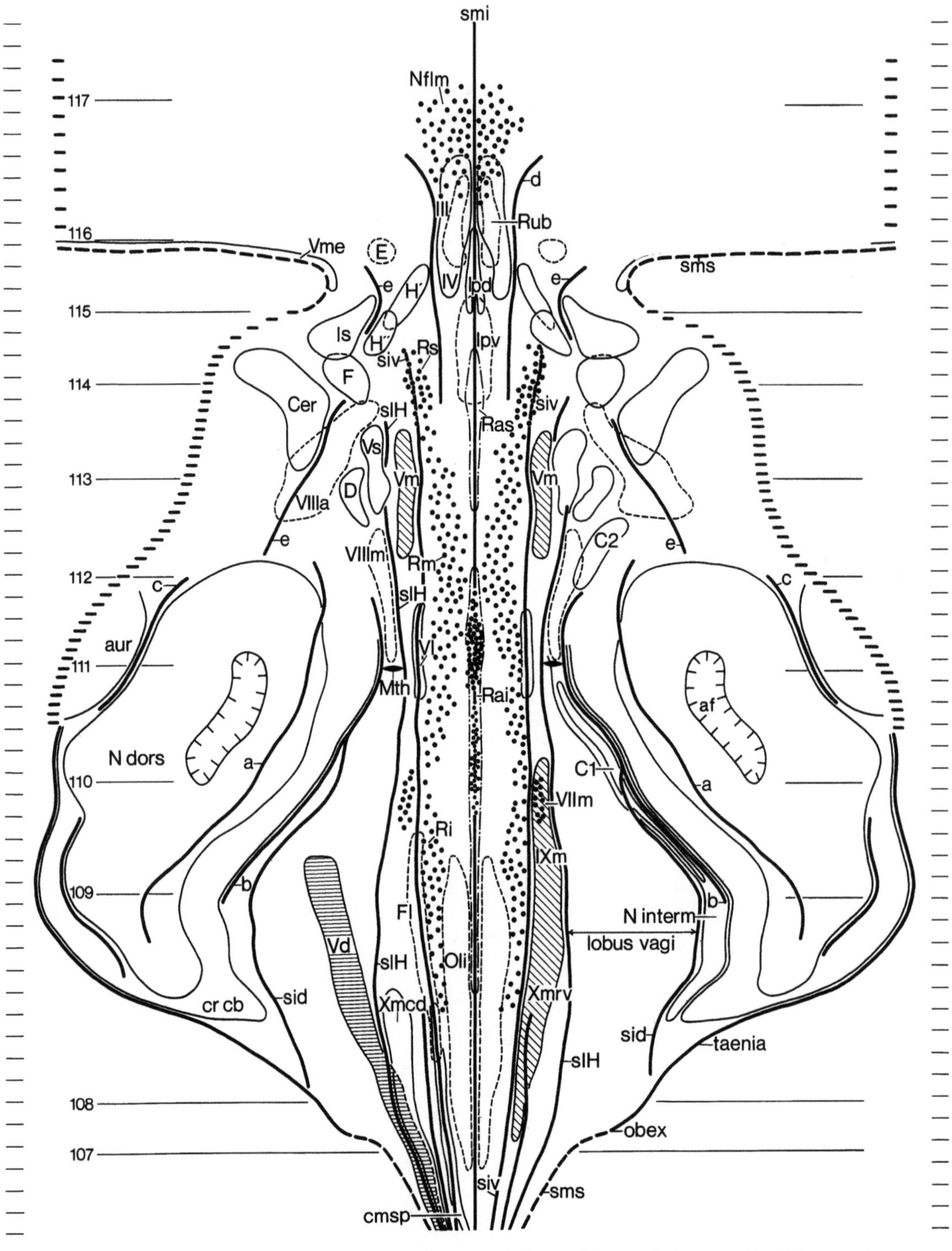

Fig. 101. Topological reconstruction of the brain stem of the ratfish *Hydrolagus collei.* Distance between two sections, indicated by *small horizontal bars,* equals 300 μm. For further explanation see Fig. 2

Fig. 102. Topographical reconstruction showing the cell masses of the brain stem of *Hydrolagus collei* as projected upon a horizontal plane. Distance between two sections, indicated by *small horizontal bars,* equals 300 μm. For further explanation see Fig. 3

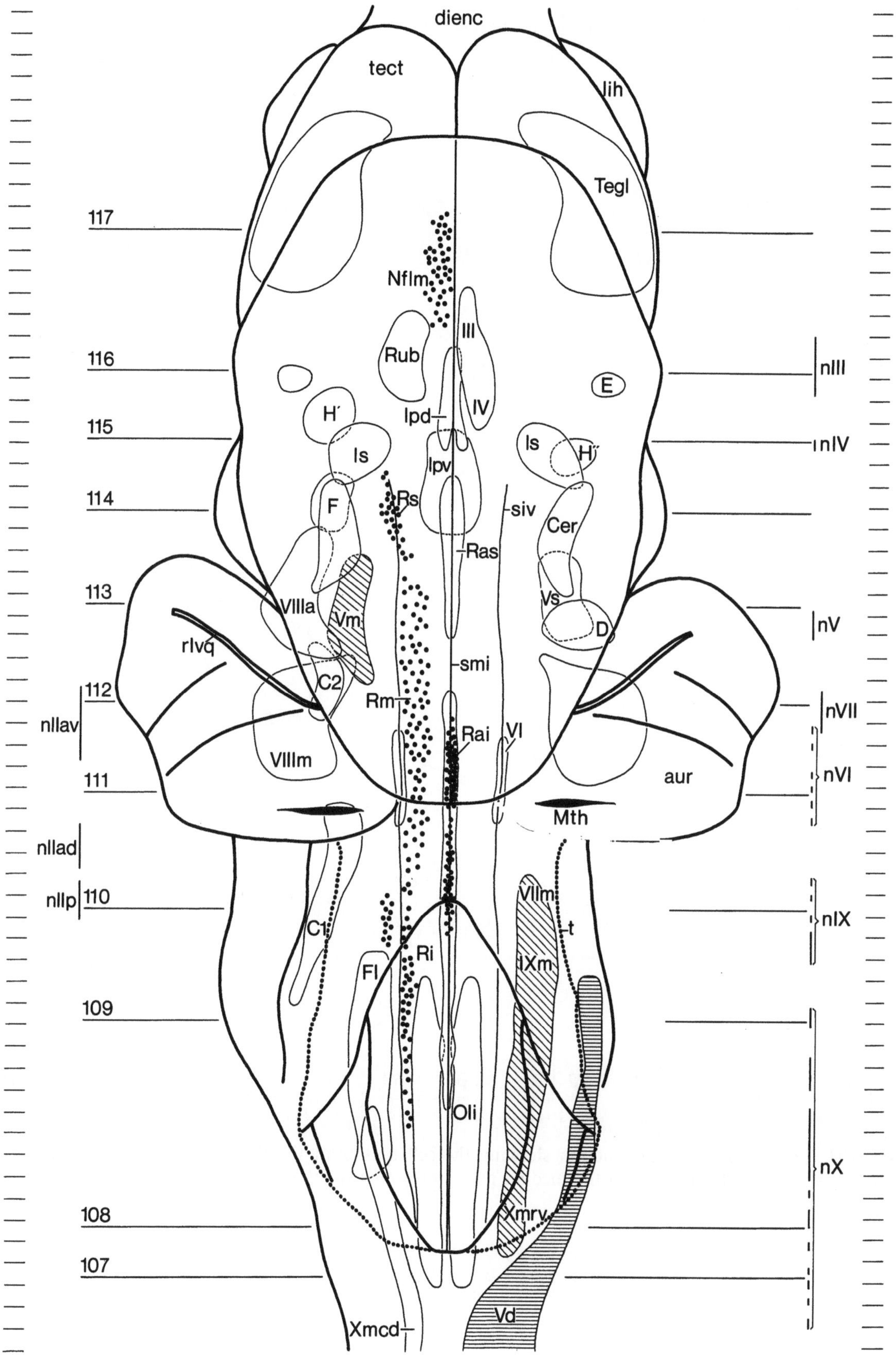
dienc
tect
lih
Tegl
117
Nflm
III
Rub
116
nIII
E
H′
lpd
IV
115
nIV
Is
lpv
H″
114
F
Rs
siv
Cer
Ras
113
VIIIa
Vm
Vs
D
nV
rlvq
smi
C2
112
Rm
nVII
nllav
Rai
VI
VIIIm
111
aur
nVI
Mth
nllad
nllp
110
VIIm
nIX
t
C1
Ri
IXm
Fl
109
Oli
nX
108
Xmrv
107
Vd
Xmcd

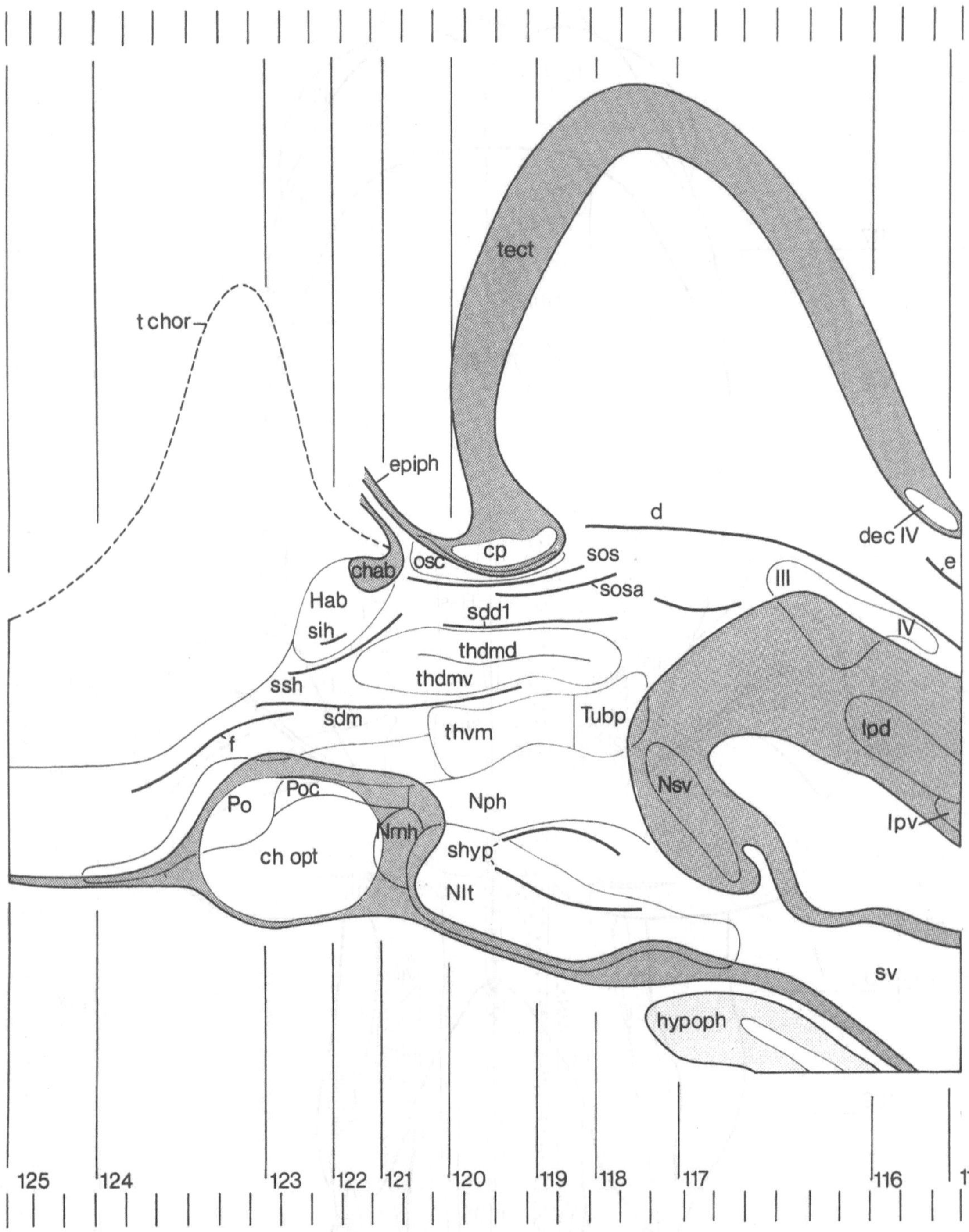

Fig. 103. Topographical reconstruction showing the cell masses of the diencephalon of *Hydrolagus collei* lying periventricularly or in the median plane, projected upon a midsagittal plane. For further explanation see Fig. 4

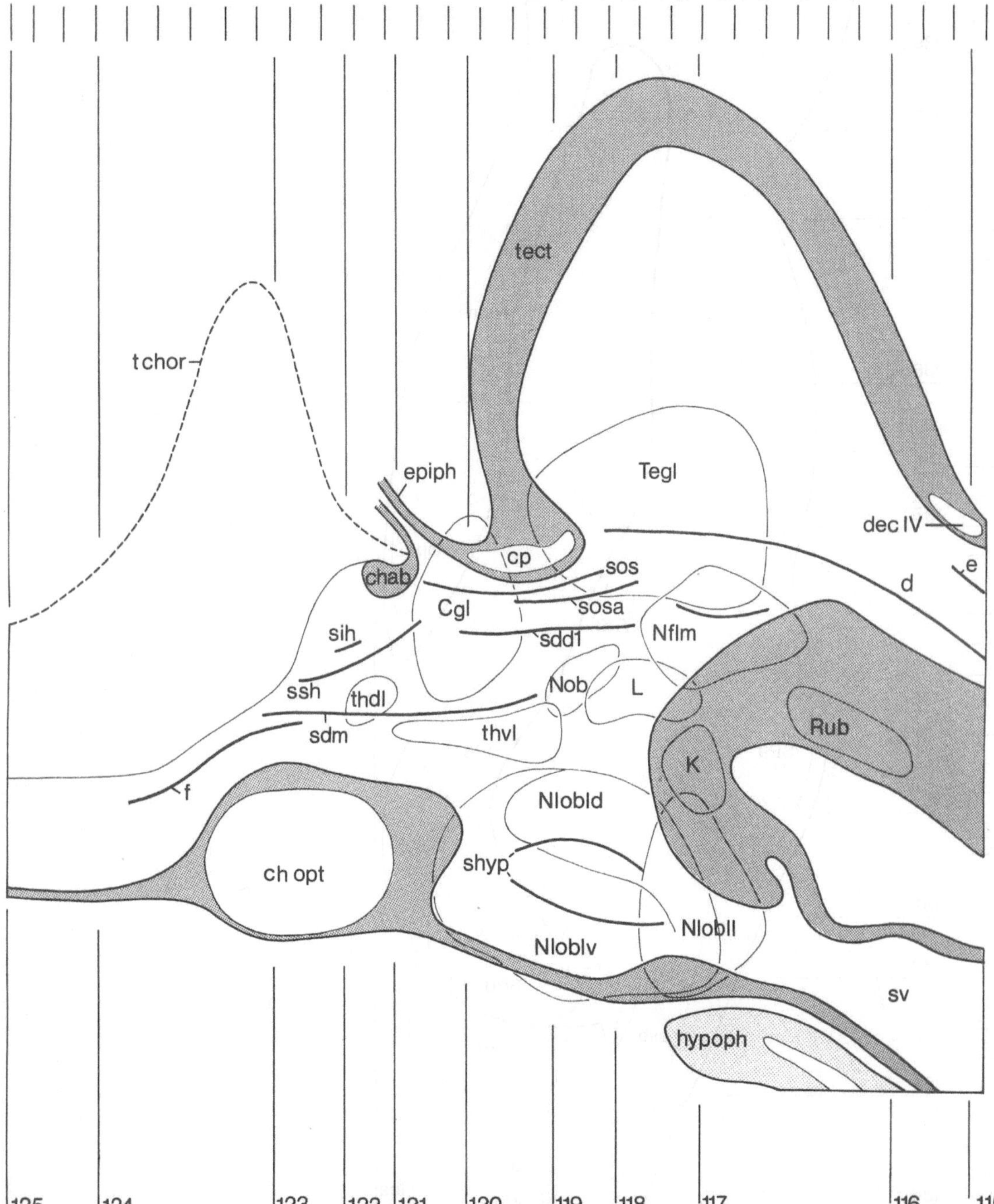

Fig. 104. Topographical reconstruction showing the migrated cell masses of the diencephalon of *Hydrolagus collei* as projected upon a midsagittal plane. For further explanation see Fig. 4

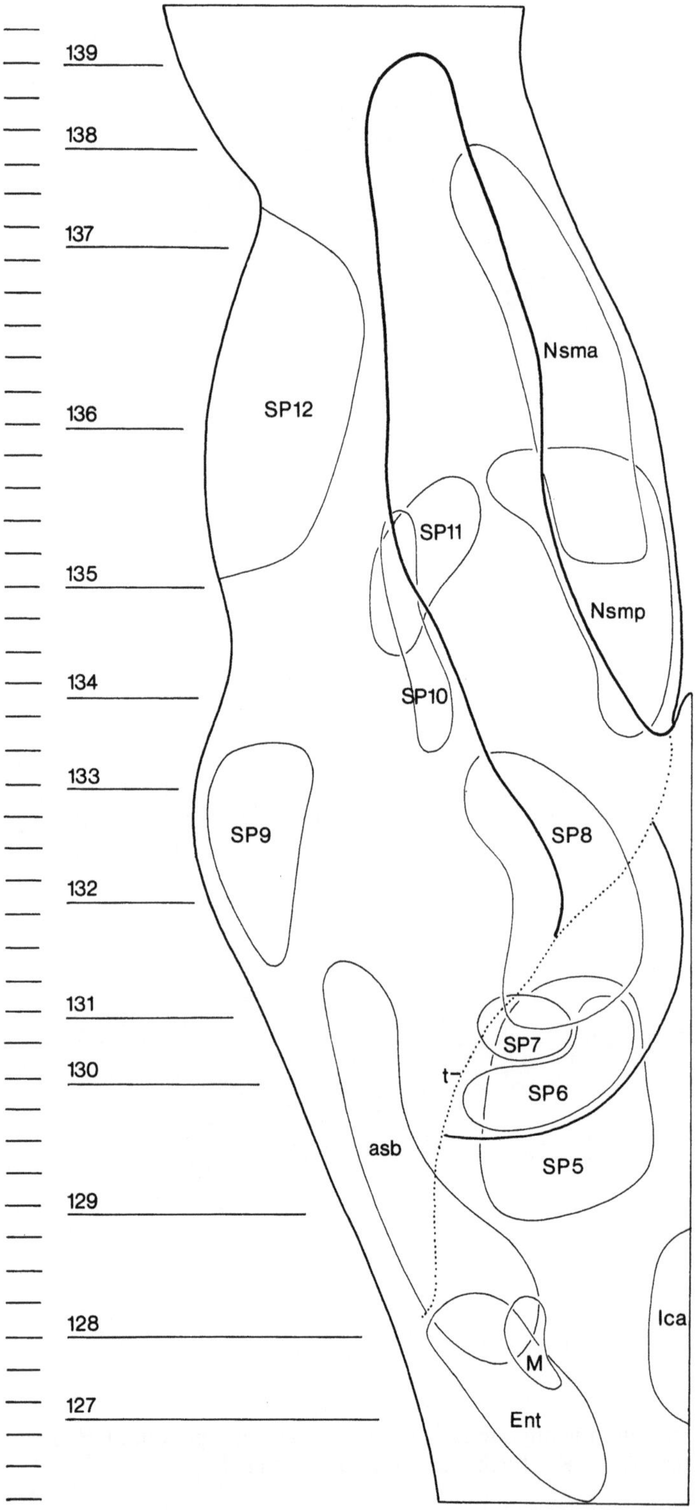

Fig. 105. Topographical reconstruction showing the cell masses of the subpallium and of the telencephalon impar of *Hydrolagus collei* as projected upon a horizontal plane. For further explanation see Fig. 6

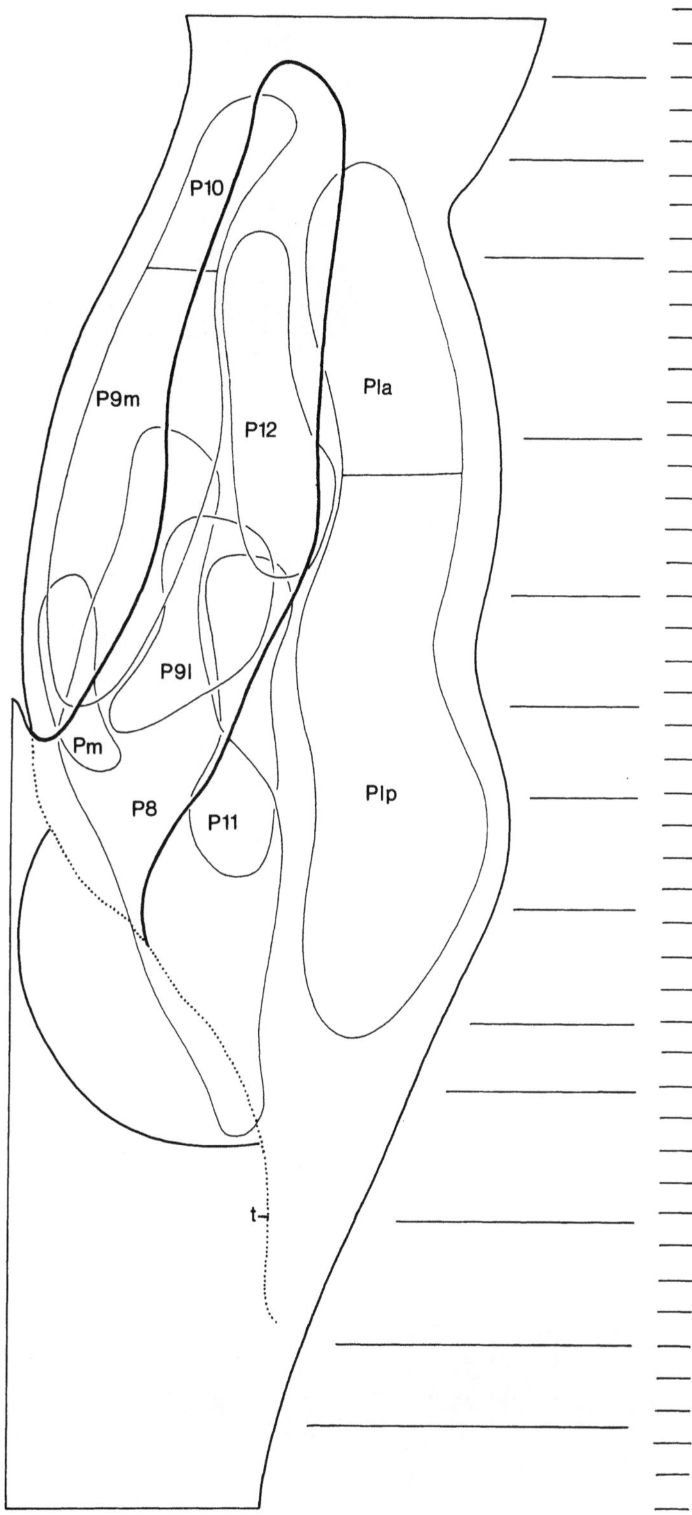

Fig. 106. Topographical reconstruction showing the pallial cell masses of the telencephalon of *Hydrolagus collei* as projected upon a horizontal plane. For further explanation see Fig. 6

Figs. 107–139. Photomicrographs of transverse sections of the brain of *Hydrolagus collei*. For further explanation see Fig. 8–33

107

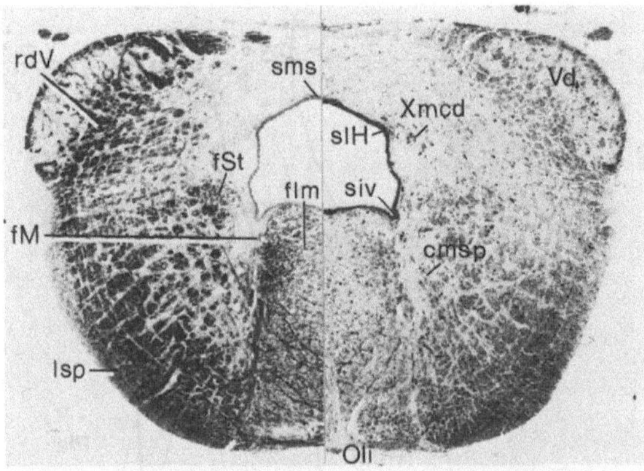
rdV
sms
Vd
Xmcd
slH
fSt
flm
siv
fM
cmsp
lsp
Oli

108

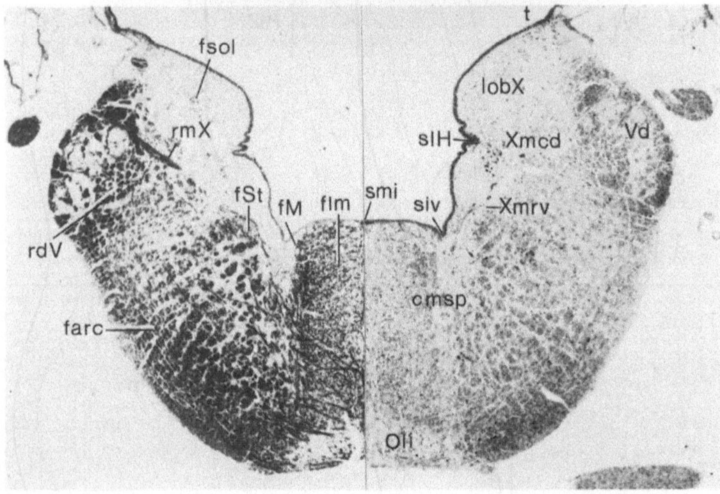
t
fsol
lobX
rmX
slH
Xmcd
Vd
fSt
fM
flm
smi
siv
Xmrv
rdV
cmsp
farc
Oli

109

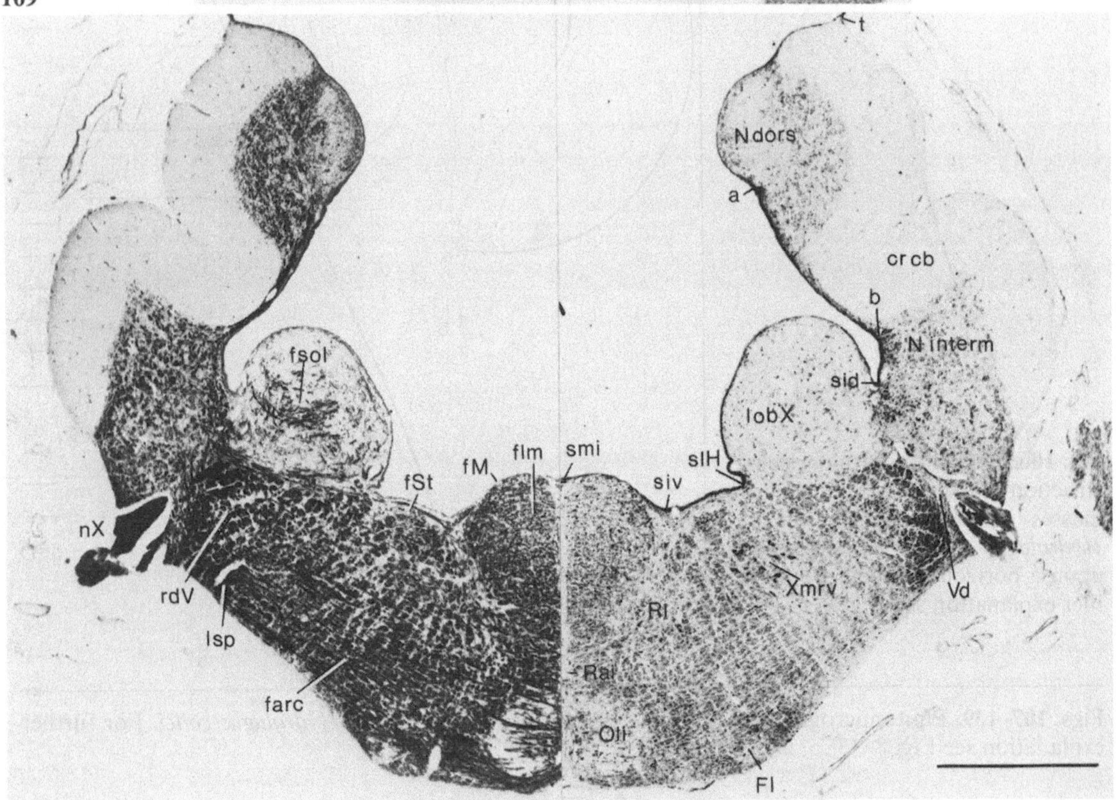
t
Ndors
a
crcb
b
N interm
fsol
sid
lobX
fM
flm
smi
slH
fSt
siv
nX
Xmrv
Vd
rdV
RI
lsp
Ral
farc
Oli
FI

110

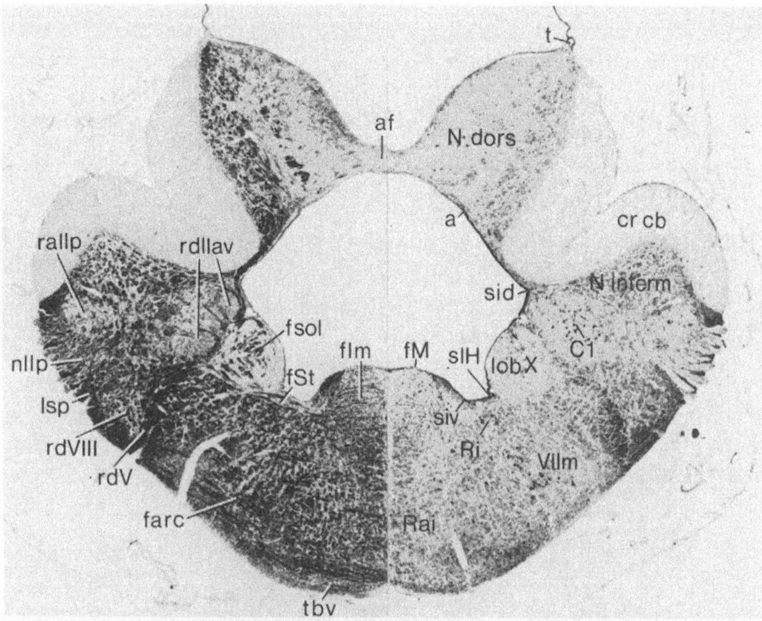

t
af
N dors
a
cr cb
rallp
rdllav
sid
N interm
fsol
C1
nllp
flm
fM
slH
lobX
fSt
lsp
siv
rdVIII
Ri
rdV
Vllm
farc
Rai
tbv

111

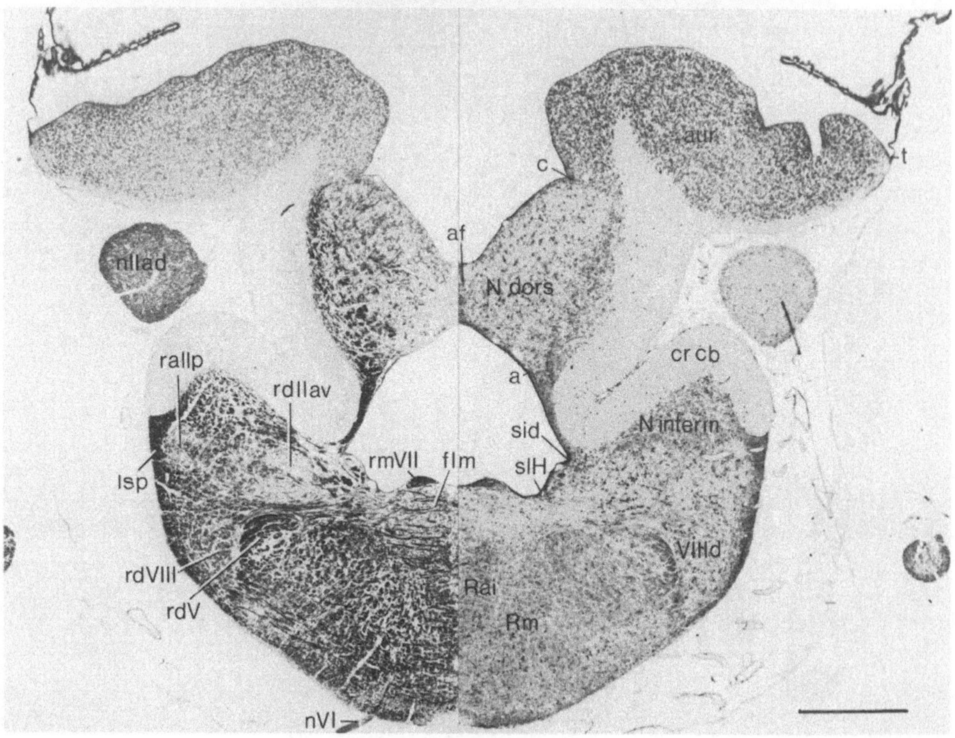

aur
c
t
af
nllad
N dors
rallp
cr cb
a
rdllav
sid
N interm
rmVII
flm
slH
lsp
VIIId
rdVIII
Rai
rdV
Rm
nVI

112

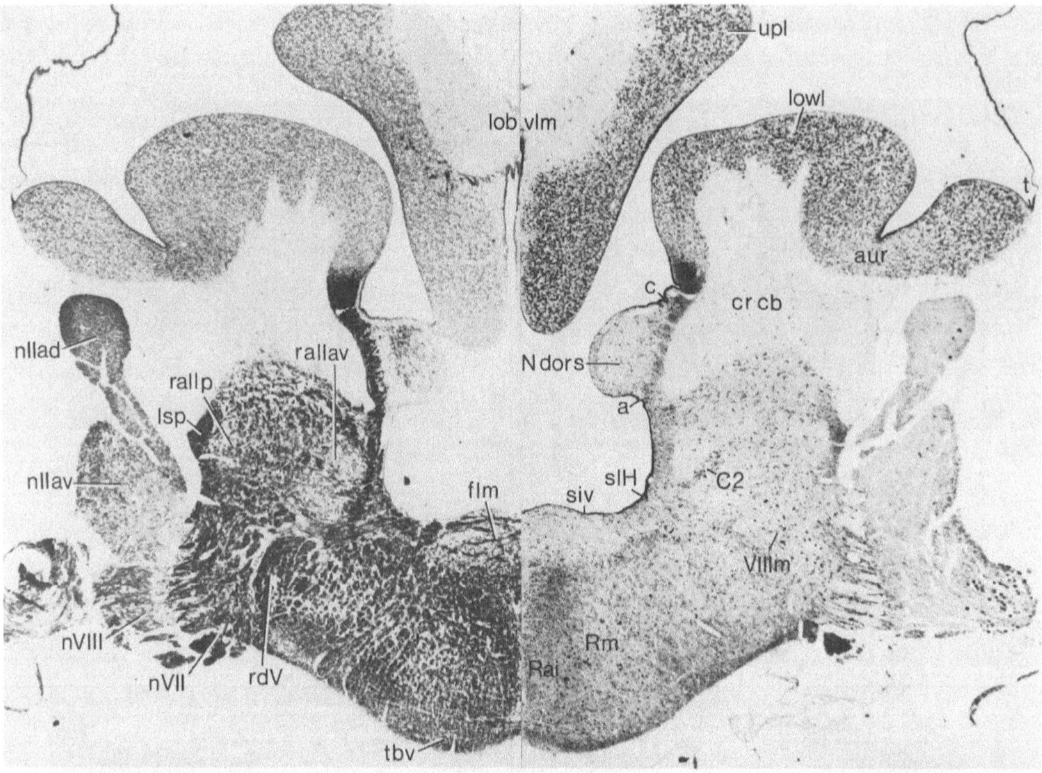
upl
lowl
lob vlm
t
aur
c
cr cb
nllad
rallav
Ndors
rallp
a
lsp
nllav
flm
siv
slH
C2
VIIIm
nVIII
Rm
nVII
rdV
Rai
tbv

113

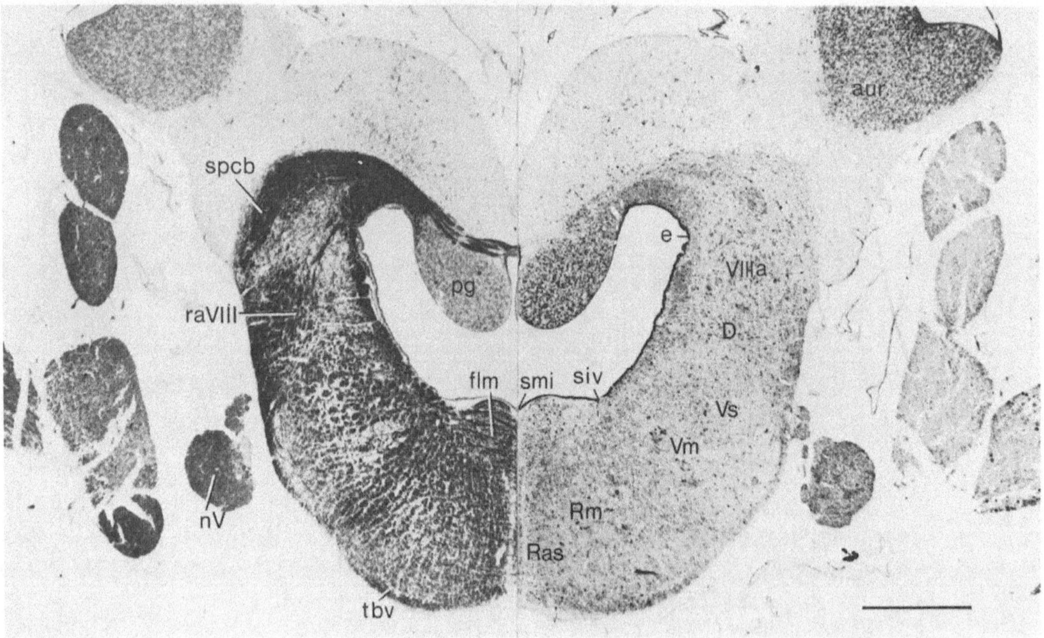
aur
spcb
e
VIIIa
pg
raVIII
D
flm
smi
siv
Vs
Vm
nV
Rm
Ras
tbv

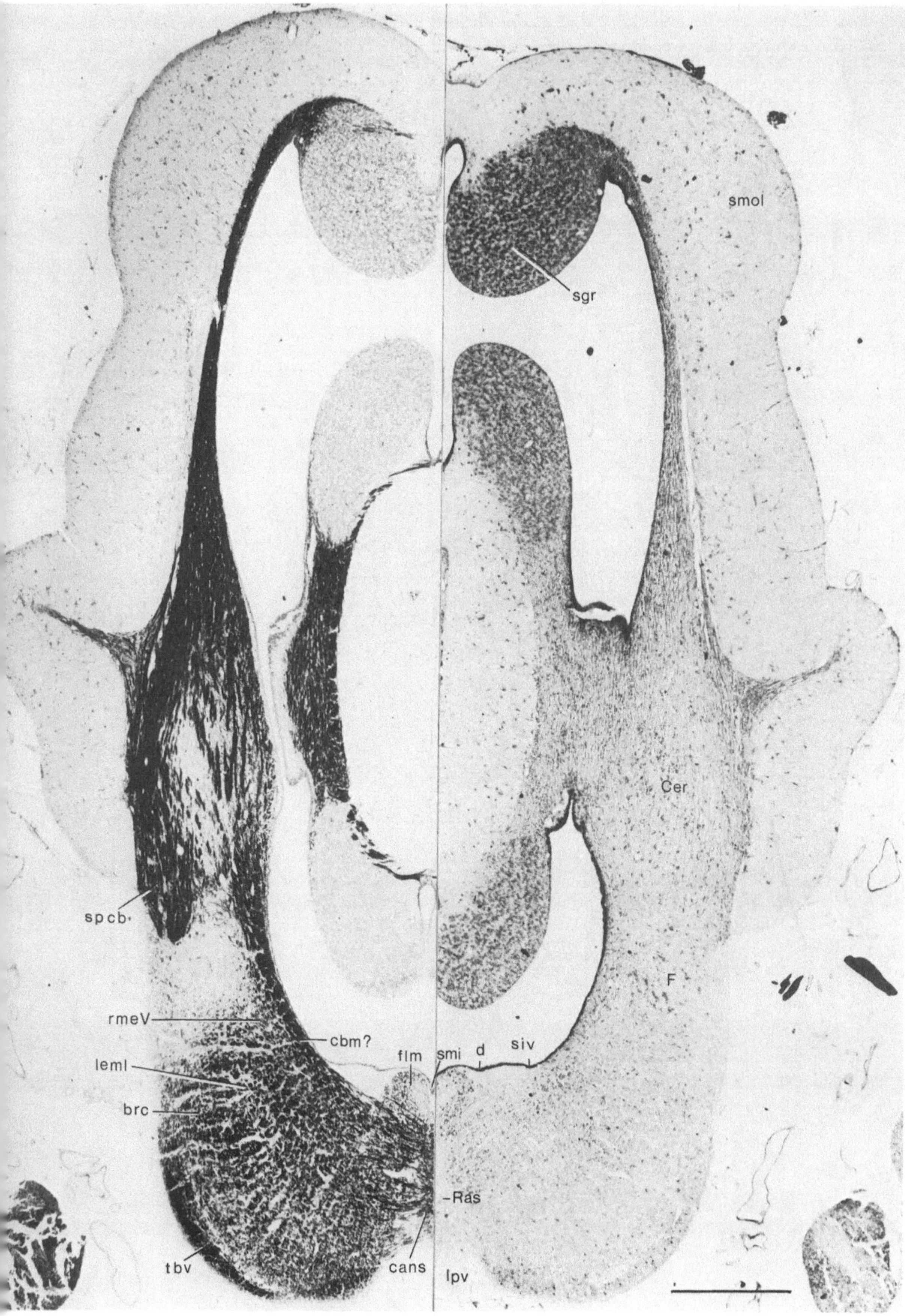
smol
sgr
Cer
F
sp cb
rmeV
cbm?
leml
brc
flm
smi
d
siv
-Ras
tbv
cans
lpv

115

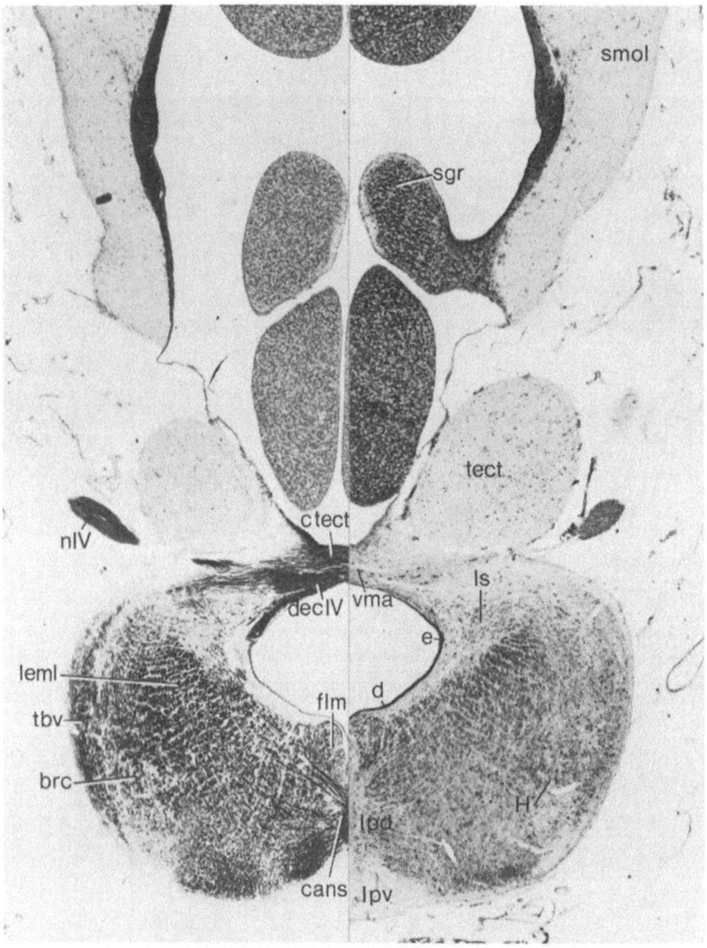
smol
sgr
tect
nIV
ctect
decIV
vma
ls
e
leml
tbv
brc
flm
d
H
lpd
cans
lpv

116

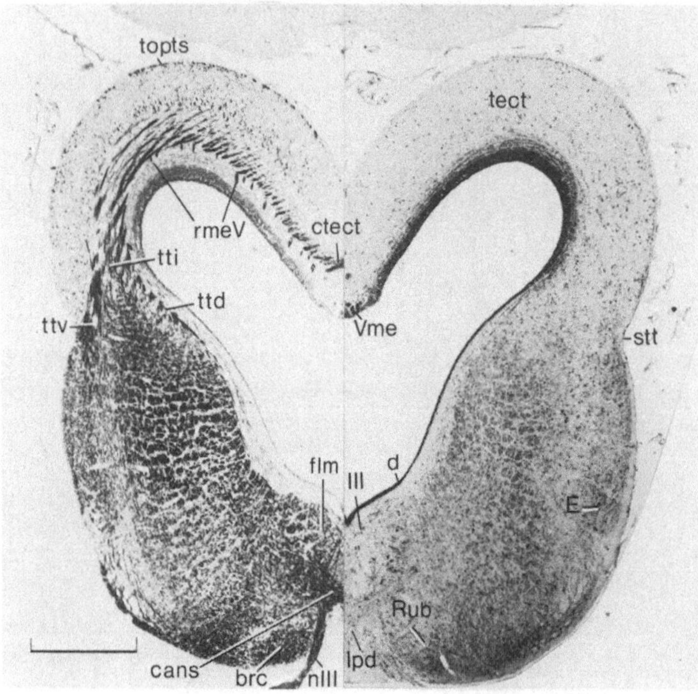
topts
tect
rmeV
ctect
tti
ttd
ttv
Vme
stt
flm
d
III
E
Rub
cans
brc
nIII
lpd

117

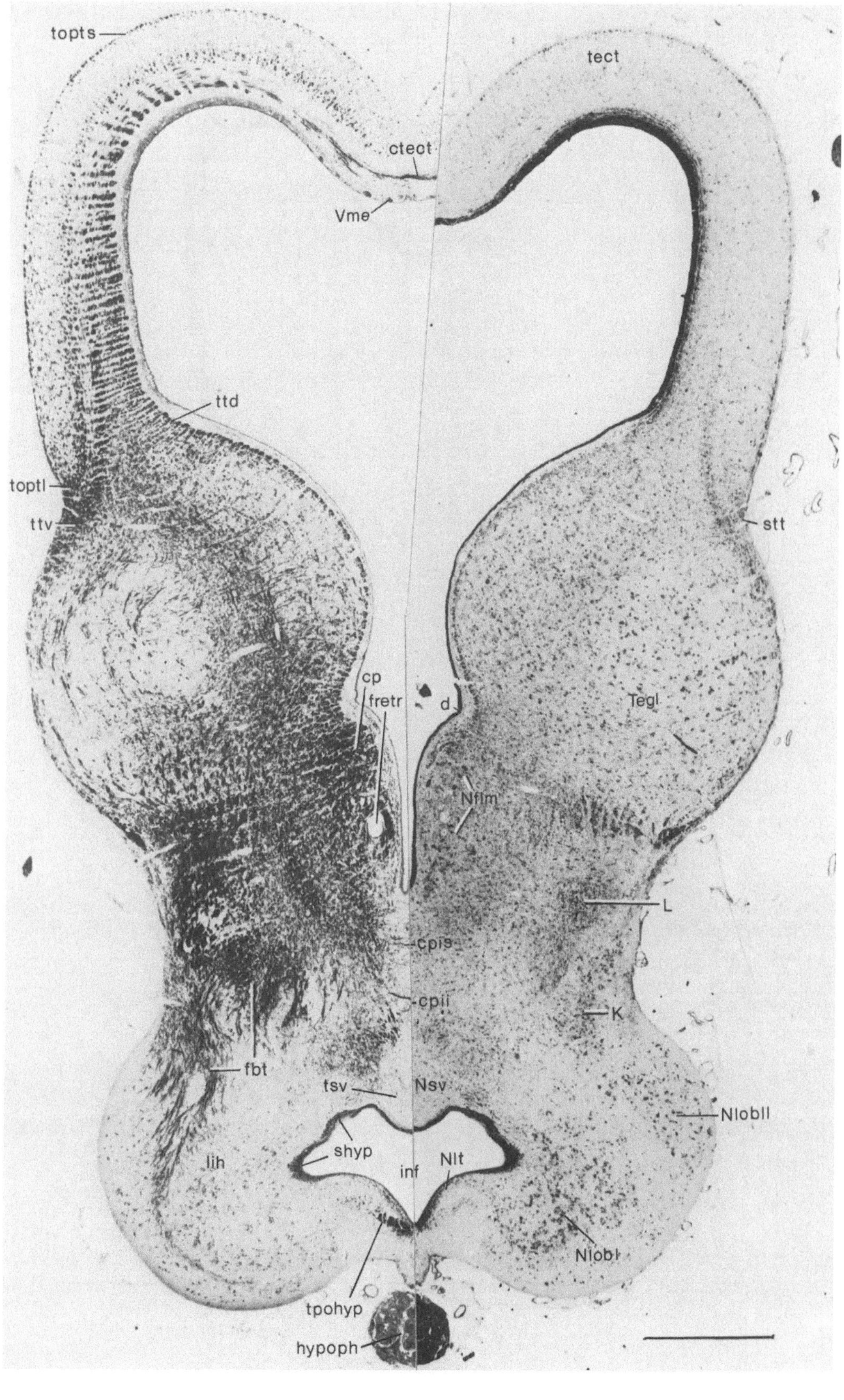
topts
tect
ctect
Vme
ttd
toptl
ttv
stt
cp
fretr
d
Tegl
Nflm
L
cpis
cpii
K
fbt
tsv
Nsv
Nlobll
shyp
lih
inf
Nlt
Nlobl
tpohyp
hypoph

118

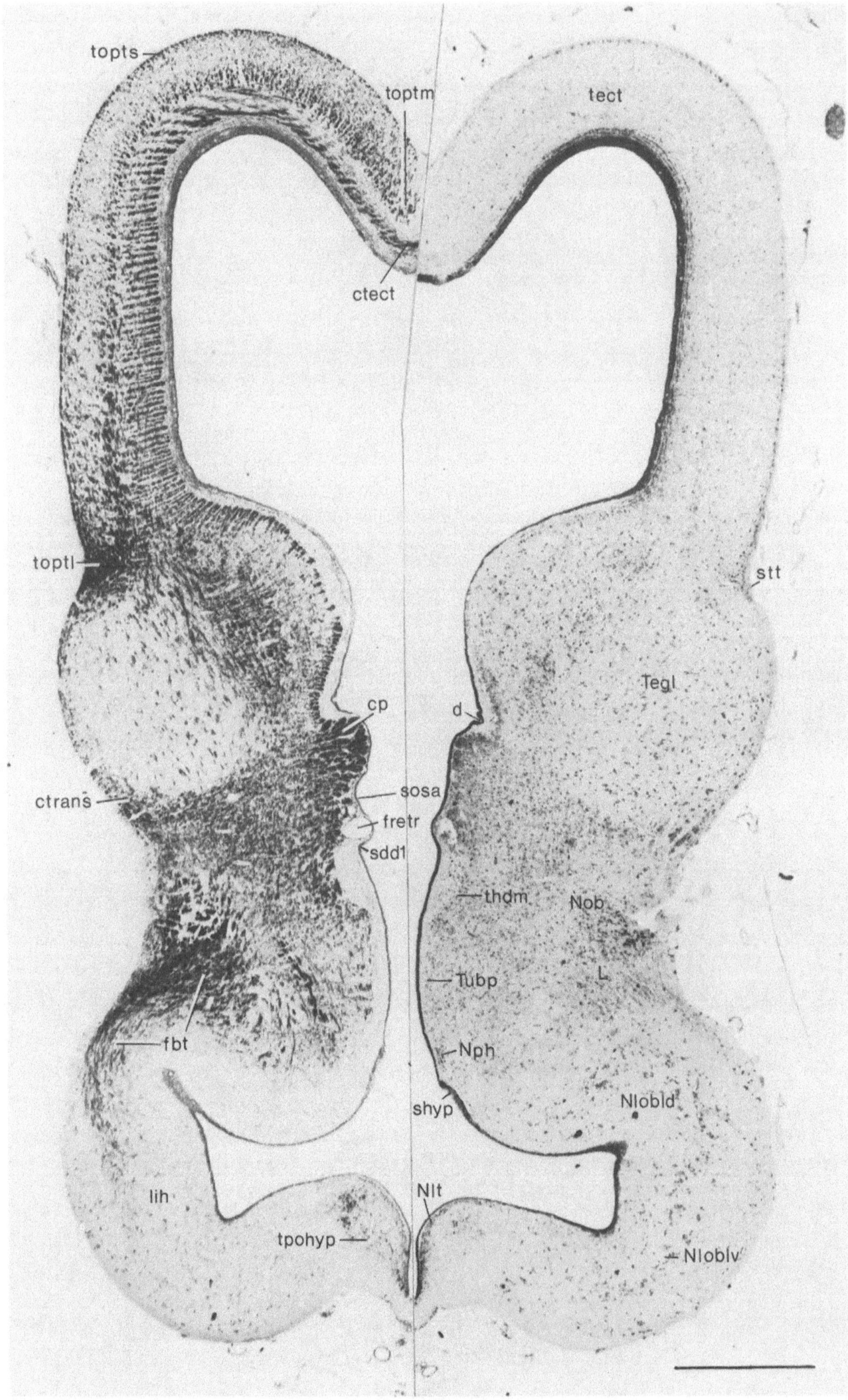

topts
toptm
tect
ctect
toptl
stt
Tegl
cp
d
ctrans
sosa
fretr
sdd1
thdm
Nob
L
Tubp
fbt
Nph
shyp
Nlobld
lih
Nlt
tpohyp
Nloblv

119

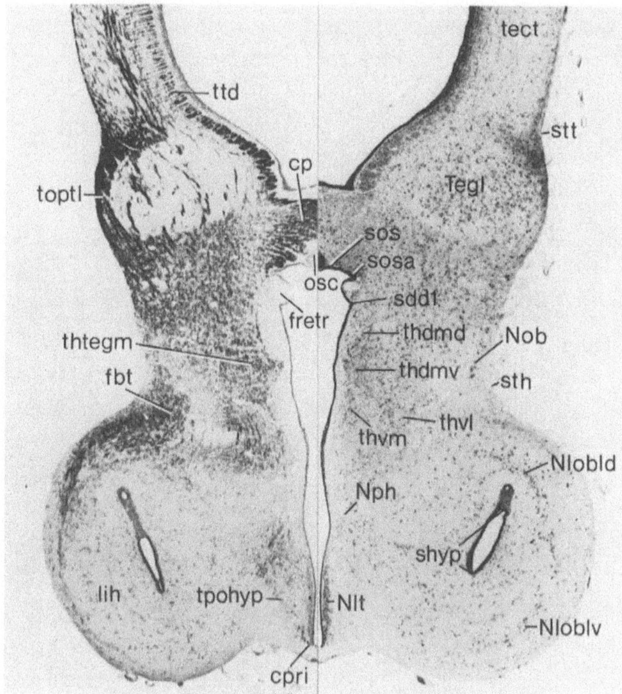
tect
ttd
stt
cp
toptl
Tegl
sos
sosa
osc
sdd1
fretr
thdmd
Nob
thtegm
thdmv
sth
fbt
thvl
thvm
Nlobld
Nph
shyp
lih
tpohyp
Nlt
Nloblv
cpri

120

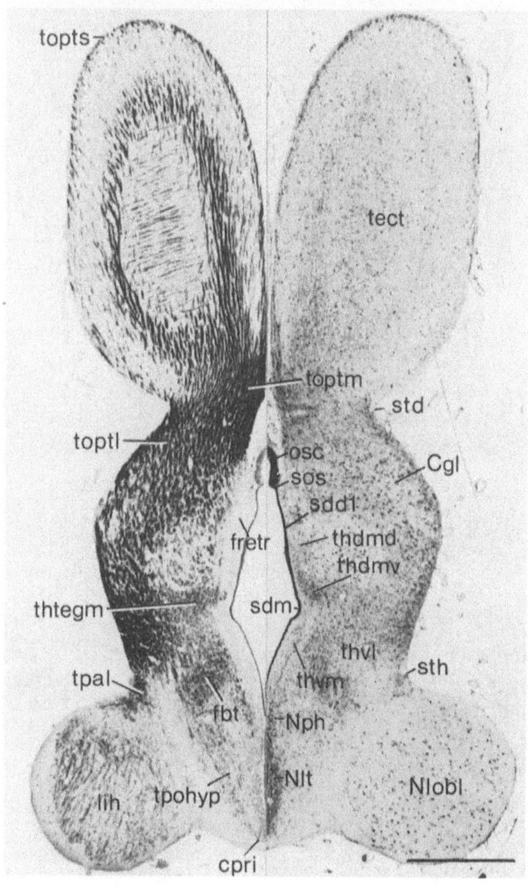
topts
tect
toptm
std
toptl
osc
Cgl
sos
sdd1
fretr
thdmd
thdmv
thtegm
sdm
thvl
sth
tpal
thvm
fbt
Nph
Nlt
lih
tpohyp
Nlobl
cpri

121

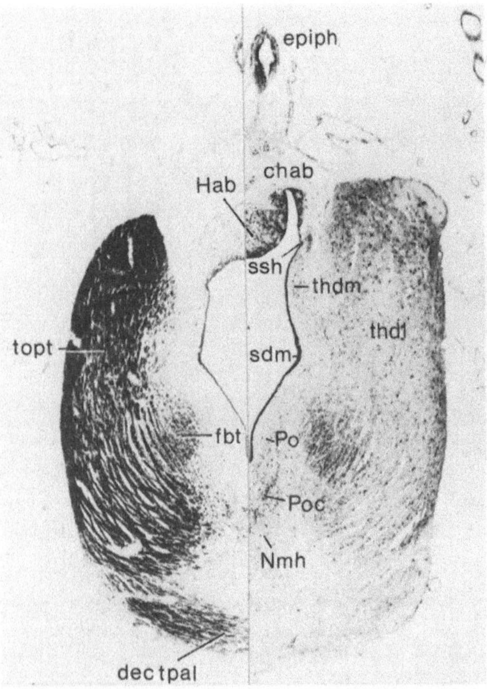

122

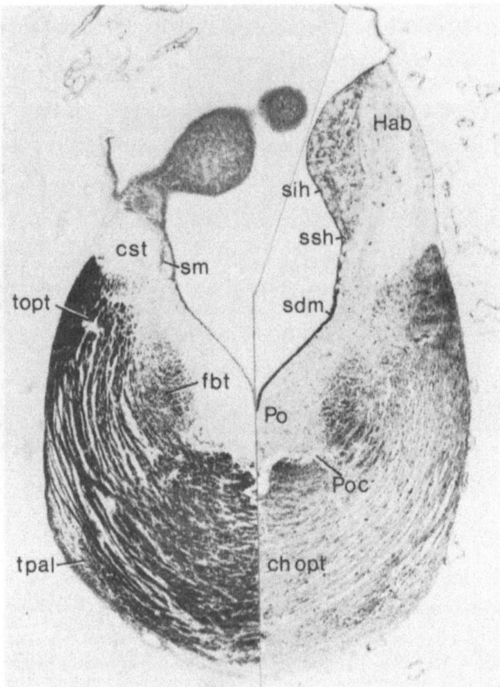

123

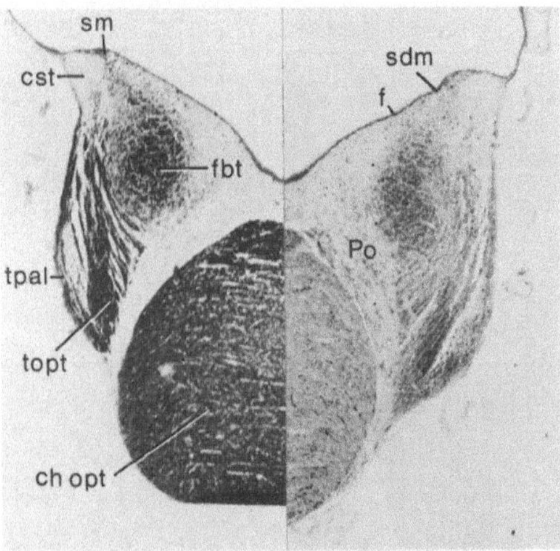

124

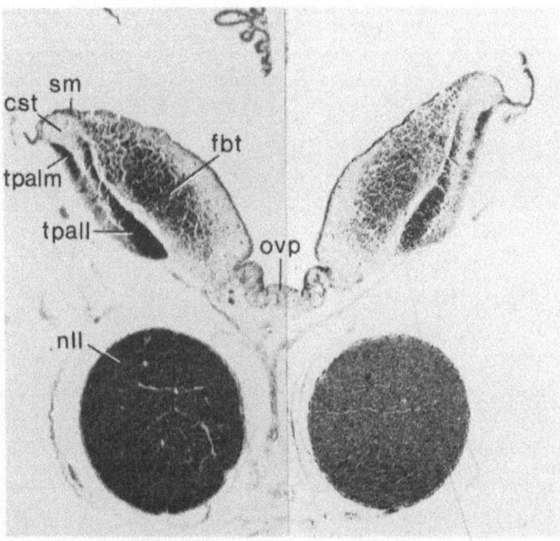

125

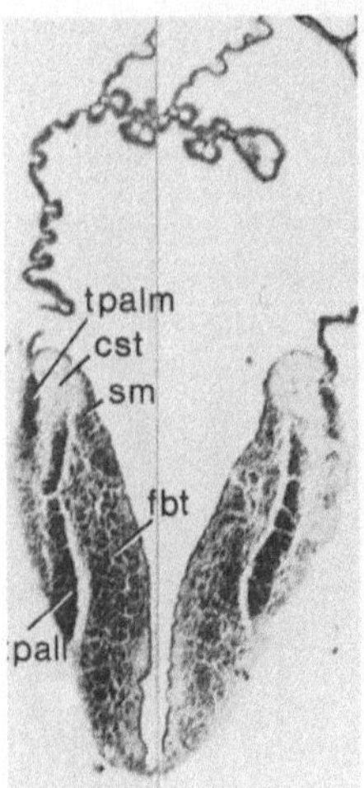

126

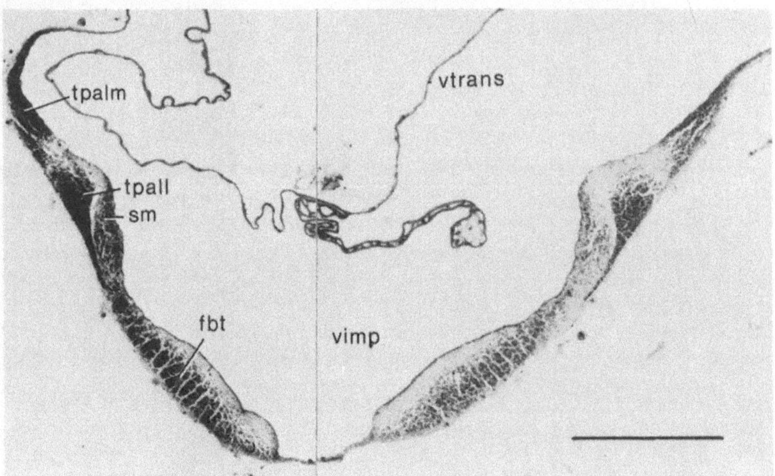

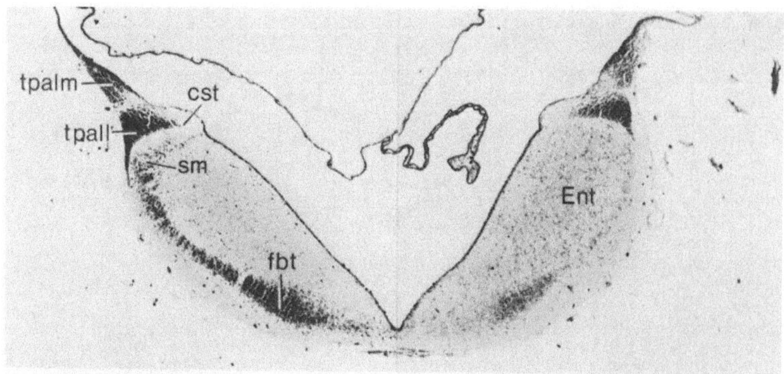
tpalm
cst
tpall
sm
Ent
fbt

127

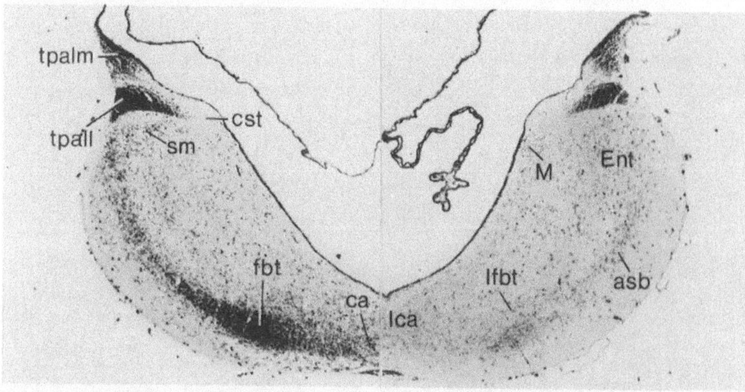
tpalm
tpall
cst
sm
M
Ent
fbt
lfbt
asb
ca
lca

128

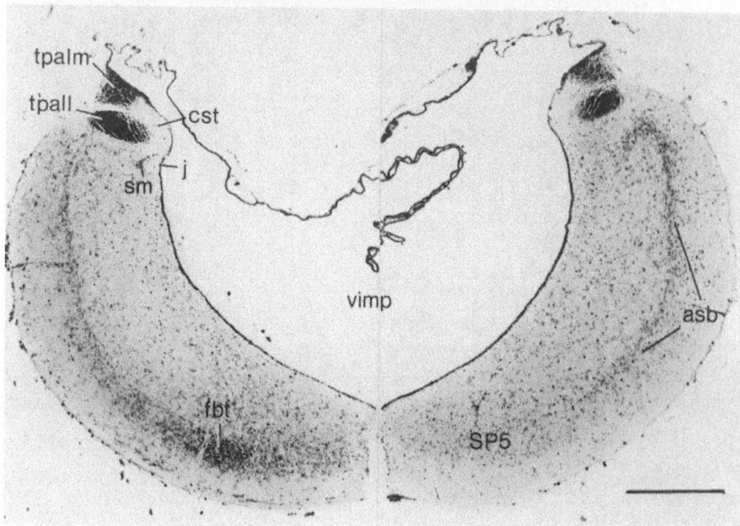
tpalm
tpall
cst
sm
j
vimp
asb
fbt
SP5

129

130

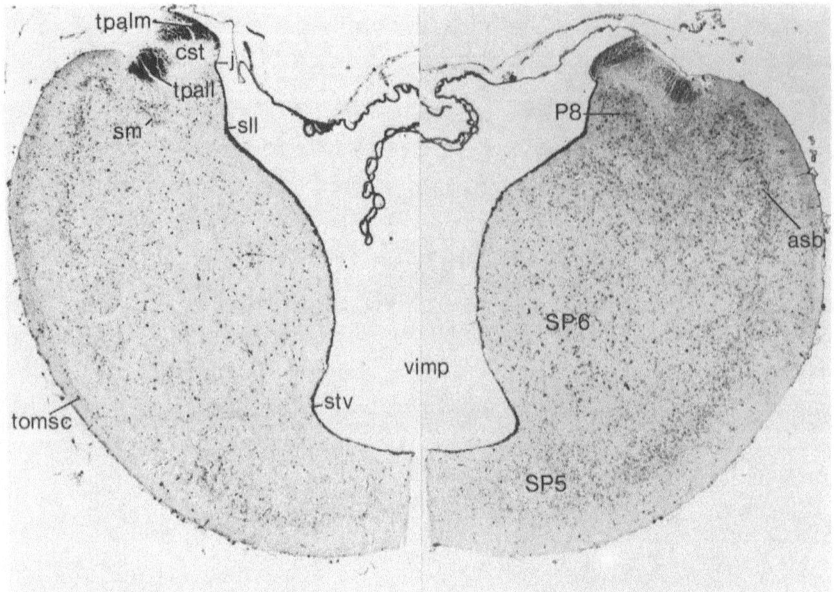
tpalm
cst
j
tpall
sm
sll
P8
asb
SP6
vimp
stv
tomsc
SP5

131

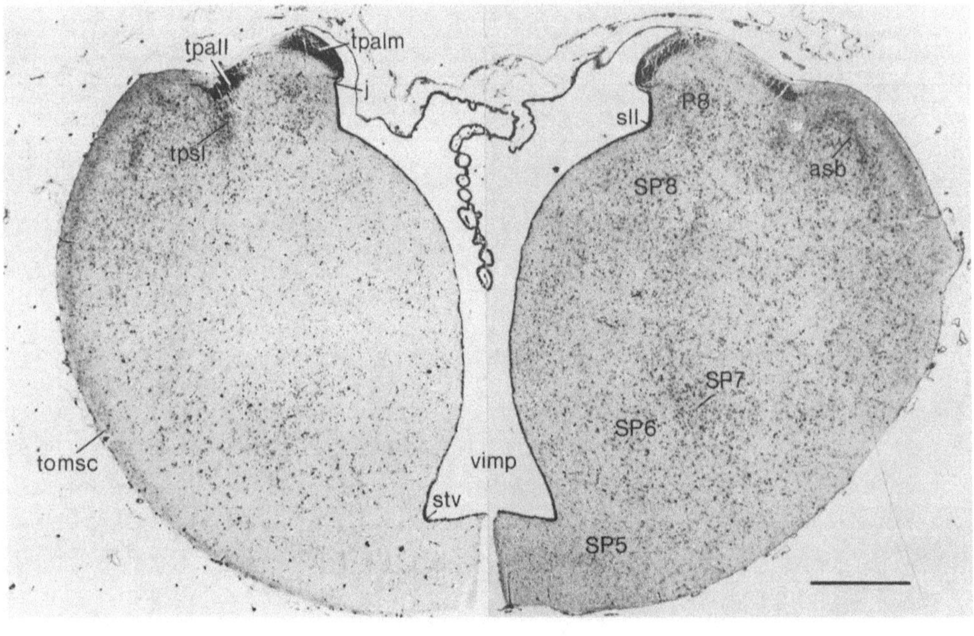
tpall
tpalm
j
sll
P8
tpsl
asb
SP8
SP7
SP6
tomsc
vimp
stv
SP5

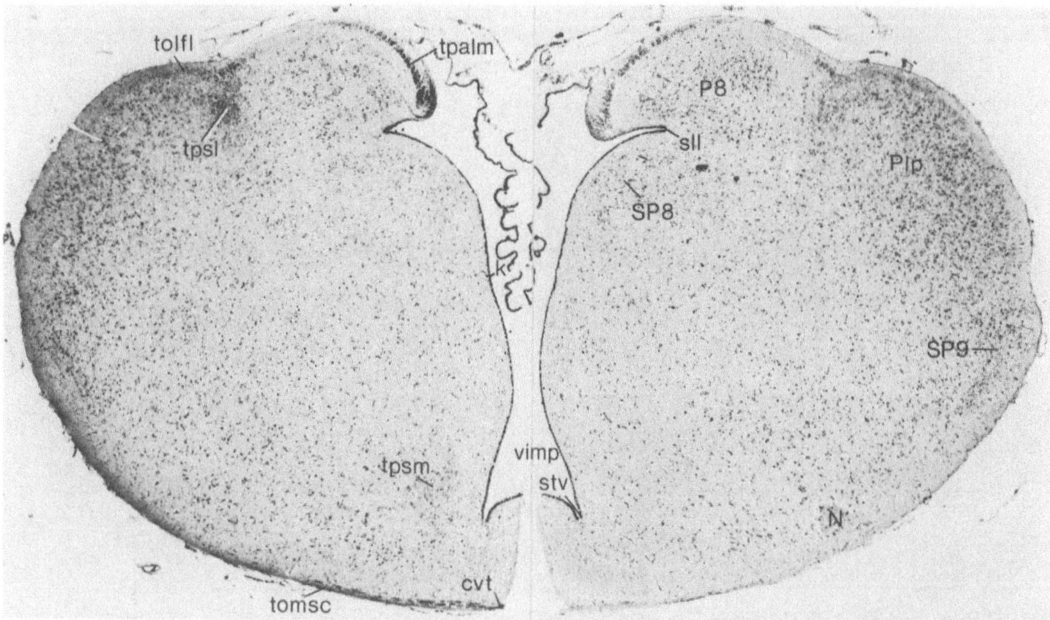
tolfl
tpalm
P8
tpsl
sll
Plp
SP8
k
SP9
vimp
tpsm
stv
N
cvt
tomsc

132

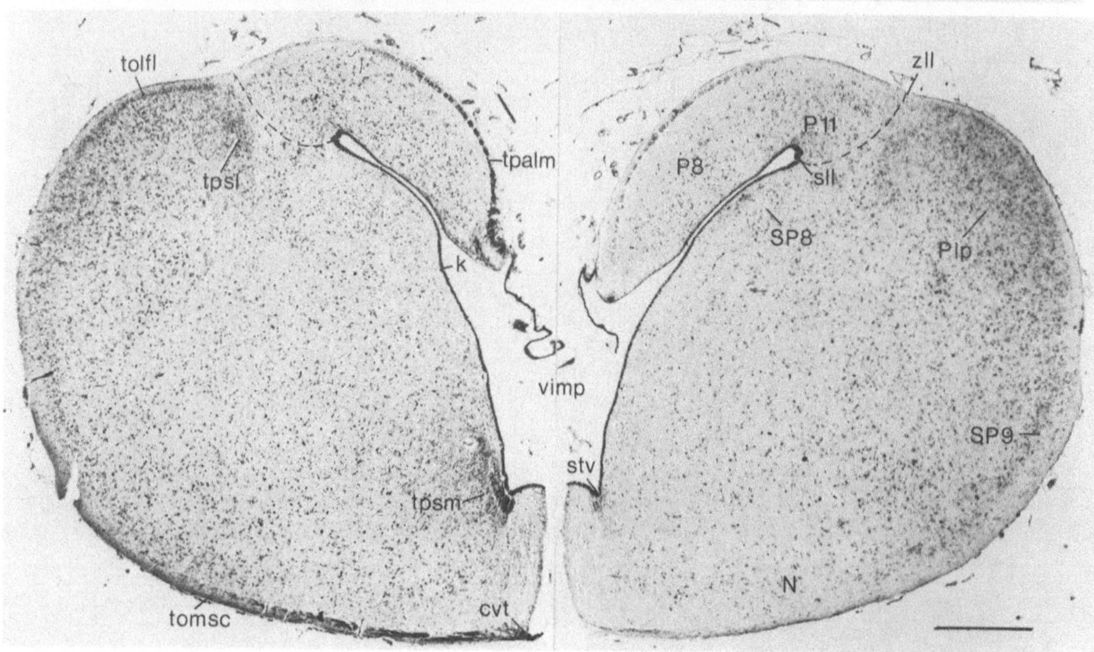
tolfl
zll
P11
tpalm
P8
tpsl
sll
SP8
Plp
k
vimp
SP9
stv
tpsm
N
tomsc
cvt

133

134

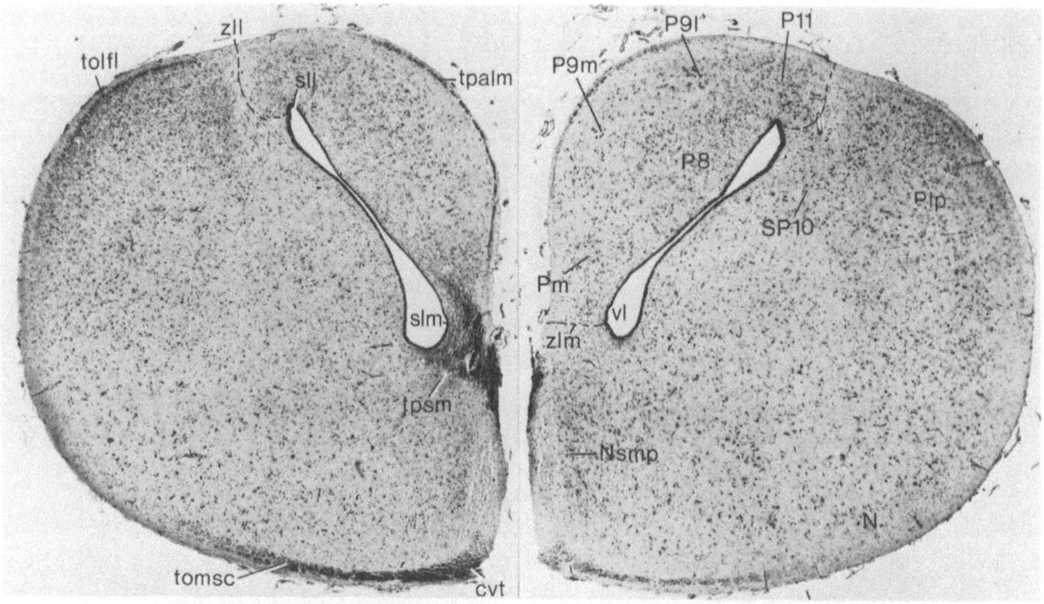
zll
tolfl
sll
tpalm
slm
tpsm
tomsc
cvt
P9l
P11
P9m
P8
Plp
SP10
Pm
vl
zlm
Nsmp
N

135

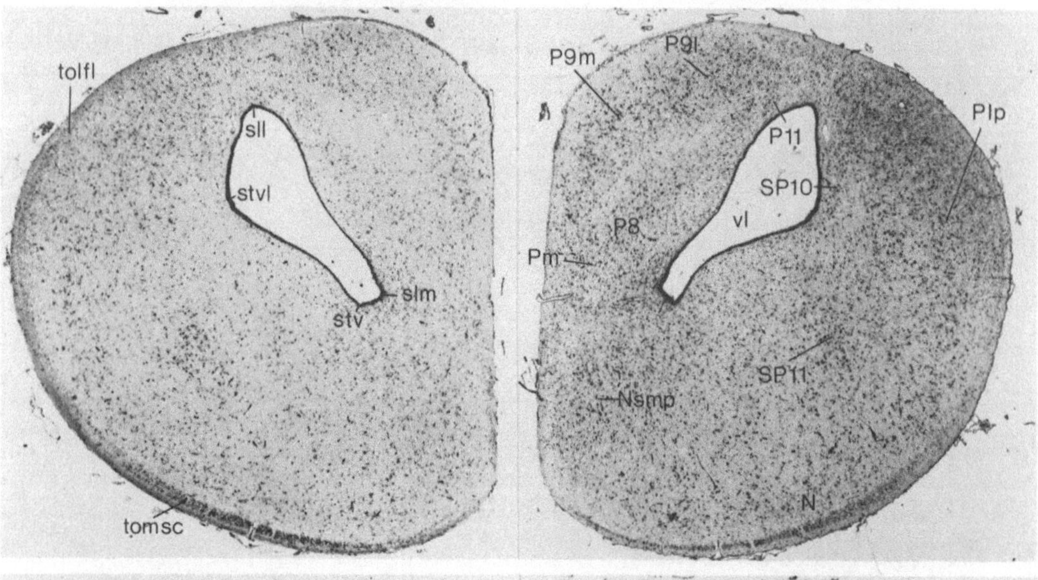
tolfl
sll
stvl
slm
stv
tomsc
P9m
P9l
Plp
P11
SP10
vl
P8
Pm
SP11
Nsmp
N

136

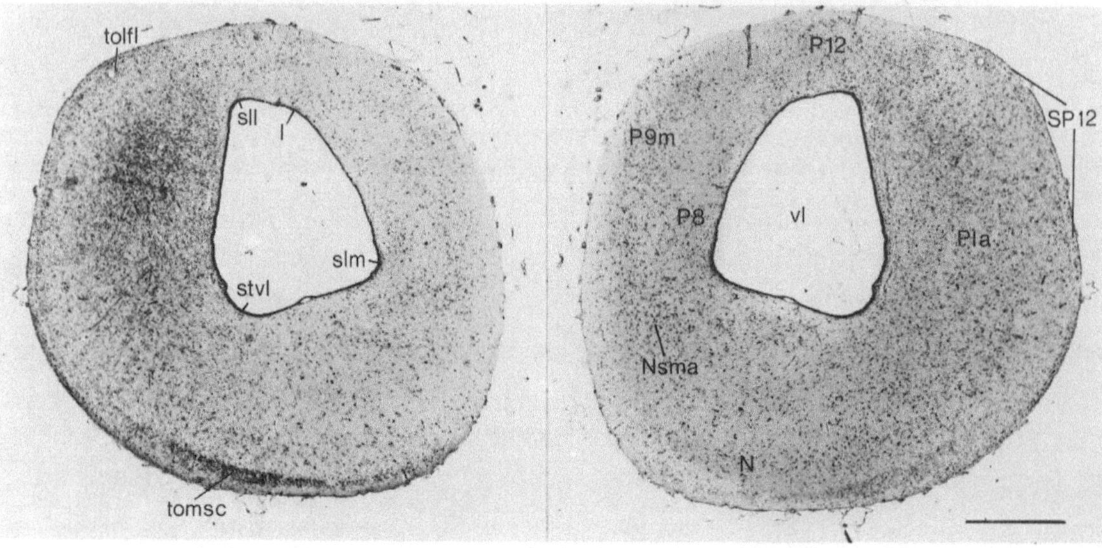
tolfl
sll
l
slm
stvl
tomsc
P12
SP12
P9m
P8
vl
Pla
Nsma
N

137

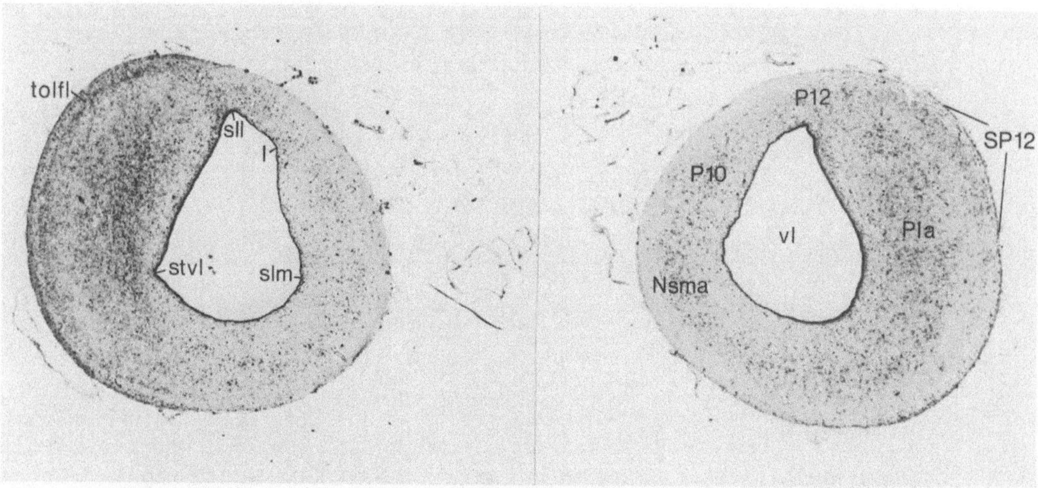
tolfl
sll
l
stvl
slm
P12
SP12
P10
vl
Pla
Nsma

138

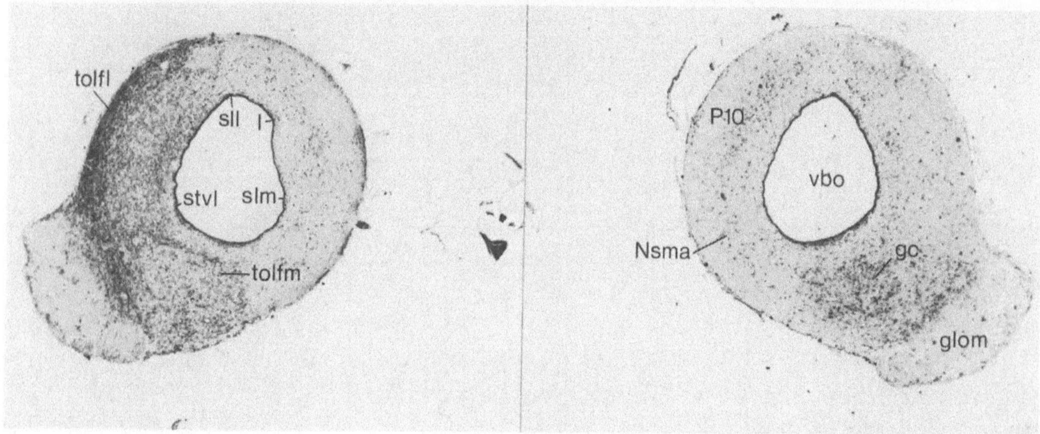
tolfl
sll
l
stvl
slm
tolfm
P10
vbo
Nsma
gc
glom

139

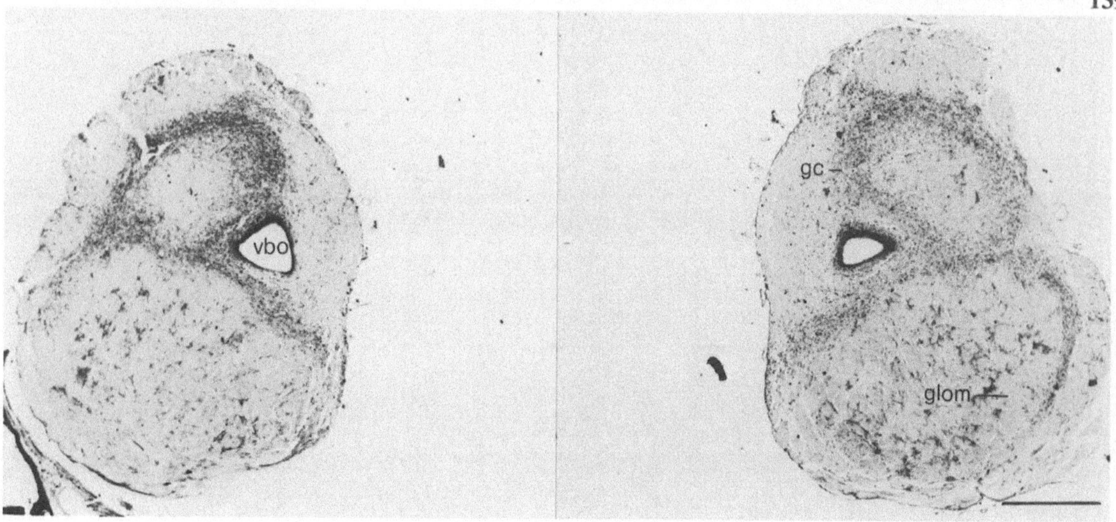
vbo
gc
glom

References

Addens JL (1933) The motor nuclei and roots of the cranial and first spinal nerves of vertebrates. I. Introduction and cyclostomes. Z Anat Entwicklungsgesch 101:307–410

Addens JL (1945) The Nucleus of Bellonci and adjacent cell groups in Selachians. I. The ganglion habenulae and its connections. Proc K Ned Akad Wet 48:345–359

Allis EP (1901) The lateral sensory canals, the eye-muscles, and the peripheral distribution of certain of the cranial nerves of Mustelus laevis. Q J Microsc Sci 45:87–236

Alluchon-Gérard M-J (1971) Types cellulaires et étapes de la différenciation de l'adénohypophyse chez l'embryon de roussette (Scyllium canicula, Chondrichthyens). Z Zellforsch Mikrosk Anat 120:525–545

Alluchon-Gérard M-J (1978a) Ultrastructure of the dogfish adenohypophysis. Cell Tissue Res 193:139–154

Alluchon-Gérard M-J (1978b) Electron microscopic study of the post-thyroidectomy changes in the ventral lobe of the adenohypophysis in very young spotted dogfish (Scyllium canicula, Chondrichthyes). Gen Comp Endocrinol 36: 585–597

Altner H (1964a) Vergleichende Untersuchungen über Cytologie und Vascularisation des Saccus vasculosus der Chondrichthyes. Z Zellforsch Mikrosk Anat 64:570–592

Altner H (1964b) Die Bedeutung von Lipiden im Epithel des Saccus vasculosus des Selachiergehirns. Naturwissenschaften 51:225–226

Altner H (1965) Über Aktivitätsphasen im Epithel des Saccus vasculosus von Etmopterus spinax. Zur Frage der Funktion des Saccus vasculosus der Haie. Z Zellforsch Mikrosk Anat 66: 663–672

Andres KH (1970) Anatomy and ultrastructure of the olfactory bulb in fish, amphibia, reptiles, birds and mammals. Ciba Found Symp Taste and Smell in Vertebrates, p 177–196

Andres KH (1975) Neue morphologische Grundlagen zur Physiologie des Riechens und Schmeckens. Arch Oto-Rhino-Laryngol 210: 1–41

Andrianov GN, Brown HR, Ilyinsky OB (1974) Responses of central neurons to electrical and magnetic stimuli of the ampullae of Lorenzini in the black sea skate. J Comp Physiol 93:287–299

Ariëns Kappers CU (1906) The structure of the teleostean and selachian brain. J Comp Neurol 16:1–112

Ariëns Kappers CU (1920) Die vergleichende Anatomie des Nervensystems der Wirbeltiere und des Menschen I. Bohn, Haarlem, p 624

Ariëns Kappers CU (1921) Die vergleichende Anatomie des Nervensystems der Wirbeltiere und des Menschen II. Bohn, Haarlem, p 704

Ariëns Kappers CU (1947) Anatomie comparée du système nerveuse. Bohn, Haarlem, p 754

Ariëns Kappers CU, Carpenter FW (1911) Das Gehirn von Chimaera monstrosa. Folia Neurobiol 5:127–160

Ariëns Kappers CU, Huber GC, Crosby EC (1936) The comparative anatomy of the nervous system of vertebrates, including man, vol 1. Macmillan, New York, p 864

Ariëns Kappers CU, Theunissen WF (1907) Zur vergleichenden Anatomie des Vorderhirns der Vertebraten. Anat Anz 30:496–509

Azzena GB, Desole C, Palmieri G (1970) Cerebellar projections of the masticatory and extraocular muscle proprioception. Exp Neurol 27:151–161

Bäckström K (1924) Contributions to the forebrain morphology in selachians. Acta Zool 5:123–240

Ball JN, Baker BI (1969) The pituitary gland: Anatomy and histophysiology. In: Hoar WS, Randall DJ (eds) Fish physiology, vol 2. Academic Press, New York, p 1–111

Ballintijn CM, Roberts JL (1976) Neural control and proprioceptive load matching in reflex respiratory movements of fishes. Fed Proc 35:1983–1991

Bargmann W (1953) Über das Zwischenhirn-Hypophysensystem von Fischen. Z Zellforsch Mikrosk Anat 38:275–298

Bargmann W (1954) Über Feinbau und Funktion des Saccus vasculosus. Z Zellforsch Mikrosk Anat 40:49–74

Barnard JW (1936) A phylogenetic study of the visceral afferent areas associated with the facial, glossopharyngeal, and vagus nerves, and their fiber connections. The efferent facial nucleus. J Comp Neurol 65:503–602

Bartelmez GW (1915) Mauthner's cell and the nucleus motorius tegmenti. J Comp Neurol 25:87–128

Bartels W (1971) Die Ontogenese der aminhaltigen Neuronensysteme im Gehirn von Rana temporaria. Z Zellforsch Mikrosk Anat 116: 94–118

Bauchot R, Platel R, Ridet J-M (1976) Brain-body weight relationships in Selachii. Copeia 1976: 305–309

Baumgarten HG (1972) Biogenic monoamines in the cyclostome and lower vertebrate brain. Prog Histochem Cytochem 4:1–90

Beard J (1896) The history of a transient nervous apparatus in certain Ichthyopsida. An account of the development and degeneration of ganglion cells and nerve-fibres, part I Raja batis. Zool Jahrb Anat 9:319–426

Beccari N (1930) I centri tegmentali dell'asse cerebrale dei Selaci. Arch Zool Ital 14:411–429

Belbenoit P, Bauer R (1972) Video recordings of prey capture behaviour and associated electric organ discharge of Torpedo marmorata. Marine Biol 17:93–99

Bennett MVL (1968) Neural control of electric organs. In: Ingle D (ed) The central nervous system and fish behaviour. University Press, Chicago, p 147–169

Bennett MVL (1971) Electric organs. In: Hoar WS, Randall DJ (eds) Fish physiology, vol 5. Academic Press, New York, p 347–491

Bergeijk WA van (1966) Evolution of the sense of hearing in vertebrates. Am Zool 6:371–377

Bergquist H (1932) Zur Morphologie des Zwischenhirns bei niederen Wirbeltieren. Acta Zool 13:57–303

Bern H (1969) Urophysis and caudal neurosecretory system. In: Hoar WS, Randall DJ (eds) Fish physiology, vol 2. Academic Press, New York, p 399–416

Bern HA, DeRoos CC, Biglieri EG (1962) Aldosterone and other corticosteroids from chondrichthyean interrenal glands. Gen Comp Endocrinol 2:490–494

Bernstein JJ, Gelderd JB (1973) Synaptic reorganization following regeneration of goldfish spinal cord. Exp Neurol 41:402–410

Bethe A (1899) Die Lokomotion des Haifisches (Scyllium) und ihre Beziehung zu den einzelnen Gehirnteilen und zum Labyrinth. Pflügers Arch Gesamte Physiol Menschen Tiere 76:470–493

Black D (1917) The motor nuclei of the cerebral nerves in phylogeny. I. Cyclostomi and pisces. J Comp Neurol 27:467–564

Bodznick D, Northcutt RG (1979) Some connections of the lateral olfactory area of the horn shark (Heterodontus francisci). Soc Neurosci Symp 5:139

Bodznick D, Northcutt RG (1980) Segregation of electro- and mechanoreceptive inputs to the elasmobranch medulla. Brain Res 195:313–321

Bone Q (1977) Mauthner neurons in elasmobranchs. J Mar Biol Assoc UK 57:253–259

Bone Q (1978) Locomotor muscle. In: Hoar WS, Randall DK (eds) Fish physiology, vol 7. Academic Press, New York, p 361–424

Bone Q, Roberts BL (1969) The density of elasmobranchs. J Mar Biol Assoc UK 49:913–937

Boord RL (1977) Auricular projections in the clearnose skate, Raja eglanteria. Am Zool 17:887

Boord RL, Campbell CBG (1977) Structural and functional organization of the lateral line system of sharks. Am Zool 17:431–441

Boord RL, Northcutt RG (1979) Ascending projections of anterior and posterior lateral line lobes of the clearnose skate, Raja eglanteria. Anat Rec 193:487

Boord RL, Northcutt RG (1982) Ascending lateral line pathways to the midbrain of the clearnose skate, Raja eglanteria. J Comp Neurol 207:274–282

Boord RL, Roberts BL (1980) Medullary and cerebellar projections of the statoacoustic nerve of the dogfish Scyliorhinus canicula. J Comp Neurol 193:57–68

Bortolami R, Callegari E, Lucchi ML (1972) Anatomical relationship between mesencephalic trigeminal nucleus and cerebellum in the duck. Brain Res 47:317–329

Braak H (1963) Das Ependym der Hirnventrikel von Chimaera monstrosa (mit besonderer Berücksichtigung des Organon vasculare praeopticum). Z Zellforsch Mikrosk Anat 60:582–608

Braford MR Jr (1972) Ascending efferent tectal projections in the south american spectacled caiman. Anat Rec 172:275–276

Brauth SE, Karten HJ (1977) Direct accessory optic projections to the vestibulocerebellum: A possible channel for oculomotor control systems. Exp Brain Res 28:73–84

Brecha N, Karten HJ, Hunt SP (1980) Projections of the nucleus of the basal optic root in the pigeon: An autoradiographic and horseradish peroxidase study. J Comp Neurol 189:615–670

Breder CM (1926) The locomotion of fishes. Zoologica NY 4:159–256

Brodal A, Saugstad LF (1965) Retrograde cellular changes in the mesencephalic nucleus in the cat following cerebellar lesions. Acta Morphol Neerl-Scand 6:147–159

Bruckmoser P (1973) Beziehungen zwischen Struktur und Funktion in der Evolution des Telencephalon. Verh Dtsch Zool Ges 66: 219–229

Bruckmoser P, Dieringer N (1973) Evoked potentials in the primary and secondary olfactory projection areas of the forebrain in Elasmobranchia. J Comp Physiol 87:65–74

Bullock TH (1979) Processing of ampullary input in the brain: Comparison of sensitivity and evoked responses among elasmobranch and siluriform fishes. J Physiol (Paris) 75:397–400

Bullock TH (1981) Comparisons of the electric and acoustic senses and their central processing. In: Tavolga WN, Popper AN, Fay RR (eds) Hearing and sound communication in fishes. Springer, Berlin Heidelberg New York, S 525–571

Bullock TH, Corwin JT (1979) Acoustic evoked activity in the brain of sharks. J Comp Physiol 129:223–234

Butler AB, Northcutt RG (1971) Ascending tectal afferent projections in the lizard Iguana iguana. Brain Res 35:597–601

Campbell B (1946) Morphology and conduction of bipolar dorsal root ganglion cells of selachian fishes. J Neurophysiol 9:173–179

Campbell B (1951) The locomotor behaviour of spinal elasmobranchs with an analysis of stinging in Urobatis. Copeia 1951:277

Catois EM (1901) Recherches sur l'histologie et l'anatomie microscopique de l'encephale chez les poissons. Bull Sci France Belgique 36:81–83

Chadwick A (1966) Prolactin-like activity in the pituitary gland of fishes and amphibians? J Endocrinol 35:75–81

Charlton HH (1928) A gland-like ependymal structure in the brain. Proc K Ned Akad Wet Amsterdam 31:823–836

Charlton HH (1932) Comparative studies on the nucleus preopticus pars magnocellularis and the nucleus lateralis tuberis in fishes. J Comp Neurol 54:237–276

Chevins PFD (1968) The anatomy and physiology of the pituitary complex in the genus Raja, Elasmobranchii. Ph D Thesis, University of Leeds

Chevins PFD, Dodd JM (1970) Pituitary innervation and control of colour change in skates, Raja naevus, R. clavata, R. montagui, and R. radiata. Gen Comp Endocrinol 15:232–241

Claas B, Münz M (1980) Bony fish lateral line efferent neurons identified by retrograde axonal transport of horseradish peroxidase (HRP). Brain Res 193:249–253

Coggeshall RE, Leonard RB, Applebaum ML, Willis WD (1978) Organization of peripheral nerves and spinal roots of the Atlantic stingray, Dasyatis sabina. J Neurophysiol 41:97–107

Cohen AH, Wällen P (1980) The neuronal correlate of locomotion in fish. Exp Brain Res 41:11–18

Cohen DH, Duff TA, Ebbesson SOE (1973) Electrophysiological identification of a visual area in shark telencephalon. Science 182:492–494

Compagno LJV (1977) Phyletic relationships of living sharks and rays. Am Zool 17:303–322

Corner EDS, Denton EJ, Forster GR (1969) On the buoyancy of some deep-sea sharks. Proc R Soc Lond B Biol Sci 171:415–429

Corwin JT (1981a) Peripheral auditory physiology in the lemon shark; evidence of parallel otolithic and non-otolithic sound detection. J Comp Physiol A 142:379–390

Corwin JT (1981b) Audition in elasmobranchs. In: Tavolga WN, Popper AN, Fay RR (eds) Hearing and sound communication in fishes. Springer, Berlin Heidelberg New York, S 81–105

Corwin JT, Bullock TH, Schweitzer J (to be published) The auditory brainstem response in five vertebrate classes. J Comp Physiol

Corwin JT, Northcutt RG (1982) Auditory centers in the elasmobranch brain stem: Deoxyglucose autoradiography and evoked potential recording. Brain Res 236:261–273

Cowan WM (1973) Neuronal death as a regulative mechanism in the control of all numbers in the nervous system. In: Rodstein M (ed) Development and aging in the nervous system. Academic Press, New York, p 19–41

Crow CH (1927) The spinal reflexes of the skate. J Physiol (Lond) 63:61–67

Cruce WLR, Nieuwenhuys R (1974) The cell masses in the brain stem of the Turtle Testudo hermanni; a topographical and topological analysis. J Comp Neurol 156:277–306

Dacey DM (1982) Axon morphology of mesencephalic trigeminal neurons in a snake, Thamnophis sirtalis. J Comp Neurol 204:268–279

Dahlgren U (1914) On the electric motor nerve centers in the skates (Rajidae). Science 40: 862–863

Dammerman KW (1910) Der Saccus vasculosus der Fische, ein Tiefeorgan. Z Wiss Zool 96: 654–726

Dart RA (1920) A contribution to the morphology of the corpus striatum. J Anat (Lond) 55:1–26

Deery DJ (1974) Hypothalamic regulation of the dogfish pituitary. In: Knowles F, Vollrath L (eds) Neurosecretion – the final neuroendocrine pathway, VI Int Symp Neurosecr, London 1973. Springer, Berlin Heidelberg New York, p 299–300

Deery DJ, Jones AC (1975) Effects of hypothalamic extracts, neurotransmitters and synthetic hypothalamic releasing hormones on the adenyl cyclase activity of the lobes of the pituitary of the dogfish (Scyliorhinus canicula L.). J Endocrinol 64:49–57

Demski LS (1977) Electrical stimulation of the shark brain. Am Zool 17:487–500

DeRoos R, DeRoos CC (1967) Presence of corticotropin activity in the pituitary gland of chondrichthyan fish. Gen Comp Endocrinol 9:267–275

Dijkgraaf S (1949) Lokalisationsversuche am Fischgehirn. Experientia 5:44–47

Dijkgraaf S (1963) The functioning and significance of the lateral line organs. Biol Rev 38:51–105

Dijkgraaf S, Kalmijn AJ (1962) Verhaltensversuche zur Funktion der Lorenzinischen Ampullen. Naturwissenschaften 49:400

Dobson S, Dodd JM (1977a) Endocrine control of the testis in the dogfish Scyliorhinus canicula L. I Effects of partial hypophysectomy on gravimetric, hormonal and biochemical aspects of testis function. Gen Comp Endocrinol 32:41–52

Dobson S, Dodd JM (1977b) Endocrine control of the testis in the dogfish Scyliorhinus canicula L. II Histological and ultrastructural changes in the testis after partial hypophysectomy (ventral lobectomy). Gen Comp Endocrinol 32:53–71

Dodd JM, Dodd MHJ, Jenkins N (1980) Presence of a gonadotropin in the Rachendachhypophyse of the pituitary gland of the rabbitfish Hydrolagus collei (Chondrichthyes: Holocephali). Gen Comp Endocrinol 40:342–343

Dodd JM, Evennett PJ, Goddard CK (1960) Reproductive endocrinology in cyclostomes and elasmobranchs. Symp Zool Soc Lond 1:77–103

Donkelaar HJ ten (1976a) Descending pathways from the brain stem to the spinal cord in some reptiles. I. Origin. J Comp Neurol 167:421–442

Donkelaar HJ ten (1976b) Descending pathways from the brain stem to the spinal cord in some reptiles. II. Course and site of termination. J Comp Neurol 167:443–464

Donkelaar HJ ten, de Boer-van Huizen R, Schouten FTM, Eggen SJH (1981) Cells of origin of descending pathways to the spinal cord in the clawed toad (Xenopus laevis). Neurosci 6:2297–2312

Donkelaar HJ ten, Kusuma A, de Boer-van Huizen R (1980) Cells of origin of pathways descending to the spinal cord in some quadrupedal reptiles. J Comp Neurol 192:827–851

Donkelaar HJ ten, Nieuwenhuys R (1979) The brain stem. In: Gans C (ed) Biology of the reptilia. Academic Press, London, p 133–200

Dowling JE, Ripps H (1970) Visual adaptation in the retina of the skate. J Gen Physiol 56:491–520

Droge MH, Leonard RB (1979) Fictive locomotion in the stingray Dasyatis sabina. Neurosci Abstr 5:368

Eaton RC, Lavender WA, Wieland CM (1981) Identification of Mauthner initiated response pattern in goldfish: Evidence from simultaneous cinematography and electrophysiology. J Comp Physiol 144:521–531

Ebbesson SOE (1968) Retinal projections in two teleost fishes (Opsanus tau and Gymnothorax funebris). An experimental study with silver impregnation methods. Brain Behav Evol 1:134–154

Ebbesson SOE (1972) New insights into the organization of the shark brain. Comp Biochem Physiol A 42:121–129

Ebbesson SOE (1980a) The parcellation theory and its relation to interspecific variability in brain organization, evolutionary and ontogenetic development, and neuronal plasticity. Cell Tissue Res 213:179–212

Ebbesson SOE (1980b) On the organization of telencephalon in elasmobranchs. In: Ebbesson SOE (ed) Comparative neurology of the telencephalon. Plenum Press, New York London, p 1–16

Ebbesson SOE (1981) Projections of the optic tectum and the mesencephalic nucleus of the trigeminal nerve in the tegu lizard (Tupinambis nigropunctatus). Cell Tissue Res 216:151–165

Ebbesson SOE, Campbell CBG (1973) On the organization of cerebellar efferent pathways in the nurse shark (Ginglymostoma cirratum). J Comp Neurol 152:233–254

Ebbesson SOE, Heimer L (1970) Projections of the olfactory tract fibers in the nurse shark (Ginglymostoma cirratum). Brain Res 17:47–55

Ebbesson SOE, Hodde KC (1981) Ascending spinal systems in the nurse shark, Ginglymostoma cirratum. Cell Tissue Res 216:313–331

Ebbesson SOE, Meyer DL (1980) The visual system of the guitarfish (Rhinobatos productus). Cell Tissue Res 206:243–250

Ebbesson SOE, Meyer DL (1982) Efferents to the retina have multiple sources in teleost fish. Science 214:924–926

Ebbesson SOE, Northcutt RG (1976) Neurology of anamniotic vertebrates. In: Masteron RB, Bitterman ME, Campbell CBG, Hotton N (eds) Evolution of brain and behavior in vertebrates. Erlbaum, Hillsdale NJ, p 115–146

Ebbesson SOE, Ramsey JS (1968) The optic tracts of two species of sharks (Galeocerdo cuvieri and Ginglymostoma cirratum). Brain Res 8: 36–53

Ebbesson SOE, Schroeder D (1971) Connections of the nurse shark's telencephalon. Science 173:254–256

Ebbesson SOE, Vanegas H (1976) Projections of the optic tectum in two teleost species. J Comp Neurol 165:161–180

Eccles JC, Táboriková H, Tsukahara N (1970) Responses of the granule cells of the selachian cerebellum (Mustelus canis). Brain Res 17:87–102

Edinger L (1888) Untersuchungen über die vergleichende Anatomie des Gehirns, 1. das Vorderhirn. Abhandl Senckenberg Naturforsch Ges 15:91–119

Edinger L (1892) Untersuchungen über die vergleichende Anatomie des Gehirns, 2. das Zwischenhirn, I Teil: Das Zwischenhirns der Selachier und der Amphibien. Abhandl Senckenberg Naturforsch Ges 18:1–55

Edinger L (1901) Das Cerebellum von Scyllium canicula. Arch Mikrosk Anat 58:661–678

Edinger L (1908) Vorlesungen über den Bau der nervösen Zentralorgane des Menschen und der Tiere. Vogel, Leipzig

Ewart JC (1890) The cranial nerves of the Torpedo. Proc R Soc Lond B 47:290–291

Ewart JC (1893) The electric organ of the skate: Note on an electric centre in the spinal cord. Proc R Soc Lond B 53:388–391

Faber DS, Korn M (1978) Neurobiology of the Mauthner Cell. Raven, New York

Fahrenholz C (1928) Eine Rachendachhypophyse bei Chimaera monstrosa. Anat Anz 66:342–348

Farner H-P (1978) Untersuchungen zur Embryonalentwicklung des Gehirns von Scyliorhinus canicula (L.). I Bildung der Hirngestalt, Migrations modi und -phasen, Bau des Zwischenhirns. J Hirnforsch 19:313–332

Faucette JR (1969a) The olfactory bulb and medial hemisphere wall of the rat-fish, Chimaera. J Comp Neurol 137:377–406

Faucette JR (1969b) The acessory olfactory bulbs and the lateral telencephalic wall of the rat-fish, Chimaera. J Comp Neurol 137:407–432

Fay RR, Kendall JI, Popper AN, Tester AL (1974) Vibration detection by the Macula neglecta of sharks. Comp Biochem Physiol A 47:1235–1240

Ferguson JL, Mulvanny PJ, Brauth SE (1978) Distribution of neurons projecting to the retina of Caiman crocodilus. Brain Behav Evol 15:294–306

Fields RD, Lange GD (1980) Electroreception in the ratfish (Hydrolagus colliei). Science 207: 547–548

Finger TE (1978a) Gustatory pathways in the bullhead catfish. II Facial lobe connections. J Comp Neurol 180:691–706

Finger TE (1978b) Cerebellar afferents in teleost catfish (Ictaluridae). J Comp Neurol 181: 173–182

Finger TE (1978c) Efferent neurons of the teleost cerebellum. Brain Res 153:608–614

Finger TE, Karten HJ (1978) The accessory optic system in teleosts. Brain Res 153:144–149

Finstad WD, Nelson DR (1975) Circadian activity rhythm in the hornshark, Heterodontus francisci: Effect of light intensity. Bull South Calif Acad Sci 74:20–26

Firth AJ, Vollrath L (1973) Determination of the distribution of luteinising hormone-like activity within the dogfish pituitary gland by means of the Xenopus oocyte meiosis assay. J Endocrinol 58:347–348

Foster RE Hall WC (1975) The connections and laminar organization of the optic tectum in a reptile (Iguana iguana). J Comp Neurol 163: 397–426

Fox GQ (1977) The morphology of the oval nuclei of neonatal Torpedo marmorata. Cell Tissue Res 178:155–167

Fox GQ, Richardson GP (1979) The developmental morphology of Torpedo marmorata: Electric organ-electrogenic phase. J Comp Neurol 185:293–316

Fritzch R, Himstedt W (1981) Pretectal neurons project to the salamander retina. Neurosci Lett 24:13–17

Gabriel KH (1970) Vergleichend-histologische Studien am Subcommissuralorgan. Anat Anz 127:129–170

Gaskell WH (1886) On the structure, distribution and function of the nerves which innervate the visceral and vascular system. J Physiol Lond 7:1–81

Gaskell WH (1889) On the relation between the structure, function, distribution and origin of the cranial nerves; together with a theory of the origin of the nervous system of vertebrates. J Physiol Lond 10:153–211

Gelderd JB (1979) Quantitative anatomical and behavioral analyses of regeneration and collateral sprouting following spinal cord transection in the nurse shark (Ginglymostoma cirratum). Acta Neurobiol Exp Suppl 39:121–142

Gerlach J (1947) Beiträge zur vergleichenden Morphologie des Selachierhirnes. Anat Anz 96: 79–165

Glasser S, Ingle D (1978) The nucleus isthmus as a relay station in the ipsilateral visual projection to the frogs optic tectum. Brain Res 159:214–218

Glees P (1940) Der periphere und zentrale Anteil des sympathischen Nervensystems der Selachier. Acta Neerl Morph 3:209–248

Goldstein MH, Jacobs VL (1969) The central projections of the mesencephalic root of the trigeminus in a lizard (Lacerta viridis). Brain Res 14:307–320

Graeber RC (1978) Behavioral studies correlated with central nervous system integration of vision in sharks. In: Hodgson ES, Mathewson RF (eds) Sensory biology of Sharks, Skates, and Rays. US Gov't Printing, Washington DC, p 195–225

Graeber RC (1980) Telencephalic function in elasmobranchs. A behavioral perspective. In: Ebbesson SOE (ed) Comparative neurology of the Telencephalon. Plenum Press, New York London, p 17–39

Graeber RC, Ebbesson SOE (1972) Retinal projections in the lemon shark (Negaprion brevirostris) Brain Behav Evol 5:461–477

Graeber RC, Ebbesson SOE, Jane JA (1973) Visual discrimination in sharks without optic tectum. Science 180:413–415

Graeber RC, Schroeder DM, Jane JA, Ebbesson SOE (1978) Visual discrimination following partial telencephalic ablations in nurse sharks (Ginglymostoma cirratum). J Comp Neurol 180:325–344

Grillner S, Perret C, Zangger P (1976) Central generation of locomotion in the spinal dogfish. Brain Res 109:255–269

Grover BG, Sharma SC (1979) Tectal projections in the goldfish (Carassius auratus). A degeneration study. J Comp Neurol 184:435–454

Grover BG, Sharma SC (1981) Organization of extrinsic tectal connections in goldfish (Carassius auratus). J Comp Neurol 196:471–488

Gruber SH, Myrberg AA (1977) Approaches to the study of the behavior of sharks. Am Zool 17:471–486

Gruberg ER, Udin SB (1978) Topographic projections between the nucleus isthmi and the tectum of the frog Rana pipiens. J Comp Neurol 179:487–500

Haller B (1898) Vom Bau des Wirbeltiergehirns. I. Salmo und Scyllium. Morphol Jahrb 26:345–641

Haller von Hallerstein V (1929) Die Gliederung des Zwischen- und Mittelhirns der Wirbeltiere. Morphol Jahrb 63:359–407

Halpern M, Wang RT, Colman DR (1976) Centrifugal fibers to the eye in a non-avian vertebrate: Source revealed by horseradish peroxidase studies. Science 194:1185–1188

Hamasaki DI, Streck P (1971) Properties of the epiphysis cerebri of the small-spotted dogfish shark, Scyliorhinus caniculus L. Vision Res 11:189–198

Harrach M, Graf von (1970) Elektronenmikroskopische Beobachtungen am Saccus vasculosus einiger Knorpelfische. Z Zellforsch 105: 188–209

Harris AJ (1965) Eye movements of the dogfish Squalus acanthias L. J Exp Biol 43:107–130

Harris JE (1936) The role of the fins in the equilibrium of the swimming fish. J Exp Biol 13: 476–494

Harris JE, Whiting HP (1954) Structure and function in the locomotory system of the dogfish embryo, the myogenic stage of movement. J Exp Biol 31:501–524

Harrison F, Corbin KB (1942) The central pathway for the jaw jerk. Am J Physiol 123: 439–445

Hayle TH (1973a) A comparative study of spinal projections to the brain (except cerebellum) in three classes of poikilothermic vertebrates. J Comp Neurol 149:463–476

Hayle TH (1973b) A comparative study of spinocerebellar systems in three classes of poikilothermic vertebrates. J Comp Neurol 149: 477–495

Healey EG (1957) The nervous system. In: Brown ME (ed) The physiology of fishes. Academic Press, London New York, p 1–119

Heimer L (1969) The secondary olfactory connections in mammals, reptiles and sharks. Ann NY Acad Sci 167:129–147

Herrick CJ (1910) The morphology of the forebrain in amphibia and reptilia. J Comp Neurol 20:413–547

Herrick CJ (1913) Anatomy of the brain. In: The reference handbook of the medical sciences, vol 2. Wood, New York, p 274–342

Herrick CJ (1943) The cranial nerves. A review of fifty years. J Sci Lab Denison Univ 38: 41–51

Herrick CJ (1948) The Brain of the Tiger Salamander. Chicago University Press, Chicago, p 407

Hodgson ES, Mathewson RF (1978) Sensory biology of Sharks, Skates, and Rays. US Gov't Print Off, Washington DC, p 666

Hoëvell JJLD van (1911) Remarks on the reticular cells of the oblongata in different vertebrates. Proc Acad Sci (Amsterdam) 13:1047–1065

Hoëvell JJLD van (1916) The phylogenetic development of the cerebellar nuclei. Proc Acad Sci (Amsterdam) 18:1421–1435

Hogben L (1936) The pigmentary effector system. VII. The chromatic function in elasmobranch fishes. Proc R Soc Lond B 120:142–158

Holmgren N (1922) Points of view concerning forebrain morphology in lower vertebrates. J Comp Neurol 34:391–459

Hoogland PV, Welker E (1981) Telencephalic projections to the eye in Python reticulatus. Brain Res 213:173–176

Horst CJ van der (1918) Die motorischen Kerne und Bahnen in dem Gehirn der Fische, ihr taxonomischer Wert und ihre neurobiotaktische Bedeutung. Tijdschr Ned Dierk Ver 16:168–270

Horstmann E (1954) Die Faserglia des Selachiergehirns. Z Zellforsch 30:588–617

Houser GL (1901) The neurones and supporting elements of the brain of a selachian. J Comp Neurol 11:65–175

Hugosson R (1955) Studien über die Entwicklung der longitudinalen Zellsäulen und der Anlagen der Gehirnnervenkerne in der Medulla oblongata bei verschiedenen Vertebraten. Z Anat Entwicklungesch 118:543–566

Hugosson R (1957) Morphologic and experimental studies on the development and significance of the rhombencephalic longitudinal cell columns. Thesis Lund, p 133

Hyde IH (1904) Localization of the respiratory centre in the skate. Am J Physiol 10:236–258

Itaya SK (1980) Retinal efferents from the pretectal area in the rat. Brain Res 201:436–441

Ito H, Kishida R (1977) Tectal afferent neurons identified by the retrograde HRP method in the carp telencephalon. Brain Res 130:142–145

Ito H, Sakamoto N, Takatsuji K (1982) Cytoarchitecture, fiber connections, and ultrastructure of nucleus isthmi in a teleost (Navodon modestus) with a special reference to degenerating isthmic afferents from optic tectum and nucleus pretectalis. J Comp Neurol 205:299–311

Ito H, Tanaka H, Sakamoto N, Morita Y (1981) Isthmic afferent neurons identified by the retrograde HRP method in a teleost, Navadon modestus. Brain Res 207:163–169

Jackson RG, Sage M (1973a) Regional distribution of thyroid stimulating hormone activity in the pituitary gland of the atlantic stingray, Dasyatis sabina. Fish Bull 71:93–97

Jackson RG, Sage M (1973b) A comparison of the effects of mammalian TSH on the thyroid glands of the teleost Galeichthys felis and the elasmobranch Dasyatis sabina. Comp Biochem Physiol A 44:867–870

Jaeger RJ, Benevento LA (1980) The innervation of the internal structures of the eye in the Macaque monkey: A horseradish peroxidase study. Anat Rec 196:87A

Jansen WF (1969) The action absorbing and transporting function of the saccus vasculosus. In: Sterba G (ed) Zirkumventrikuläre Organe und Liquor, Symposium Reinhardsbrunn Mai 1968. Fischer, Jena, p 123–126

Jasinski A (1969) Vascularization of the hypophyseal region in lower vertebrates (Cyclostomes and fishes). Gen Comp Endocrinol [Suppl] 2:510–521

Jasinski A, Gorbman A (1966) Hypothalamo-hypophyseal vascular and neurosecretory links in the ratfish, Hydrolagus collei (Lay and Bennett). Gen Comp Endocrinol 6:476–490

Jenkins N (1978) The endocrine control of reproduction in the dogfish (Scyliorhinus canicula L.), with special reference to the hypothalamus. Ph D Thesis, University of Wales

Jenkins N, Joss JP, Dodd JM (1980) Biochemical and autoradiographic studies on the oestradiol-concentrating cells in the diencephalon and pituitary gland of the female dogfish (Scyliorhinus canicula L.). Gen Comp Endocrinol 40: 211–219

Jerison HJ (1973) Evolution of the brain and intelligence. Academic Press, New York London, p 482

Johnston I (1981) Structure and function of fish muscles. Symp Zool Soc Lond 48:71–113

Johnston JB (1901) The brain of Acipenser. A contribution to the morphology of the vertebrate brain. Zool Jahrb Abt Anat Ontog 15: 59–260

Johnston JB (1902) The brain of Petromyzon. J Comp Neurol 7:2–82

Johnston JB (1905) The radix mesencephalica trigemini. The Ganglion isthmi. Anat Anz 87: 364–379

Johnston JB (1909) The morphology of the forebrain vesicles in vertebrates. J Comp Neurol 19:457–539

Johnston JB (1911) The telencephalon of selachians. J Comp Neurol 21:1–113

Joseph BS, King RB, Whitlock DG (1968) Central distribution of trigeminal primary afferent fibers in anuran amphibians. Anat Rec 160: 719–728

Källén B (1951) Embryological studies on the nuclei and their homologization in the vertebrate forebrain. K Fysiogr Sallsk Lund Handl NF 62:1–36

Kalmijn AJ (1971) The electric sense of sharks and rays. J Exp Biol 55:371–383

Kalmijn AJ (1974) The detection of electric fields from inamimate and animate sources other than electric organs. In: Fessard A (ed) Handbook of sensory physiology, vol III/3. Springer, Berlin Heidelberg New York, p 147–200

Kalmijn AJ (1978) Electric and magnetic sensory world of sharks, skates, and rays. In: Hodgson ES, Mathewson RF (eds) Sensory biology of Sharks, Skates, and Rays. US Gov't Print Off Washington DC, p 507–528

Kamer JC van de, Verhagen TG (1954) The cytology of the neurohypophysis, the saccus vasculosus and the recessus posterior in Scyliorhinus caniculus. Proc K Ned Akad Wet Ser C Biol Med Sci 57:358–364

Kamer JC van de, Wilschut IJC, Heussen AMA (1973) On the presence and localization of glycogen accumulations inside coronet cells of the saccus vasculosus of the dogfish (Scyliorhinus caniculus). Z Zellforsch 140:277–290

Kashin SM, Feldman AG, Orlovsky GN (1974) Locomotion of fish evoked by electrical stimulation of the brain. Brain Res 82:41–47

Keenan E (1928) The phylogenetic development of the substantia gelatinosa Rolandi. Proc K Ned Akad Wet 31:837–854

Kemali M, Miralto A (1979) Light and electron-microscopic structure of cells protruding into the mesencephalic ventricle of Scyllium stellare (Elasmobranchii, Selachii). Cell Tissue Res 200:153–157

Kemali M, Miralto A, Sada E (1980) Asymetry of the habenulae in the elasmobranch Scyllium stellare. I. Light microscopy. Z Mikrosk Anat Forsch 94:794–800

Kishida R (1979) Comparative study on the teleostean optic tectum: Lamination and cytoarchitecture. J Hirnforsch 20:57–67

Klatzo I (1967) Cellular morphology of the lemon shark brain. In: Gilbert PW, Mathewson RF, Rall DP (eds) Sharks, Skates and Rays. Hopkins Maryland, p 341–359

Klausewitz W (1962) Wie schwimmen Haifische? Nat Mus Frankf 92:219–226

Klesch WL, Sage M (1973) The control of the interrenal by the pituitary in the elasmobranch, Dasyatis sabina. Comp Biochem Physiol A 45:961–967

Klesch WL, Sage M (1975) The stimulation of corticosteroidogenesis in the interrenal of the elasmobranch Dasyatis sabina by mammalian ACTH. Comp Biochem Physiol A 52:145–146

Knowles F (1965) Evidence for a dual control, by neurosecretion, of hormone synthesis and hormone release in the pituitary of the dogfish, Scyliorhinus stellaris. Philos Trans R Soc Lond B Biol Sci 249:435–456

Knowles F, Vollrath L, Meurling P (1975) Cytology and neuroendocrine relations of the pituitary of the dogfish, Scyliorhinus canicula. Proc R Soc Lond Biol Sci 191:507–525

Knowles F, Weatherhead B, Martin R (1970) The ultrastructure of neurosecretory fibre terminals after zinc-iodine-osmium impregnation. In: Bargmann W, Scharrer B (eds) Aspects of Neurosecretion. Springer, Berlin Heidelberg New York, p 159–165

Koester DM (1981a) Afferent and efferent connections of the posterior lateral line nerve of the clearnose skate, Raja eglanteria. Anat Rec A 199:146

Koester DM (1981b) Central connections of the ocatvolateralis nerves of the clearnose skate, Raja eglanteria. PhD Thesis, University of Delaware

Koester DM, Boord RL (1978) The central projections of first order anterior lateral line neurons of the clearnose skate, Raja eglanteria. Am Zool 18:587

Kolmer W (1923) Über das Vorkommen eines Knochenmarkähnlichen Gewebes bei einem Selachier (Knorpelmark von Chimaera monstrosa). Anat Anz 56:529–534

Kooy FH (1917) The inferior olive in vertebrates. Folia Neurobiol (Leipz) 10:205–369

Korn H, Bennett MVL (1975) Vestibular nystagmus and teleost oculomotor neurons: Functions of electrotonic coupling and dendritic impulse initiation (Spheroides). J Neurophysiol 38: 430–451

Kremers JWPM, Nieuwenhuys R (1979) Topological analysis of the brain stem of the crossopterygian Fish Latimeria chalumnae. J Comp Neurol 187:613–638

Kuchnow KP (1971) The elasmobranch pupillary response. Vision Res 11:1395–1406

Kuhlenbeck H (1929a) Über die Grundbestandteile des Zwischenhirnbauplans der Anamnier. Morphol Jahrb 63:50–95

Kuhlenbeck H (1929b) Die Grundbestandteile des Endhirns im Lichte der Bauplanlehre. Anat Anz 67:1–51

Kuhlenbeck H (1970) Structural elements: Biology of nervous tissue. In: The central nervous system of vertebrates, vol 3/I. Karger, Basel New York, p 818

Kuhlenbeck H (1973) Overall morphologic pattern. In: The central nervous system of vertebrates, vol 3/II. Karger, Basel New York, p 950

Kuhlenbeck H (1977) Derivatives of the Prosencephalon: Diencephalon and Telencephalon. In: The central nervous system of vertebrates, vol 5/I. Karger, Basel New York, p 888

Kuhlenbeck H, Niimi K (1969) Further observations on the morphology of the brain in the Holocephalian Elasmobranchs Chimaera and Callorhynchus. J Hirnforsch 11:267–314

Landgrebe FW, Waring H (1941) Intermediate lobe pituitary hormone. Q J Exp Physiol 31: 31–62

Larsell O (1923) The cerebellum of the frog. J Comp Neurol 36:89–112

Larsell O (1967) The comparative anatomy and histology of the cerebellum from Myxinoids through birds. University Minnesota Press, Minneapolis, p 291

Leghissa S (1962) La struttura della corteccia mesencefalica dei ciclostomi selaci ed urodeli. Accad Sci Del' Inst Bologna Classe Sci Fisiche 9:123–152

Lenhossék M von (1892) Beobachtungen an den Spinalganglion und dem Rückenmark von Pristiurusembryonen. Anat Anz 7:519–539

Lenhossék M von (1894) Zur Kenntnis des Rückenmarkes der Rochen. Beiträge zur Histologie des Nervensystems und der Sinnesorgane. Bergman, Wiesbaden

Leonard RB, Coggeshall RE, Willis WD (1978a) A documentation of an age related increase in neuronal and axonal numbers in the Stingray, Dasyatis sabina. Lesener. J Comp Neurol 179:13–22

Leonard RB, Rudomín P, Willis WD (1978b) Central effects of volleys in sensory and motor components of peripheral nerve in the stingray, Dasyatis sabina. J Neurophysiol 41:108–125

Leonard RB, Willis WD (1977) Rhombencephalic cells of origin of descending spinal systems in the atlantic stingray. Soc Neurosci 7th Ann Meeting, Anaheim CA, p 503

Leonard RB, Willis WD (1979) The organization of the electromotor nucleus and extraocular motor nuclei in the stargazer (Astroscopus y-graecum). J Comp Neurol 183:397–414

Locy WA (1905) On a newly recognized nerve connected with the forebrain of selachians. Anat Anz 26:33–36

Lowenstein O (1974) Comparative morphology and physiology. In: Kornhuber MH (ed) Handbook of sensory physiology, vol 6. Springer, Berlin Heidelberg New York, p 75–120

Lowenstein O, Compton GJ (1978) A comparative study of the responses of isolated first-order semicircular canal afferents to angular and linear acceleration, analysed in the time and frequency domains. Proc R Soc 202:313–338

Lowenstein O, Roberts TDM (1949) The equilibrium function of the otolith organs of the thornback ray (Raja clavata). J Physiol (Lond) 110:349–415

Lowenstein O, Roberts TDM (1951) The localization and analysis of the responses to vibration from the isolated elasmobranch labyrinth. A contribution to the problem of the evolution of hearing in vertebrates. J Physiol (Lond) 114:471–489

Lowry PJ, Bennett HPJ, McMartin C, Scott AP (1974) The isolation and amino acid sequence of an adrenocorticotrophin from the pars distalis and a corticotrophin-like intermediate lobe peptide from the neurointermediate lobe of the pituitary of the dogfish Squalus acanthias. Biochem J 141:427–437

Lowry PJ, Chadwick A (1970) Purification and amino acid sequence of melanocyte stimulating hormone from the dogfish Squalus acanthias. Biochem J 118:713–718

Luiten PGM (1975) The horseradish peroxidase technique applied to the teleostean nervous system. Brain Res 89:181–186

Luiten PGM (1981a) Two visual pathways to the telencephalon in the nurse shark (Ginglymostma cirratum). I Retinal projections. J Comp Neurol 196:531–538

Luiten PGM (1981b) Two visual pathways to the telencephalon in the nurse shark (Ginglymostoma cirratum). II Ascending thalamo-telencephalic connections. J Comp Neurol 196: 539–548

Luiten PGM (1981c) Afferent and efferent connections of the optic tectum in the carp (Cyprinus carpio L.). Brain Res 220:51–65

Luiten PGM, Dijkstra-de Vlieger HP (1978) Extraocular muscle representation in the brain stem of the carp. J Comp Neurol 179:669–676

Macchi IA, Rizzo F (1962) In vitro effect of mammalian adrenocorticotropin on secretion of skate (Raja erinacea) interrena tissue. Proc Soc Exp Biol Med 110:433–436

MacDonnell MF (1980a) Mesencephalic trigeminal nucleus in sharks. A light microscopic study. Brain Behav Evol 17:152–163

MacDonnell MF (1980b) Cerebrospinal fluid contacting and supra-ependymal mesencephalic trigeminal cells in the blue and mako sharks. A scanning electron microscopic study. Brain Behav Evol 17:164–177

Manni E, Bortolami R, Azzena GB (1965) Jaw muscle proprioception and mesencephalic trigeminal cells in birds. Exp Neurol 12:320–328

Marón K (1963) Regeneration of the spinal cord in shark embryos (Scyliorhinus canicula) Folia Biol (Krakow) 11:269–275

Martin GF (1969) Efferent tectal pathways of the Opossum (Didelphis virginiana). J Comp Neurol 135:209–244

Matesz C (1981) Peripheral and central distribution of fibres of the mesencephalic trigeminal root in the rat. Neurosci Lett 27:13–17

Matesz C, Székely G (1978) The motor column and sensory projections of the branchial cranial nerves in the frog. J Comp Neurol 178:157–176

Maxwell SS (1923) Labyrinth and equilibrium. Lippincolt, Philadelphia, p 160

McClung JR, Gelderd JB, Whitman RA (1978) Axosomatic synapses in the spinal cord of the shark. Exp Neurol 61:609–614

McCormick CA (1981) Central projections of the lateral line and eighth nerve in the bowfin, Amia calva. J Comp Neurol 197:1–15

McCready PJ, Boord RL (1976) The topography of the superficial roots and ganglia of the anterior lateral line nerve of the smooth dogfish, Mustelus canis. J Morphol 150:527–538

Meek J, Schellart NAM (1978) A Golgi study of goldfish optic tectum. J Comp Neurol 182: 89–122

Mellinger J (1962a) Cytologie hypophysaire de Scyliorhinus caniculus (L.) et d'autres poissons Elasmobranchs. Microscopie optique et microscopie électronique. CR Acad Sci [D] (Paris) 255:2294–2296

Mellinger J (1962b) Existence de plusieurs systèmes neurosécrétoires hypothalamohypophysaires chez les poissons élasmobranches Scyliorhinus caniculus et Sc. stellaris. Microscopie ordinaire et microscopie electronique. CR Acad Sci [D] Paris 255:1789–1791

Mellinger JCA (1963) Étude histophysiologique du système hypothalamo-hypophysaire de Scyliorhinus caniculus (L.) en état de melanodispersion permanente. Gen Comp Endocrinol 3: 26–45

Mellinger J (1964) Les relations neurovasculoglandulaires dans l'appareil hypophysaire de la roussette, Scyliorhinus caniculus (L.) (Poissons Elasmobranches). Arch Anat Histol Embryol (Strasb) 47:1–201

Mellinger J (1972) Types cellulaires et fonctions de l'adénohypophyse de la Torpille (Torpedo marmorata). Gen Comp Endocrinol 18:608

Mellinger J (1976) Croissance et reproduction de la Torpille (Torpedo marmorata). IV Croissance, dimorphisme sexuel et variation de la glande interrénale, de la thyroide et de l'hypophyse. Bull Biol Fr Belg 110:333–386

Mellinger J, Dubois MP (1973) Confirmation, par l'immunofluorescence, de la fonction corticotrope du lobe rostral et de la fonction gonadotrope du lobe ventral de l'hypophyse d'un poisson cartilagineux, la Torpille marbrée (Torpedo marmorata). CR Acad Sci [D] Paris 276: 1879–1881

Meredith GE, Butler AB (1982) Central connections of the inner ear in a cichlid fish. Soc. Neurosci Abstr 8:625

Meurling P (1960) Presence of a pituitary portal system in elasmobranches. Nature 187:336–337

Meurling P (1967) The vascularization of the pituitary in Elasmobranchs. Sarsia 28:1–104

Meurling P (1972) Control of pars intermedia in large embryos of the spiny dogfish, Squalus acanthias. Gen Comp Endocrinol 18:609

Meurling P, Björklund A (1970) The arrangment of neurosecretory and catecholamine fibres in relation to the pituitary intermedia cells of the skate, Raja radiata. Z Zellforsch Mikrosk Anat 108:81–93

Meurling P, Fremberg M, Björklund A (1969) Control of MSH release in the intermediate lobe of Raja radiata (Elasmobranchii). Gen Comp Endocrinol 13:520

Meurling P, Klefbohm B (1974) Control of pars intermedia activity in late embryos of the spiny dogfish, Squalus acanthias. Cell Tissue Res 155:221–229

Meyer DL, Fiebig E, Ebbesson SOE (1981) A note on the reciprocal connections between the retina and the brain in the puffer fish Tetraodon fluviatilis. Neurosci Lett 23:111–115

Miralto A, Kemali M (1980) Asymetry of the habenulae in the Elasmobranch Scyllium stellare. II Electron microscopy. Z Mikrosk Anat Forsch 94:801–813

Montgomery JC (1977) Vestibular nuclei of the dogfish Scyliorhinus canicula. J Physiol 269: 40–41P

Montgomery JC (1978) Dogfish vestibular system. Ph D Thesis Univ Bristol

Montgomery JC (1980) Dogfish horizontal canal system: Responses of primary afferent, vestibular and cerebellar neurons to rotational stimulation. Neurosci 5:1761–1769

Montgomery JC (1981) Origin of the parallel fibers in the cerebellar crest overlying the intermediate nucleus of the elasmobranch hindbrain. J Comp Neurol 202:185–191

Montgomery JC, Roberts BL (1979) Organization of vestibular afferents to the vestibular nuclei of the Dogfish. Brain Behav Evol 16:81–98

Mori S, Nishimura H, Aoki M (1980) Brain stem activation of the spinal stepping generator. In: Hobson JA, Brazier MAB (eds) The reticular formation revisited: Specifying function for a nonspecific system. Raven, New York, p 241–260

Morita Y, Ito H, Masai H (1980) Central gustatory paths in the crucian carp Carassius carassius. J Comp Neurol 191:119–132

Münz H, Claas B (1981) Centrifugal innervation of the retina in cichlid and poecilid fishes. A horseradish peroxidase study. Neurosci Lett 22:223–226

Münz H, Claas B, Stumpf WE, Jennes L (1982) Centrifugal innervation of the retina by luteinizing hormone releasing hormone (LHRH)-immunoreactive telencephalic neurons in teleostean fishes. Cell Tissue Res 222:313–323

Myrberg AA Jr (1978) Underwater sound – its effect on the behavior of sharks. In: Hodgson ES, Mathewson RF (eds) Sensory biology of Sharks, Skates, and Rays. US Gov't Print Off, Washington, DC, p 391–417

Nakajima Y (1970) Fine structure of the medullary command nucleus of the electric organ of the skate. Tissue Cell 2:47–59

Nelson DR, Johnson RH (1970) Diel activity rhythms in the nocturnal bottom-dwelling sharks, Heterodontus francisci and Cephaloscyllium ventriosum. Copeia 1970:732–739

Nicholson C, Llinás R (1969) Inhibition of Purkinje cells in the cerebellum of elasmobranch fishes. Brain Res 12:477–481

Nicholson C, Llinás R, Precht W (1969) Neural elements of the cerebellum in elasmobranch fishes; structural and functional characteristics. In: Llinás R (ed) Neurobiology of cerebellar evolution and development. Am Med Assoc, Chicago, p 215–243

Nieuwenhuys R (1964) Comparative anatomy of the spinal cord. In: Eccles JC, Schadé JP (eds) Progress in brain research, vol 11. Elsevier, Amsterdam, p 1–57

Nieuwenhuys R (1967a) Comparative anatomy of the cerebellum. In: Fox CA, Snider RS (eds) Progress in brain research, vol 25. Elsevier, Amsterdam, p 1–93

Nieuwenhuys R (1967b) Comparative anatomy of olfactory centres and tracts. In: Zotterman Y (ed) Progress in brain research, vol 23. Elsevier, Amsterdam, p 1–64

Nieuwenhuys R (1969) A survey of the structure of the forebrain in higher bony fishes (Osteichthyes). Ann NY Acad Sci 167:31–64

Nieuwenhuys R (1972) Topological analysis of the brain stem of the Lamprey Lampetra fluviatilis. J Comp Neurol 145:165–177

Nieuwenhuys R (1974) Topological analysis of the brain stem: A general introduction. J Comp Neurol 156:255–276

Nieuwenhuys R, Bodenheimer TS (1966) The diencephalon of the primitive bony fish Polypterus in the light of the problem of homology. J Morphol 118:415–450

Nieuwenhuys R, Opdam P (1976) Structure of the brain stem. In: Llinás R, Precht W (eds) Neurobiology of the Frog. Springer, Berlin Heidelberg New York, p 811–855

Norris HW, Hughes SP (1920) The cranial, occipital and anterior spinal nerves of the dogfish, Squalus acanthias. J Comp Neurol 31: 293–402

Northcutt RG (1977) Elasmobranch central nervous system organization and its possible evolutionary significance. Am Zool 17:411–429

Northcutt RG (1978) Brain organization in the cartilaginous fishes. In: Hodgson ES, Mathewson RF (eds) Sensory biology of Sharks, Skates, and Rays. US Gov't Print Off, Washington DC, p 117–193

Northcutt RG (1979a) Retinofugal pathways in fetal and adult spiny dogfish, Squalus acanthias. Brain Res 162:219–230

Northcutt RG (1979b) Experimental determination of the primary trigeminal projections in lampreys. Brain Res 163:323–327

Northcutt RG (1980) Central auditory pathways in anamniotic vertebrates. In: Popper AN, Fay RR (eds) Proceedings in life science: Comparative studies of hearing in vertebrates. Springer, New York, p 79–118

Northcutt RG (1981) Evolution of the telencephalon in nonmammals. Ann Rev Neurosci 4:301–350

Northcutt RG, Boord RL (1981) Efferent projections of the optic tectum in the clearnose skate, Raja eglanteria. Anat Rec 199:185A

Northcutt RG, Butler AB (1976) Retinofugal pathways in the longnose Gar Lepisosteus osseus (Linnaeus). J Comp Neurol 166:1–16

Northcutt RG, Butler AB (1980) Projections of the optic tectum in the longnose gar, Lepisosteus osseus. Brain Res 190:333–346

Northcutt RG, Wathey JC (1979) Some connections of the skate dorsal and medial pallia. Soc Neurosci Symp 5:145

Northcutt RG, Wathey JC (1980) Guitarfish possess ipsilateral as well as contralateral retinofugal projections. Neurosci Lett 20:237–242

O'Leary DP, Dunn RF, Honrubia V (1976) Analysis of afferent responses from isolated semicircular canal of the guitarfish using rotational acceleration white-noise inputs. I Correlation of response dynamics with receptor innervation. J Neurophysiol 39:631–644

Oliva O (1967) On topography of eye muscles of several elasmobranchs with regard to their life habits. Vestn Cesk Spol Zool 31:51–67

Opdam P, Kemali M, Nieuwenhuys R (1976) Topological analysis of the brain stem of the frogs Rana esculenta and Rana catesbeiana. J Comp Neurol 165:307–332

Opdam P, Nieuwenhuys R (1976) Topological analysis of the brain stem of the Axolotl Ambystoma mexicanum. J Comp Neurol 165:285–306

Panneton WM, Martin GF (1979) Midbrain projections to the trigeminal, facial and hypoglossal nuclei in the opossum. A study using axonal transport techniques. Brain Res 168:493–511

Papez JW (1929) Comparative neurology. Crowell, New York, p 518

Parent A (1973) Distribution of monoamine-containing neurons in the brain stem of the frog, Rana temporaria. J Morphol 139:67–78

Parent A, Dube L, Braford MR Jr, Northcutt RG (1978) The organization of monoamine-containing neurons in the brain of the sunfish (Lepomis gibbosus) as revealed by fluorescence microscopy. J Comp Neurol 182:495–516

Paul DH (1969) Parallel fibre response in the elasmobranch cerebellum. J Physiol 202:110–112

Paul DH (1982) The cerebellum of fishes: A comparative neurophysiological and neuroanatomical review. Adv Biochem Physiol 8: 111–177

Paul DH, Roberts BL (1975) Connections between the cerebellum and the reticular formation in the dogfish Scyliorhinus canicula. J Physiol 249:62–63P

Paul DH, Roberts BL (1976) Response of cerebellar nuclear neurones in the elasmobranch fish Scyliorhinus canicula. J Physiol 257:43–44P

Paul DH, Roberts BL (1977a) Studies on a primitive cerebellar cortex. I The anatomy of the lateral-line lobes of the dogfish, Scyliorhinus canicula. Proc R Soc Lond B Biol Sci 195: 453–466

Paul DH, Roberts BL (1977b) Studies on a primitive cerebellar cortex. II The projection of the posterior lateral-line nerve to the lateral-line lobes of the dogfish brain. Proc R Soc Lond B Biol Sci 195:467–478

Paul DH, Roberts BL (1977c) Studies on a primitive cerebellar cortex. III The projection of the anterior lateral-line nerve to the lateral-line lobes of the dogfish hindbrain. Proc R Soc Lond B Biol Sci 195:479–496

Paul DH, Roberts BL (1977d) The location and properties of the efferent neurons of the head lateral-line organs of dogfish. J Comp Physiol 116:117–127

Paul DH, Roberts BL (1978) Organization of the reticular formation in the dogfish Scyliorhinus canicula. J Physiol 280:71–72P

Paul DH, Roberts BL (1979) The significance of cerebellar function for a reflex movement of the dogfish. J Comp Physiol 134:69–74

Paul DH, Roberts BL (1981) The activity of cerebellar neurones of an elasmobranch fish (Scyliorhinus canicula) during a reflex movement of a fin. J Physiol 321:369–383

Paul DH, Roberts BL, Ryan KP (1977) Comparison between the lateral-line lobes of the dogfish and the cerebellum: An ultrastructural study. J Hirnforsch 18:335–343

Perks AM (1969) The neurohypophysis. In: Hoar WS, Randall DJ (eds) Fish physiology, vol II. The Endocrine System. Academic Press London New York, p 112–205

Petrovicky P (1966) Reticular formation of the pigeon. Fol Morphol 14:334–345

Pinganaud G, Clairambault P (1979) The visual system of the trout Salmo irideus Gibb. A degeneration and radioautographic study. J Hirnforsch 20:413–431

Platt CJ, Bullock TH, Czéh G, Kovačević N, Konjević D, Gojković M (1974) Comparison of electroreceptor, mechanoreceptor and optic evoked potentials in the brain of some rays and sharks. J Comp Physiol 95:323–355

Polimanti O (1911) Contributi alla fisiologia des sistema nervosa centrale e del movimento dei pesci. I. Selacoidei. Zool Jahrb 30:473–716

Polimanti O (1913) Contributions a la physiologie du système nerveaux central et du mouvement des poissons. Arch Ital Biol 59:393

Poloumordwinoff D (1898) Recherches sur les terminaisons nerveuses sensitives dans les muscles striés volontaires. Trav Soc Sci Arcachon 3: 73–79

Pouwels E (1978) On the development of the cerebellum of the trout, Salmo gairdneri IV. Development of the pattern of connectivity. Anat Embryol 153:55–65

Rafols JA, Matzke HA (1970) Efferent projections of the superior colliculus in the opossum. J Comp Neurol 138:147–160

Ramón-Moliner E (1967) La différentiation morphologique des neurones. Arch Ital Biol 105:149–188

Ramón-Moliner E (1969) The leptodendritic neuron: Its distribution and significance. Ann NY Acad Sci 167:65–70

Ramón-Moliner E, Nauta WJH (1966) The isodendritic core of the brain stem. J Comp Neurol 126:311–335

Reiner A, Karten HJ (1978) A bisynaptic retinocerebellar pathway in the turtle. Brain Res 150:163–169

Repérant J, Lemire M, Miceli D, Peyrichoux J (1976) A radioautographic study of the visual system in fresh water teleost following intraocular injection of tritiated fucose and proline. Brain Res 118:123–131

Repérant J, Peyrichoux J, Weidner C, Miceli D, Rio J-P (1980) The centrifugal visual system in Vipera aspis. An experimental study using retrograde axonal transport of HRP and (^{3}H) adenosine. Brain Res 183:435–441

Retzius G (1881) Das Gehörorgan der Wirbeltiere. Samson & Wallin, Stockholm

Retzius G (1895) Über den Bau des Rückenmarkes der Selachier. Biol Unters 7:34–39

Rijnberk GA van (1904) Beobachtungen über die Pigmentation der Haut bei Scyllium catulus und canicula, und ihre Zuordnung zu der segmentalen Hautinnervation dieser Thiere. Petrus Camper Ned Bijdr Anat 3:137–173

Rizzolo A (1929) A study of equilibrium in the smooth dogfish (Galeus canis Mitchill) after removal of different parts of the brain. Biol Bull 57:245

Roberts BL (1969) The spinal nerves of the dogfish, Scyliorhinus. J Mar Biol Assoc UK 49:51–75

Roberts BL (1978) Mechanoreceptors and the behaviour of elasmobranch fishes with special reference to the acoustico-lateralis system. In: Hodgson ES, Mathewson RF (eds) Sensory biology of Sharks, Skates and Rays. US Gov't Print Off, Washington DC, p 331–390

Roberts BL (1981a) The organization of the nervous system of fishes in relation to locomotion. Symp Soc Zool Soc 48:115–136

Roberts BL (1981b) Central processing of acousticolateralis signals in elasmobranchs. In: Tavolga WN, Popper AN, Fay RR (eds) Hearing and sound communication in fishes. Springer, New York, p 357–373

Roberts BL, Ryan KP (1975) Cytological features of the giant neurons controlling electric discharge in the ray Torpedo. J Mar Biol Assoc UK 55:123–131

Roberts BL, Williamson RM (to be published) Interactions between long descending pathways and cord neurons in the spinal dogfish. J Physiol

Roberts BL, Williamson RM (to be published) Pattern formation in the dogfish spinal cord. In: Roberts A, Roberts BL (eds) Neural origins of rhythmic movements. Cambridge University Press, Cambridge

Roberts BL, Witkovsky P (1975) A functional analysis of the mesencephalic nucleus of the fifth nerve in the selachian brain. Proc R Soc Lond B 190:473–495

Rohon JV (1878) Das Centralorgan des Nervensystems der Selachier. Denkschr Kaiserl Akad Wiss, Math-Naturwiss Kl 38:43–109

Rohon JV (1884) Zur Histiogenese des Rückenmarkes der Forelle. Sitzg Ber Math Phys Kl Königl Bayr Akad Wiss 14:39–56

Rosiles JR, Leonard RB (1980) The organization of the extraocular motor nuclei in the atlantic stingray, Dasyatis sabina. J Comp Neurol 193:677–687

Rosiles JR, Leonard RB, Willis WD (1977) Organization of the cranial motor nuclei in the stingray, Dasyatis sabina. Anat Rec 187: 699

Rubinson K (1968) Projections of the tectum opticum of the frog. Brain Behav Evol 1: 529–561

Rubinson K (1970) Connections of the mesencephalic nucleus of the trigeminal nerve in the frog. An experimental study with silver impregnation methods. Brain Res 19:3–14

Rüdeberg C (1969) Light and electron microscopic studies on the pineal organ of the dogfish, Scyliorhinus canicula (L.). Z Zellforsch 96: 548–581

Rüdeberg S-I (1961) Morphogenetic studies on the cerebellar nuclei and their homologization in different vertebrates including man. Lund, Thesis, p 148

Sage M, Bern HA (1972) Assay of prolactin in vertebrate pituitaries by its dispersion of xantophore pigment in the teleost Gillichthys mirabilis. J Exp Zool 180:169–174

Saito N (1966) Spike potential of the electro-motoneuron of the electric skate, Narke japonica. Jpn J Physiol 16:509–519

Saito T (1930) Über die retikulären Zellen im Gehirn des japanischen Dornhaies (Acathias mitsukurii (Jordan et Fowler). Folia Anat Jap 8:323–343

Sanders FK (1940) Second-order olfactory and visual learning in the optic tectum of the goldfish. J Exp Biol 17:416–434

Sathyanesan AG (1965) The hypophysis and hypothalamo-hypophyseal system in the chimaeroid fish Hydrolagus collei (Lay and Bennett) with a note on their vascularization. J Morphol 166:413–449

Sauerbeck E (1896) Beiträge zur Kenntnis vom feineren Bau des Selachierhirns. Anat Anz 12: 41–52

Sawyer WH (1965) Active neurohypophysial principles from a cyclostome (Petromyzon marinus) and two cartilaginous fishes (Squalus acanthias and Hydrolagus collei (sic)). Gen Comp Endocrinol 5:427–439

Scanes CG, Dobson S, Follett BK, Dodd JM (1972) Gonadotrophic activity in the pituitary gland of the dogfish (Scyliorhinus canicula). J Endocrinol 54:343–344

Schaper A (1898) The fine structure of the selachian cerebellum (Mustelus vulgaris) as shown by chrome-silver impregnation. J Comp Neurol 8:1–20

Scharrer E (1952) Das Hypophysen-Zwischenhirnsystem von Scyllium stellare. Z Zellforsch 37:196–204

Schepman AHM (1918) De octavo-laterale zintuigen en hun verbindingen in de hersenen der vertebraten. Dissertation, Univ Amsterdam, p 146

Schroeder DM (1981) Tectal projections of an infrared sensitive snake, Crotalus viridis. J Comp Neurol 195:477–500

Schroeder DM, Ebbesson SOE (1974) Nonolfactory telencephalic afferents in the nurse shark (Ginglymostoma cirratum). Brain Behav Evol 9:121–155

Schroeder DM, Ebbesson SOE (1975) Cytoarchitecture of the optic tectum in the nurse shark. J Comp Neurol 160:443–461

Schroeder DM, Vanegas H, Ebbesson SOE (1980) Cytoarchitecture of the optic tectum of the squirrelfish, Holocentrus. J Comp Neurol 191:337–351

Sharma SC (1972) The retinal projections in the goldfish: An experimental study. Brain Res 39:213–223

Short S, Butler PJ, Taylor EW (1977) The relative importance of nervous, humoral and intrinsic mechanisms in the regulation of heart rate and

stroke volume in the dogfish (Scyliorhinus canicula). J Exp Biol 70:77–92
Sligar CM, Voneida TJ (1976) Tectal efferents in the blind cave fish Astyanax hubbsi. J Comp Neurol 165:107–124
Smeets WJAJ (1981a) Retinofugal pathways in two Chondrichthyans, the Shark Scyliorhinus canicula and the Ray Raja clavata. J Comp Neurol 195:1–11
Smeets WJAJ (1981b) Efferent tectal pathways in two chondrichthyans, the shark Scyliorhinus canicula and the ray Raja clavata. J Comp Neurol 195:13–23
Smeets WJAJ (1982) The afferent connections of the tectum mesencephali in two chondrichthyans, the shark Scyliorhinus canicula and the ray Raja clavata. J Comp Neurol 205: 139–152
Smeets WJAJ (1983) The secondary olfactory connections in two chondrichthyans, the shark Scyliorhinus canicula and the ray Raja clavata. J Comp Neurol (in press)
Smeets WJAJ, Nieuwenhuys R (1976) Topological analysis of the brain stem of the sharks Squalus acanthias and Scyliorhinus canicula. J Comp Neurol 165:333–368
Smeets WJAJ, Timerick SJB (1981) Cells of origin of pathways descending to the spinal cord in two chondrichthyans, the shark Scyliorhinus canicula and the ray Raja clavata. J Comp Neurol 202:473–491
Solger B (1899) Mauthner'sche Fasern bei Chimaera. Morphol Jahrb 27:322–324
Somana R, Kotchabhakdi N, Walberg F (1980) Cerebellar afferents from the trigeminal sensory nuclei in the cat. Exp Brain Res 38:57–64
Speidel CC (1922) Further comparative studies in other fishes of cells that are homologous to the large irregular glandular cells in the spinal cord of the skates. J Comp Neurol 34:303–317
Stell WK (1972a) The morphological organization of the vertebrate retina. In: Fourtes MGF (ed) The Handbook of Sensory Physiology, vol VII/2. Springer, Berlin Heidelberg New York, p 111–213
Stell WK (1972b) The structure and morphologic relations of rods and cones in the retina of the spiny dogfish Squalus. Comp Biochem Physiol 42A:141–151
Stell WK, Witkovsky P (1973a) Retinal structure in the smooth dogfish, Mustelus canis: General description and light microscopy of giant ganglion cells. J Comp Neurol 148:1–32
Stell WK, Witkovsky P (1973b) Retinal structure in the smooth dogfish Mustelus canis: Light microscopy of photoreceptor and horizontal cells. J Comp Neurol 148:33–46
Sterling P (1977) Anatomy and physiology of goldfish oculomotor system. I Structure of abducens nucleus. J Neurophysiol 40:557–572
Sterzi G (1900) Ricerche intomo all'anatomia comparata ed all'ontogenesi delle meningi. Atti Ist Veneto Lett Arti CL Sci Mat Nat Sci 60:1101
Sterzi G (1909) Il sistema nervoso centrale dei vertebrate, vol 2, Pesci I. Selaci. Draghi, Padova
Sterzi G (1912) Il sistema nervoso centrale dei Vertebrati, Pesci, vol II. Draghi, Padova, p 473
Stieda L (1873) Über den Bau des Rückenmarkes der Rochen und der Haie. Z Wiss Zool 23: 435–442
Strutz J, Schmidt CL, Sturmer C (1980) Origin of efferent fibers of the vestibular apparatus in goldfish. A horseradish peroxidase study. Neurosci Lett 18:5–9
Sumpter JP (1976) An annual cycle of plasma testosterone and oestradiol, and pituitary gonadotrophin, in the female dogfish (Scyliorhinus canicula L.). Gen Comp Endocrinol 29:268–269
Sumpter JP, Follett BK, Jenkins N, Dodd JM (1978a) Studies on the purification and properties of gonadotrophin from ventral lobes of the pituitary gland of the dogfish (Scyliorhinus canicula L.). Gen Comp Endocrinol 36:264–274
Sumpter JP, Jenkins N, Dodd JM (1978b) Gonadotrophic hormone in the pituitary gland of the dogfish (Scyliorhinus canicula L.). Distribution and physiological significance. Gen Comp Endocrinol 36:275–285
Sumpter JP, Jenkins N, Duggan RT, Dodd JM (1980) The steroidogenic effects of pituitary extracts from a range of elasmobranch species on isolated testicular cells of quail, and their neutralisation by a dogfish anti-gonadotrophin. Gen Comp Endocrinol 40:331
Szabo T (1955) Quelques precisions sur le noyau de commande centrale de la decharge electrique chez la Raie (Raja clavata). J Physiol (Paris) 47:283–285
Szabo T, Libouban S (1979) On the course and origin of cranial nerves in the teleost fish Gnathonemus determined by ortho- and retrograde horseradish peroxidase axonal transport. Neurosci Lett 11:265–270
Székely G, Lázár G (1976) Cellular and synaptic architecture of the optic tectum. In: Llinás R, Precht W (eds) Neurobiology of the Frog. Springer, Berlin Heidelberg New York, p 811–855
Szentagothai J (1948) Anatomical considerations of monosynaptic reflex arcs. J Neurophysiol 11:445–453
Taber E (1961) The cytoarchitecture of the brain stem of the cat. I. Brain stem nuclei of cat. J Comp Neurol 116:27–69

Ten Cate J (1927) On the segmental innervation of the skin of rays. Proc Acad Sci (Amsterdam) 30:734–739

Ten Cate J (1930) Contribution à la physiologie comparée du cervelet. II Le cervelet des Plagiostomes. Arch Néerl Physiol 15:479–528

Teräväinen H, Rovainen CM (1971) Electrical activity of myotomal muscle fibers, motoneurons, and sensory dorsal cells during spinal reflexes in lampreys (Petromyzon Lampetra). J Neurophysiol 34:999–1009

Thomson KS, Simanek DE (1977) Body form and locomotion in sharks. Am Zool 17:343–354

Thors F, Nieuwenhuys R (1979) Topological analysis of the brain stem of the lungfish Lepidosiren paradoxa. J Comp Neurol 187:589–612

Tilney F (1923) Genesis of cerebellar functions. Arch Neurol Psychiatry 9:137–169

Timerick SJB (1982) The response of the pectoral fin muscles of the dogfish (Scyliorhinus canicula) to stimulation of the labyrinth and long descending pathways. J Physiol 327:64

Vanegas H, Amat J, Essayag-Millán E (1973) Electrophysiological evidence of tectal efferents to the fish eye. Brain Res 54:309–313

Vanegas H, Ebbesson SOE (1973) Retinal projections in the Perch-like teleost Eugerres plumieri. J Comp Neurol 151:331–358

Vesselkin NP, Ermakova TV, Repérant J, Kosareva AA, Kenigfest NB (1980) The retinofugal and retinopetal systems in Lampetra fluviatilis. An experimental study using radioautographic and HRP methods. Brain Res 195:453–460

Vesselkin NP, Kovačević N (1973) Nonolfactory telencephalic afferent projections in elasmobranch fishes. Zh evol ion biokhim fiziol (Tome) 9:585–592

Voneida TJ, Sligar CM (1976) A comparative neuroanatomic study of retinal projections in two fishes: Astyanax hubbsi (the Blind Cave Fish), and Astyanax mexicanus. J Comp Neurol 165: 89–106

Voorhoeve JJ (1917) Over den Bouw van de kleine Hersenen der Plagiostomen. Thesis, Amsterdam, p 88

Wallenberg A (1907) Beiträge zur Kenntnis des Gehirns der Teleostier und Selachier. Anat Anz 31:369–399

Waring H (1942) The coordination of vertebrate melanophore responses. Biol Rev 17:120–150

Watanabe A (1966) The saccus vasculosus of the ray (Dasyatis akajei). Arch Histol Jpn 27: 427–449

Weinberg E (1928) The mesencephalic root of the fifth nerve. A comparative study. J Comp Neurol 46:249–405

Wilczynski W, Northcutt RG (1977) Afferents to the optic tectum of the leopard frog: An HRP study. J Comp Neurol 173:213–230

Williams HT (1973) The telencephalon of the newborn dogfish shark, Squalus acanthias. J Hirnforsch 14:261–285

Williamson RM, Roberts BL (1980) The timing of motoneuronal activity in the swimming spinal dogfish. Proc R Soc Lond B Biol Sci 211:119–133

Wilson JF, Dodd JM (1973a) The role of the pineal complex and lateral eyes in the colour change response of the dogfish, Scyliorhinus canicula L. J Endocrinol 58:591–598

Wilson JF, Dodd JM (1973b) Effects of pharmacological agents on the in vivo release of melanophore – stimulating hormone in the dogfish, Scyliorhinus canicula. Gen Comp Endocrinol 20:556–566

Wilson JF, Dodd JM (1973c) Distribution of monoamines in the diencephalon and pituitary of the dogfish, Scyliorhinus canicula L. Z Zellforsch 137:451–469

Wilson JF, Goos HJT, Dodd JM (1974) An investigation of the neural mechanisms controlling the colour range responses of the dogfish Scyliorhinus caynicula L. by mesencephalic and diencephalic lesions. Proc R Soc Lond B Biol Sci 187:171–190

Wingstrand KG (1966) Comparative anatomy and evolution of the hypophysis. In: Harris GW, Donovan BT (eds) The Pituitary gland, vol 1. Butterworths, London, p 58–126

Wintrebert P (1920) La contraction rhythmic aneurale des myotomes chez les embryons de Selaciens. I. Obervation de Scyliorhinus canicula L. Arch Zool Exp Gén 60:222–459

Witkovsky P (1971) Synapses made by myelinated fibers running to teleost and elasmobranch retinas. J Comp Neurol 142:205–221

Witkovsky P, Powell CC, Brunken WJ (1980) Some aspects of the organization of the optic tectum of the skate Raja. Neurosci 5:1989–2002

Witkovsky P, Roberts BL (1975) The light microscopical structure of the mesencephalic nucleus of the fifth nerve in the selachian brain. Proc R Soc Lond B Biol Sci 190:457–471

Witkovsky P, Roberts BL (1976) Electron microscopic observations of the mesencephalic nucleus of the fifth nerve in the Selachian brain. J Neurocytol 5:643–660

Witkovsky P, Stell WK (1973) Retinal structure in the smooth dogfish Mustelus canis: Light microscopy of bipolar cells. J Comp Neurol 148:47–60

Wunderer H (1908) Über Terminal Körperchen der Anamnien. Arch Mikrosk Anat 71:504–569

Young JZ (1933a) The autonomic nervous system of Selachians. Q J Microsc Sci 75:571–624

Young JZ (1933b) Comparative studies on the physiology of the iris. I. Selachians. Proc R Soc Lond B 112:228–241

Young JZ (1980) Nervous control of stomach movements in dogfishes and rays. J Mar Biol Assoc UK 60:1–17

Ziegels J (1976) The vertebrate subcommissural organ: A structural and functional review. Arch Biol 87:429–476

Ziegels J (1979) The subcommissural organ of submammalian vertebrates: A histochemical study. J Hirnforsch 20:11–18

Abbreviations

A	nucleus A
a	sulcus a
af	area of fusion of both rhombencephalic alar plates
apdl	area periventricularis dorsolateralis
app	area periventricularis pallialis
apvl	area periventricularis ventrolateralis
asb	area superficialis basalis
aur	auricula cerebelli
B	nucleus B
b	sulcus b
b′	sulcus b′
brc	brachium conjunctivum
C1	nucleus C1
C2	nucleus C2
c	sulcus c
ca	commissura anterior
cans	commissura ansulata
cbb	tractus cerebellovestibularis et cerebellobulbaris rectus
cbm	tractus cerebellomotorius
cc	canalis centralis medullae spinalis
ccp	corpus commune posterius
Cer	nucleus cerebelli
Cgl	corpus geniculatum laterale
ch opt	chiasma opticum
chab	commissura habenulae
cmsp	columna motoria spinalis
co cb	corpus cerebelli
colfi	commissura olfactoria inferior
colfs	commissura olfactoria superior
cp	commissura posterior
cpii	commissura postinfundibularis, pars inferior
cpis	commissura postinfundibularis, pars superior
cpo	commissura postoptica
cpri	commissura preinfundibularis
cr cb	crista cerebellaris
cst	commissura superior telencephali
ctect	commissura tecti mesencephali
ctrans	commissura transversa
cv	cornu ventrale
cvt	commissura ventralis telencephali
D	nucleus D
d	sulcus d
dd	diverticulum dorsale telencephali
ddp	decussatio dorsalis pallii
dec tpal	decussatio tractus pallii
dec IV	decussatio nervi trochlearis
dienc	diencephalon
E	nucleus E
e	sulcus e
e_1	sulcus e_1
e_2	sulcus e_2
emth	eminentia thalami
emvcb	eminentia ventralis cerebelli
Ent	nucleus entopeduncularis
epiph	epiphysis
EW	nucleus of Edinger-Westphal
F	nucleus F
f	sulcus f
farc	fibrae arcuatae
fbt	fasciculus basalis telencephali
Fd	nucleus funiculi dorsalis
fd	funiculus dorsalis
fi	foramen interventriculare (Monroi)
Fl	nucleus funiculi lateralis
fl	funiculus lateralis
flm	fasciculus longitudinalis medialis
fM	fibre of Mauthner
fpd	fasciculus predorsalis
fretr	fasciculus retroflexus
fsol	fasciculus solitarius
fSt	fasciculus medianus of Stieda
fstr	fibrae striatae
fv	funiculus ventralis
fvm	fasciculus ventromedialis of Schroeder and Ebbesson
G	nucleus G
g	sulcus g
Gc	griseum centrale rhombencephali
gc	granular cells of the bulbus olfactorius
glom	glomeruli olfactorii
H, H′, H″	nucleus H, H′, H″
h	sulcus h
Hab	ganglion habenulae
hypoph	hypophysis

Ic	nucleus intercollicularis
Ica	nucleus interstitialis commissurae anterioris
Ifbt	nucleus interstitialis fasciculi basalis telencephali
inf	infundibulum
Ipd	nucleus interpeduncularis, pars dorsalis
Ipv	nucleus interpeduncularis, pars ventralis
Is	nucleus isthmi
J	nucleus J
j	sulcus j
K	nucleus K
k	sulcus k
L	nucleus L
l	sulcus l
lb	tractus lobobulbaris
leml	lemniscus lateralis
lemV	lemniscus trigeminalis
lih	lobus inferior hypothalami
lob vlm	lobus vestibulolateralis, pars medialis
lobX	lobus vagi
lowl	lower leaf of auricula cerebelli
lsp	lemniscus spinalis
M	nucleus M
msp	medulla spinalis
Mth	Mauthner cell
N	nucleus N
N dors	nucleus dorsalis areae octavolateralis
N interm	nucleus intermedius areae octavolateralis
Ncp	nucleus commissurae posterioris
Nflm	nucleus fasciculi longitudinalis medialis
nllad	nervus lineae lateralis anterior, pars dorsalis
nllav	nervus lineae lateralis anterior, pars ventralis
nllp	nervus lineae lateralis posterior
Nlobl	nucleus lobi lateralis
Nlobld	nucleus lobi lateralis, pars dorsalis
Nlobll	nucleus lobi lateralis, pars lateralis
Nloblm	nucleus lobi lateralis, pars medialis
Nloblv	nucleus lobi lateralis, pars ventralis
Nlt	nucleus lateralis tuberis
Nmh	nucleus medius hypothalami
Nob	nucleus opticus basalis
Nph	nucleus periventricularis hypothalami
Nsc	nucleus suprachiasmaticus
Nscd	nucleus septi caudodorsalis
Nscv	nucleus septi caudoventralis
Nsl	nucleus septi lateralis
Nsm	nucleus septi medialis
Nsma	nucleus septi medialis, pars anterior
Nsmd	nucleus septi medialis, pars dorsalis
Nsmp	nucleus septi medialis, pars posterior
Nsv	nucleus sacci vasculosi
nterm	nervus terminalis
nII	nervus opticus
nIII	nervus oculomotorius
nIV	nervus trochlearis
nV	nervus trigeminus
nVI	nervus abducens
nVII	nervus facialis
nVIII	nervus octavus
nIX	nervus glossopharyngeus
nX	nervus vagus
O	nucleus O
olcb	tractus olivocerebellaris
Oli	oliva inferior
osc	organon subcommissurale
ovp	organon vasculosum preopticum
P	nucleus P
P1-P12	area pallialis 1-12
P9l	area pallialis 9, pars lateralis
P9m	area pallialis 9, pars medialis
pal	pallium
Pdc	pallium dorsale, pars centralis
Pdcl	pallium dorsale, pars centralis lateralis
Pdcm	pallium dorsale, pars centralis medialis
Pds	pallium dorsale, pars superficialis
Pdsa	pallium dorsale, pars superficialis anterior
Pdsp	pallium dorsale, pars superficialis posterior
pg	prominentia granularis
Pl	pallium laterale
Pla	pallium laterale, pars anterior
plev	plica encephali ventralis
Plp	pallium laterale, pars posterior
Pm	pallium mediale
Po	nucleus preopticus
Poc	nucleus preopticus, pars compacta
pret	area pretectalis
Pretc	nucleus pretectalis centralis
Rai	nucleus raphes inferior
rallav	radix ascendens nervi lineae lateralis anterioris, pars ventralis
rallp	radix ascendens nervi lineae lateralis posterioris
Ras	nucleus raphes superior
raVIII	radix ascendens nervi octavi
rb	area retrobulbaris
rd	radix dorsalis nervi spinalis
rdllav	radix descendens nervi lineae lateralis anterioris, pars ventralis
rdV	radix descendens nervi trigemini

rdVIII	radix descendens nervi octavi
rhomb	rhombencephalon
Ri	nucleus reticularis inferior
Ris	nucleus reticularis isthmi
rlvq	recessus lateralis ventriculi quarti
Rm	nucleus reticularis medius
rmam	recessus mamillaris
rmeV	radix mesencephalica nervi trigemini
rmVII	radix motoria nervi facialis
rmX	radix motoria nervi vagi
rne	recessus neuroporicus externus
rni	recessus neuroporicus internus
rpo	recessus preopticus
Rs	nucleus reticularis superior
rsp	tractus rubrospinalis
rsVII	radix sensoria nervi facialis
rsX	radix sensoria nervi vagi
Rub	nucleus ruber
rv	radix ventralis nervi spinalis
sdd1	sulcus diencephalicus dorsalis 1
sdd2	sulcus diencephalicus dorsalis 2
sdm	sulcus diencephalicus medius
sdv	sulcus diencephalicus ventralis
sg	substantia gelatinosa
sgr	stratum granulare cerebelli
shyp	sulcus hypothalamicus
sid	sulcus intermedius dorsalis
sih	sulcus intrahabenularis
siv	sulcus intermedius ventralis
slH	sulcus limitans of His
sll	sulcus limitans lateralis
slm	sulcus limitans medialis
sm	stria medullaris
smi	sulcus medianus inferior
smol	stratum moleculare cerebelli
sms	sulcus medianus superior
sos	sulcus organi subcommissuralis
sosa	sulcus organi subcommissuralis accessorius
sP	stratum Purkinje
Sp1-12	area subpallialis 1-12
spal	subpallium
spcb	tractus spinocerebellaris
spo	sulcus preopticus
spth	tractus spinothalamicus
ssh	sulcus subhabenularis
std	sulcus tectodiencephalicus
stdl	sulcus telencephalicus dorsolateralis
stdm	sulcus telencephalicus dorsomedialis
sth	sulcus thalamohypothalamicus
str	striatum
stt	sulcus tectotegmentalis
stv	sulcus telencephalicus ventralis
stvl	sulcus telencephalicus ventrolateralis
stvm	sulcus telencephalicus ventromedialis
sv	saccus vasculosus
t	taenia
t chor	tela chorioidea
tb	tractus tectobulbaris
tbv	tractus tectobulbaris ventralis
tect	tectum mesencephali
tecthab	tractus tectohabenularis
teghab	tractus tegmentohabenularis
Tegl	nucleus tegmentalis lateralis
tegm	tegmentum mesencephali
telenc	telencephalon
th	thalamus
thc	tractus thalamocorticalis of Johnston
thd	thalamus dorsalis
thdl	thalamus dorsalis, pars lateralis
thdm	thalamus dorsalis, pars medialis
thdmd	thalamus dorsalis, pars dorsomedialis
thdmv	thalamus dorsalis, pars ventromedialis
thtect	tractus thalamotectalis
thtegm	tractus thalamotegmentalis
thtel	tractus thalamotelencephalicus
thv	thalamus ventralis
thvl	thalamus ventralis, pars lateralis
thvm	thalamus ventralis, pars medialis
tolf	tractus olfactorius
tolfl	tractus olfactorius lateralis
tolfm	tractus olfactorius medialis
tomsc	tractus olfactorius medialis septi cruciatus
topt	tractus opticus
toptb	tractus opticus basalis
toptl	tractus opticus lateralis
toptm	tractus opticus medialis
topts	tractus opticus superficialis
Torl	nucleus tori lateralis
Torsc	nucleus tori semicircularis
torsc	torus semicircularis
tpal	tractus pallii
tpall	tractus pallii, pars lateralis
tpalm	tractus pallii, pars medialis
tpalp	tractus pallii, pars periventricularis
tpals	tractus pallii, pars superficialis
tpohyp	tractus preopticohypophyseus
tps	tractus pallialis septi
tpsl	tractus pallialis septi, pars lateralis
tpsm	tractus pallialis septi, pars medialis
tsv	tractus sacci vasculosi
tt	fasciculus tectotegmentalis
ttd	fasciculus tectotegmentalis dorsalis
tti	fasciculus tectotegmentalis intermedius
ttv	fasciculus tectotegmentalis ventralis
Tubp	nucleus tuberculi posterioris
tubp	tuberculum posterius
upl	upper leaf of auriculae cerebelli
vbo	ventriculus bulbi olfactorii
vimp	ventriculus impar telencephali
vl	ventriculus lateralis telencephali

vma velum medullare anterius
vtrans velum transversum

zll zona limitans lateralis
zlm zona limitans medialis

III nucleus nervi oculomotorii
IV nucleus nervi trochlearis
Vd nucleus descendens nervi trigemini
Vm nucleus motorius nervi trigemini
Vme nucleus mesencephalicus nervi trigemini
Vs nucleus sensorius nervi trigemini
VI nucleus nervi abducentis
VIIm nucleus motorius nervi facialis
VIIIa nucleus octavus ascendens
VIIId nucleus octavus descendens
VIIIm nucleus octavus magnocellularis
IXm nucleus motorius nervi glossopharyngei
Xm nucleus motorius nervi vagi
Xmcd nucleus motorius nervi vagi, pars caudodorsalis
Xml nucleus motorius nervi vagi, pars lateralis
Xmm nucleus motorius nervi vagi, pars medialis
Xmrv nucleus motorius nervi vagi, pars rostroventralis

Subject Index

In this subject index the numbers in normal type refer to the figures in the atlas; the numbers in **bold face** correspond to the number of the diagrams in the text.

MIX
Papier aus verantwortungsvollen Quellen
Paper from responsible sources
FSC® C105338

If you have any concerns about our products,
you can contact us on
ProductSafety@springernature.com

In case Publisher is established outside the EU,
the EU authorized representative is:
Springer Nature Customer Service Center GmbH
Europaplatz 3, 69115 Heidelberg, Germany

Printed by Libri Plureos GmbH
in Hamburg, Germany